SHUTTLEWORTH COLLEGE
LIBRARY

Shuttleworth
101401

CROP PESTS IN THE UK

CROP PESTS IN THE UK

Collected edition of MAFF leaflets

Edited by

Marion Gratwick
M.Sc., D.I.C., C.Biol., M.I.Biol., F.R.E.S.
Ministry of Agriculture, Fisheries and Food
Harpenden Laboratory

CHAPMAN & HALL
London · Glasgow · New York · Tokyo · Melbourne · Madras

Published by Chapman & Hall, 2–6 Boundary Row, London SE1 8HN

Chapman & Hall, 2–6 Boundary Row, London SE1 8HN, UK

Blackie Academic & Professional, Wester Cleddens Road, Bishopbriggs, Glasgow G64 2NZ, UK

Van Nostrand Reinhold Inc., 115 5th Avenue New York NY10003, USA

Chapman & Hall Japan, Thomson Publishing Japan, Hirakawacho Nemoto Building, 6F, 1-7-11 Hirakawa-cho, Chiyoda-ku, Tokyo 102, Japan

Chapman & Hall Australia, Thomas Nelson Australia, 102 Dodds Street, South Melbourne, Victoria 3205, Australia

Chapman & Hall India, R. Seshadri, 32 Second Main Road, CIT East, Madras 600 035, India

First edition 1992

Typeset in Times by Falcon Graphics

Printed and bound in Singapore by Fong & Sons Printers Pte Ltd

ISBN 0 412 56260 5 0 442 31644 5 (USA)

A catalogue record for this book is available from the British Library

Library of Congress Cataloging-in-Publication data available

ERRATA

Page iv

- ISBN should read 0 412 46260 5 not 0 412 56260 5.

Page 38

- Fig. 6.4 The figure number and caption should read:
  Fig. 6.3 Rosegrain aphids: adults (x7)
- Fig. 6.3 The figure number and caption should read:
  Fig. 6.4 Grain aphids: adult and immature stages (x7)

Crop Pests in the UK: Collected edition of MAFF leaflets. Edited by Marion Gratwick. Published in 1992 by Chapman & Hall, London.

Contents

* Includes pests belonging to other groups

Acknowledgements

The Ministry of Agriculture, Fisheries and Food (MAFF) gratefully acknowledges the contributions made to the writing and revising of the leaflets by science specialists in other organizations. In the later years of the series these contributions came mainly from staff at the research institutes of the Agricultural and Food Research Council, in particular from entomologists at East Malling, Littlehampton, Wellesbourne and Broom's Barn, who gave invaluable help with the leaflets on pests of fruit, protected crops, certain vegetables and sugar beet, respectively. Assistance was also given by entomologists at the Scottish Crop Research Institute and the Scottish Colleges.

The Editor thanks all contributors, particularly the MAFF Entomologists in the Regions, for their co-operation and forbearance during her 20 years of editing the leaflets. She also acknowledges the assistance of her colleague Mr J. F. Southey with the editing of the nematode leaflets throughout this period.

In preparing this collected edition and in writing the historical introduction, the Editor has greatly appreciated the help and encouragement she has received from colleagues and former colleagues in the Entomology Discipline at Harpenden and in the Regions. Special thanks are due to Dr D. A. Cooper for valuable discussion and comment on the introduction.

Illustrations

MAFF is grateful for the generous help received over the years with the loan or gift of photographs for use in the leaflets. The individuals and organizations supplying photographs reproduced here are listed below. All these photographs have been published in one or more earlier editions of the leaflets.

Figure 16.2 is based, with permission, on a drawing by Dr T. Lewis. The map in Figure 39.4 was provided by Broom's Barn Experimental Station.

Source of photograph	*Figure number*
D.V Alford	24.1, 28.7, 29.5, 72.1
H.C. Baillie	63.1
J.A. Dunn	9.3, 13.1, 13.2, 13.3
R.A. Dunning	46.1, 46.3
C.A. Edwards	26.2
B.J. Emmett	63.3, 63.4, 63.5
P.N. Headley	26.6
D.S. Hill	16.1, 23.1, 48.1, 48.3, 57.1
D.J. Hooper	87.3
F.R. Petherbridge	4.2
Ingrid Williams	63.2
G.H. Winder	75.2
A.L. Winfield	14.1
Horticulture Research International:	
East Malling	1.2, 2.1, 7.1, 7.2, 12.1, 12.2, 12.3, 12.4, 12.5, 19.4, 19.5, 20.1, 20.2, 25.3, 25.4, 25.5, 25.6, 27.1, 27.2, 27.3, 27.4, 29.3, 29.8, 29.10, 30.2,

30.4, 44.1, 67.4, 68.1, 68.2, 68.3, 70.1, 70.2, 70.3, 70.4, 70.5, 73.3, 81.4

Littlehampton 8.1, 8.4, 53.3, 53.4, 55.1, 55.2, 55.3, 84.1

Wellesbourne 48.4, 48.5, 82.2, 90.4

Institute of Arable Crops Research:

Broom's Barn Experimental Station 39.1, 39.2, 39.3, 74.1, 74.2, 74.3, 74.4, 74.5, 74.6, 75.1, 75.3, 75.4, 75.5, 77.1, 77.2, 77.3, 80.1, 80.2, 80.3, 80.4, 87.2

Long Ashton Research Station 20.3, 27.5, 27.6, 27.7, 31.1

Rothamsted Experimental Station 22.1, 22.3, 37.3, 49.1, 49.3, 82.1, 85.3, 85.4

Institute for Grassland and Animal Production:

Hurley Research Station 38.4

Royal Horticultural Society:

Wisley 17.2, 19.6, 25.1, 29.4, 29.9, 67.1

Scottish Agricultural Research Institute:

Scottish Crop Research Institute 40.1, 40.2, 42.5, 59.1, 59.2, 59.3, 69.1, 69.2, 69.3, 69.4

Scottish Colleges:

North of Scotland College of Agriculture 26.1, 29.2, 42.1, 43.3, 47.1, 47.5, 48.2, 51.5, 52.5, 54.4, 57.2, 57.3, 66.1, 88.3

West of Scotland Agricultural College 71.5

The Ministry photographs reproduced here came from various MAFF Entomology Collections as shown below. The Harpenden Laboratory colour photographs, with a few exceptions (33.2, 33.5, 33.6 and 89.2), were taken by Mr J.R. Morrison, the Photographer at this Laboratory since 1960. Thanks are due to Mr Morrison not only for his photography of pests and plant damage throughout this long period but also for his technical advice on photographs and colour reproduction.

Source of photograph	*Figure number*
ADAS National Collection	1.4, 1.5, 1.6, 2.2, 5.1, 5.4, 6.1, 6.7, 9.1, 15.1, 15.2, 16.3, 17.4, 18.1, 18.3, 18.4, 18.7, 19.1, 19.2, 21.7, 22.2, 24.2, 26.3, 26.4, 29.1, 30.1, 34.1, 36.2, 36.5, 41.1, 43.1, 43.4, 45.1, 45.2, 47.3, 47.4, 51.4, 52.1, 52.3, 53.1, 54.1, 54.2, 55.4, 57.4, 61.2, 61.4, 62.1, 62.2, 73.2, 78.2, 83.1, 87.1, 90.1, 91.2, 91.5
Harpenden Laboratory	1.1, 2.3, 2.4, 3.1, 3.2, 4.3, 5.2, 6.2, 6.3, 6.4, 6.5, 6.6, 8.3, 9.2, 10.1, 10.2, 11.2, 17.3, 18.2, 18.6, 19.3, 20.4, 21.5, 21.9, 25.2, 25.7, 25.8, 26.5, 28.6, 29.6, 30.3, 30.5, 32.2, 32.3, 32.6, 33.1, 33.2, 33.3, 33.5, 33.6, 35.2, 35.3, 37.1, 37.2, 41.2, 41.3, 42.3, 44.3, 44.4, 46.2, 46.4, 47.2, 47.7, 49.2, 50.2, 50.3, 52.4, 53.2, 53.5, 53.6, 54.3, 56.1, 56.2, 56.3, 56.4, 58.1, 58.2, 58.3, 61.3, 63.6, 64.2, 65.1, 65.2, 65.4, 67.2, 67.3, 71.3, 71.4, 76.2, 78.1, 79.2, 83.3, 85.2, 87.4, 87.5, 88.1, 88.4, 89.1, 89.2, 89.3, 90.2, 90.3, 91.7, 91.9
ADAS Regional Laboratories:	
Bristol	28.3, 28.4, 28.5, 28.8, 52,2, 66.3, 72.2, 81.1, 81.2, 81.3, 91.1, 91.3, 91.4, 91.6
Cambridge	1.3, 4.1, 5.5, 18.5, 22.4, 35.4, 35.5, 60.2, 71.1, 76.1, 76.3, 76.4, 76.5, 78.4
Cardiff	71.2

Evesham	5.3, 17.1, 28.1, 28.2, 64.1, 64.3, 73.1, 73.4, 86.1
Kirton	65.3, 83.2, 88.2
Leeds	21.1,21.2, 21.4, 21.6, 21.8, 61.1, 61.5
Newcastle upon Tyne	32.1
Reading	36.1, 36.3, 36.4, 36.6, 78.3, 81.5
Shardlow	3.3, 21.3, 60.1, 60.3, 66.2, 91.8
Starcross	42.2,47.6
Trawsgoed	51.1, 51.6, 83.4
Wolverhampton	38.2, 38.3
Wye	8.2, 11.1, 11.3, 29.7, 35.1, 38.1, 38.5, 38.6, 44.2, 50.1, 79.1

Preface

Following the demise of the MAFF's Permanent Leaflet Series in 1985, it was suggested that the final editions of the crop pest advisory leaflets should be produced in a bound volume for the benefit of future agricultural entomologists and others interested in crop pests. This idea originated outside MAFF but was enthusiastically supported by entomologists in MAFF, the agricultural research institutes and the universities. As editor of the leaflets from 1965 until 1985, I offered to undertake the task of compilation and editing, and this offer was accepted.

To prevent the book from becoming quickly outdated, the sections on control measures have been redrafted by the Advisory Entomologists of MAFF to exclude references to individual pesticides; names of chemical groups have, however, been retained in references to pesticide resistance. Readers seeking further information on specific control measures should contact their local ADAS advisor. Certain other information likely to be ephemeral, such as the names of new varieties of crop plants, has also been omitted.

Editorial changes have included amending some of the section headings to make them more consistent throughout. The text of a few leaflets has also been slightly rearranged, but no attempt has been made to standardize the position of 'Nature of damage', which in many leaflets had been changed during their evolution. Extra information has been added to some leaflets to fill outstanding gaps. The content of three leaflets has been altered more extensively. Thus, *Chafer grubs* (Chapter 32) now incorporates material from Leaflet 449, *Japanese beetle*, which has been omitted from this collection as this pest has not become established in Europe. *Nematodes on strawberry* (Chapter 81) has been enlarged by the addition of further information on free-living nematodes from the plant pathology leaflet on *Soil-borne virus diseases of fruit plants*. The joint plant pathology/entomology leaflet on *Raspberry cane blight and midge blight* has been included, but the information has been mainly restricted to raspberry cane midge and the title changed accordingly.

The figures have been renumbered to form a continuous series but in a way that retains the individuality of the leaflets. The captions have been rewritten in a uniform style. A few new figures have been added and a few old ones replaced.

In bringing the leaflets together under one cover, the main problem was to decide on the most suitable arrangement. The existing numerical order, which bore no relationship to the subject matter, was rejected. Three other possibilities were considered: alphabetical order according to title, a crop grouping, and a systematic order of pest groups. Each had advantages and disadvantages, their relative importance depending on the reader. It was clearly impossible to satisfy everyone so, to reach a consensus, a poll was held among MAFF entomologists. They were asked which arrangement they thought would be most likely to suit the majority of the readership, bearing in mind the inherent difficulty with the polyphagous species in any crop grouping. The result was a clear vote for the systematic order of pest groups, so this arrangement was adopted with alphabetical order within each pest group. However, to aid readers with less zoological knowledge or with particular crop interests, two lists of leaflet titles have been included: an alphabetical list (Appendix A), and a list of titles under *all* the relevant major host crops (Appendix B). The former leaflet numbers appear after the titles in the contents list.

As an introduction to this collected edition, I have written a history of the leaflets. This account is based on my research into late 19th century publications on crop pests in Britain and their authors, records in the early issues of the *Journal of the Board of Agriculture*, and examination of available copies of the leaflets, and their collected editions, published since their origin in 1885.

M. Gratwick
Harpenden Laboratory

Foreword

Crop pest information has been made available through official publications in some form since 1885 and the first of the permanent series of advisory leaflets was issued by the Board of Agriculture in 1893. This well-known leaflet series, which covered a range of subjects in addition to crop pests, survived several minor changes in name and format until 1985 when a decision was taken by MAFF to discontinue the series.

For many years the leaflets have been acknowledged as a valuable source of up-to-date crop pest information for the adviser and for new entrants to the advisory service, as well as for farmers and growers and agricultural/horticultural students. Copies of the leaflets are also frequently requested by foreign research and extension workers. The carefully selected photographs of crop injury and of important stages of the pest, increasingly reproduced in colour since 1965, have proved an invaluable aid in the field. The concise and accurate descriptions of the life cycles and biology of the pests were supported by regularly updated advice on control strategies and recommended pesticides. As a former adviser in crop protection I can confirm, from my own experience, the value of these leaflets in providing comprehensive and easily accessible information on a wide range of pest problems.

Following the decision to discontinue this valuable source of crop pest information, it was suggested that the last editions of the leaflets should be produced in a bound volume as a way of preserving the information for all those concerned with crop protection in the future. This has resulted in the production of a completely unique compendium of information on the full range of invertebrate pests of agricultural and horticultural crops in the UK. To prevent the contents from becoming out of date, detailed information on specific pesticides has been omitted while retaining the advice and background biology required to plan control strategies.

The editor of this volume, Miss Marion Gratwick, has acknowledged the continuous input into the leaflets by science specialists in MAFF and other organizations and also the contribution made by present MAFF entomologists in re-casting the leaflets into a suitable form for this publication. Special thanks, however, are due to Miss Gratwick who, for 20 years, ensured by her untiring efforts in her role as Editor that the information contained in the leaflets was well presented and kept up to date. It is largely because of her dedication and determination that the considerable task of collating and editing this valuable collection has been successfully completed.

H. J. Gould
formerly Director,
Harpenden Laboratory
and
Head of Entomology,
Agricultural Development
and Advisory Service

Introduction

HISTORY OF THE LEAFLETS

Eleanor Ormerod's publications

The Destructive Insects Act 1877 gave the Privy Council powers to prevent the introduction into this country of the Colorado beetle, which was then causing great damage to the potato crop in America. In spite of this, the agricultural work of the Privy Council continued to be mainly concerned with diseases of animals. Nothing was done by any government body, until 1885, to disseminate information on pests or diseases of plants. As far as pests were concerned, the gap was filled by Miss Eleanor Ormerod, who, at the end of 1877, started issuing annual reports of observations on injurious insects. These reports were compiled from observations sent to Miss Ormerod by voluntary contributors and consisted mainly of records of pest damage, habits of the insects, and methods used to control them. Initially the information was confined to 16 insect species, but the coverage was soon extended to other injurious insects and later to other crop pests, including birds. The reports were illustrated with black and white drawings. Miss Ormerod produced these reports, and published them at her own expense, every year for 24 successive years.

In 1881 Miss Ormerod published a book entitled *A Manual of Injurious Insects, with Methods of Prevention and Remedy for their Attacks to Food Crops, Forest Trees, and Fruit*. This publication was a digest of information which had been sent to Miss Ormerod up to that time by her contributors, of whom she named 44. It contained illustrated descriptions of over 90 insect species and a few mites. An enlarged edition, which included millepedes, eelworms (nematodes) and slugs, was published in 1890. Miss Ormerod also produced and arranged free circulation of four-page leaflets on the most common farm pests.

Charles Whitehead's monographs

In 1882 the Royal Agricultural Society of England established an Entomological Department and appointed Miss Ormerod as Honorary Consulting Entomologist — a position she held for 10 years. Charles Whitehead, as a member of the Society's Council and the Chairman of the Seeds and Plant Diseases Committee, played a key role in both these events. He believed that similar action should be taken by the Agricultural Department of the Privy Council Office. To further this idea, he offered to act as honorary correspondent and to prepare a series of illustrated monographs on certain destructive insects for circulation among farmers, fruit growers and hop growers. This offer was accepted in 1884, and in August of the following year Mr Whitehead presented the first of his *Reports on Insects Injurious to Hop Plants, Corn Crops, and Fruit Crops in Great Britain*. The introduction to the first report explained their purpose. The native insects that were causing many crop losses in Britain were increasing and spreading; in addition, there was a fear of new pests being introduced from foreign countries. Mr Whitehead had, therefore, decided that it was 'most desirable to diffuse entomological information as to the habits and life-history of injurious insects in a simple and intelligible form, for the use of farmers, fruit-growers, market gardeners, and all who cultivate the land, and at the same time to give practical modes of prevention, and remedies against their attacks.' These reports were presented to Parliament and widely distributed. Thus, the series of government leaflets on crop pests was born.

Mr Whitehead's first report consisted of 10 monographs on pests of hop plants. The choice of crop to launch the series may seem surprising, but the author was an authority on hop growing. He lived in Kent and knew that hop growers in particular were anxious to learn what had been discovered about the insects liable to ruin their crops. (The 1882 crops had been devastated by aphids.) The second report, which was presented four months after the first, contained monographs on 16 pests of corn and grass, three pests of stored corn, and seven pests of pea, bean and clover crops. The third report, presented in the following year, described 29 fruit pests and included a short section on natural enemies. At the beginning of 1887, a further report was added to complete the series: this dealt with pests of root crops, forage crops and some vegetables, and described 22 species. Each of the four reports was individually priced, the price varying from 2*d*. to 7½*d*. according to length. Although the overall title of the reports specified insects, Mr Whitehead followed the tradition of previous agricultural entomologists such as John Curtis (1791–1862) by including mites, millepedes and eelworms, but he omitted slugs. Each of the 87 monographs was subdivided into four sections: (i) an introduction, which included an account of damage; (ii) life history, which included a description of the species; (iii) prevention; and (iv) remedies. The descriptions were illustrated with black and white drawings, most of which were the same as those used by Miss Ormerod in her manual. These came from several sources including Curtis' *Farm Insects*, first published in 1860, but a few were Eleanor Ormerod's own drawings.

Charles Whitehead's annual reports

Following completion of his series of monographs, Mr Whitehead was appointed Agricultural Adviser to the Lords of the Committee of Council for Agriculture. He was paid £250 per year and allotted a room in the Privy Council Office, but he chose to work at his home — Barming House, Maidstone — where he had a library and workshop and could do experiments in the field or garden. Until the end of 1889 he obtained some entomological assistance from Miss Ormerod, presumably by correspondence. Mr Whitehead had been writing to Miss Ormerod for several years as he is quoted in some of her reports and is acknowledged as a contributor to her 1881 manual.

Mr Whitehead's first annual report as Agricultural Adviser was for 1887. It was a record of the most serious attacks of insects and fungi on farm, orchard and garden crops during the year, with revised or new monographs on five of the species. This report was sold at a price of 3*d*. The second and third annual reports followed the same general pattern but contained descriptions of many more species. They also referred to the issue of special illustrated leaflets and circulars that had been distributed to relevant growers in response to severe outbreaks — of pea and bean beetle and winter moth in 1888, and of winter moth, mangel wurzel (mangold) fly and hessian fly in 1889.

By the time of his third annual report, at the end of 1889, the Board of Agriculture had been formed and Mr Whitehead became Technical Adviser to the Board. He continued to produce annual reports as before. In 1892 he recommended that the descriptions of the species should be illustrated in colour to aid identification. This recommendation was accepted and the report for that year, published in 1893, contained 10 plates of beautifully coloured figures of insects and fungi. Not surprisingly, this made the report much more expensive: two shillings compared with 2½*d*. for the 1890 report. It was, however, the last of these reports to be published, for in 1894 the first number of the *Journal of the Board of Agriculture* was issued. This was a quarterly publication containing a section on injurious insects and fungi. Detailed information was given on selected species, as in the preceding annual reports, but records of pest and disease damage to crops were not included. After a few years the section became shorter and appeared less regularly until it finally disappeared in 1899, which was the year when Mr Whitehead resigned from the Board because he could no longer use a microscope. Some years later he was knighted for his services to agriculture.

The Board of Agriculture's first leaflets

During 1890 the Board issued 16 leaflets, numbered Agr. 1/1890 to 16/1890. They were on various subjects, five being on pests or fungi. Copies of the hessian fly leaflet and the winter moth leaflet were supplied free through post offices in the districts particularly affected by these insects. The Directors of the Great Eastern Railway were also involved in the circulation of the winter moth leaflet. In 1891 and 1892, further leaflets were issued using the same unusual numbering system. These leaflets were produced in response to severe attacks of turnip diamond-back moth in 1891, and apple blossom weevil, raspberry moth and mangel wurzel fly in 1892.

Permanent series of leaflets

Although the origins of the leaflet series can be traced back to Charles Whitehead's monographs and the subsequent occasional leaflets issued to help combat serious outbreaks, it was in 1893 that the permanent series began. Eighteen leaflets were issued by the Board in that year and eight of these were on insect pests. More were produced each year and by 1915 there were 300, of which 61 were on crop pests. The change from a Board to a Ministry (of Agriculture and Fisheries), which occurred in 1919, did not affect the leaflet series.

Numbering systems The leaflets issued in 1893 were numbered A 1 to 18 and the numbers started again at the beginning of the next year. Thus, the same number was allocated to different titles while repeat titles were given different numbers. This confusing system continued until 1896 when the 'A' was replaced by 'Leaflet No.' and from then on the numbers ran continuously, with each title retaining its own number through the years and from one edition to the next. In 1917 the series was designated the Permanent Series.

The numbering system did not change again until 1930, when the name of the series was altered from 'Leaflet' to 'Advisory Leaflet' and existing leaflets were renumbered as new editions were published. The Colorado beetle leaflet proved an exception and retained its original number of 71. No more changes in nomenclature or numbering occurred for another 48 years. Then, at the end of 1978, the advisory series was expanded to include other leaflet series that had arisen independently within the Ministry and the word 'Advisory' was dropped. Thus, the series reverted, unwittingly, to its original name. At the same time an attempt was made to group the leaflets into fields of interest by giving them coded letters as well as new numbers. The coding was so complicated and the new numbers so disliked by leaflet users that the system was quickly abandoned and the old numbers reinstated — much to everyone's relief!

Layout and format The original concept of the pest series, like that of the early monographs, was that each leaflet should be confined to one species or a few closely related species. This concept was retained throughout, except for the addition, after 1970, of four crop-based leaflets that dealt with a range of pests from several groups.

The general layout adopted for the leaflets was similar to that of the monographs and has never been completely changed. After a few years the sections on 'Prevention' and

'Remedies' were combined and by 1920 had become 'Methods of Control', later amended to 'Control Measures'. A pesticide precautions panel was added in 1953. After the mid-1940s the layout of the other sections became less rigid as the number of contributors increased and the leaflets were allowed to evolve more independently. Additional sections were introduced where appropriate and in some leaflets the section on plant damage was placed after, rather than before, the description of the pest.

For many years most of the pest leaflets were confined to four pages, but six pages became more common in the 1940s and eight pages in the 1960s. Later, a few leaflets were even longer. Throughout their history the leaflets were printed on demy octavo or A5 size paper. The quality of the paper was rather poor in the early years, but the introduction of photographs (see below) necessitated the use of better-quality paper and this eventually led to a rise in the overall standard of production.

Illustrations Almost all the pest leaflets were originally illustrated with black and white drawings on the front page. The first edition of the Colorado beetle leaflet, in 1902, was exceptional in that it contained a colour plate. The leaflet was published in the year after the first Colorado beetle outbreak in this country and accurate identification was obviously deemed sufficiently important to justify the cost of colour. The plate was a fine coloured drawing which, for identification purposes, has possibly not been bettered by any of the later illustrations of this pest. However, it was not included in the reprint or in the following edition, and it was 1934 before colour, in the form of another coloured drawing, was reintroduced to this leaflet. This followed the second Colorado beetle outbreak, which occurred in 1933.

Monochrome photographs were introduced in 1905 — in a revised edition of the leaflet on goat moth and wood leopard moth — and gradually, over the years, almost all the drawings were replaced by photographs. At first the photographs were grouped together on an inside page of art paper. This had an unfortunate result, for the disappearance of drawings led to many leaflets having no front-page illustration. Eventually, by about 1950, all leaflets with photographs were printed entirely on art paper, thus allowing complete flexibility in the position of illustrations. However, it was not until the 1960s that front-page illustrations again became standard practice. This time, figures of pest damage rather than of the pests themselves were placed on the front page.

The first colour photograph appeared in the chrysanthemum eelworm leaflet of 1950, but a special case had to be made for this and it was 1965 before the cost of colour became generally acceptable. From then on, the policy was to illustrate the front page of each pest leaflet with a colour picture of damage. This objective was not finally achieved until 20 years later, when the last edition of the apple blossom weevil leaflet was published in 1985.

Pricing The leaflets were available free of charge until the end of 1921 when 'the increased cost of printing and paper', coupled with 'the need for economy in Government expenditure', led the Ministry to charge 1*d*. per leaflet (or 9*d*. per dozen) where more than one leaflet was required. This price remained unchanged for 30 years. By 1967 it had reached 4*d*. per leaflet, when it was found that the cost of collecting the moneys exceeded the income, so the leaflets were made free again.

Collected editions In 1904 the hundredth leaflet was published. The Board then issued the first 100 leaflets in book form, so that they could more easily be kept for reference purposes. These leaflets included 39 on crop pests. The first edition of the bound volume, in stiff covers, cost 6*d*. It proved to be very popular and, by 1919, 17 editions had been produced, amounting to a staggering total of 101 000 copies notwithstanding the continued availability of single leaflets. Further bound volumes followed: Leaflets 101–200, which included 13 on crop pests, in 1909, and Leaflets 201–300, which included another nine on crop pests, in 1915.

The first 200 leaflets were also published in

12 sectional volumes according to subject and priced at 1*d*. per volume. One volume contained the leaflets dealing with pests injurious to farm and garden crops, while another contained the leaflets on pests of fruit trees and bushes. In 1933 these bound volumes were discontinued in favour of portfolios of actual leaflets, which enabled out-of-date leaflets to be replaced by revised editions and new leaflets to be added. No further collected editions were published until the present volume.

Number of leaflets Between 1915 and 1965 the number of crop pest leaflets increased from 61 to 74. During that period new ones were added and a few discontinued, according to the pest status of the species concerned. The only leaflets to be discontinued in the second half of the period — in about 1960 — were those on mustard beetles, rhododendron bug and earcockles of wheat. A further seven titles have been lost since 1960, but the information has been retained by adding it to leaflets on pests with similar habits. Between 1965 and 1985, 22 new leaflets were added, bringing the final total to 92 —not very different from Charles Whitehead's original total of 87. About three-quarters of the insect species which he described in the 1885–87 monographs were also included in the 1985 leaflets. The main difference in coverage concerns mites and nematodes, which in recent years have played a prominent part in the series, together accounting for more than a quarter of the titles. Strangely enough, the only nematode species included by Mr Whitehead — earcockles of wheat — has ceased to be important and is no longer the subject of a leaflet.

Authorship All the leaflets issued by the Board of Agriculture were anonymous. It is not known how many of the earliest ones were written by Charles Whitehead as Technical Adviser. It is only known for certain that he was author of four of the fruit pest leaflets issued in 1893 and 1894. After his retirement there is a record, in 1901, of the British Museum (Natural History) being involved in revision of the leaflets. During the early years of the century, entomologists elsewhere must also have contributed, for at that time the Board received advice from authorities such as Cecil Warburton, who had succeeded Miss Ormerod at the Royal Agricultural Society, and F. V. Theobald, the zoologist at the Agricultural College at Wye (established in 1894). However, no attribution for authorship or revision has been found in the leaflets before 1928, although other publications were sometimes cited as sources of information. From 1928 onwards, some leaflets contained acknowledgements for direct help from scientists outside the Ministry, but this may reflect a change in policy of acknowledgement rather than of authorship. Ministry staff were not acknowledged until 1974 and then only for major contributions.

From the 1940s, if not before, leaflet texts were written and revised by the most appropriate entomologists or nematologists, irrespective of where they worked in the UK. Thus, scientists at the agricultural and horticultural research institutes became major contributors to leaflets on pests of particular crops. With the formation of the National Agricultural Advisory Service (NAAS) in 1946, Ministry authorship became more widely based and the advisory entomologists in each NAAS Region were invited to contribute to all new texts and revisions. In 1971 NAAS was combined with various other Services within the Ministry to form the Agricultural Development and Advisory Service (ADAS), but this did not affect the production of leaflets.

Editorship In 1918 there was a reorganization of the Plant Pathology Laboratory that had been founded four years before at the Royal Botanic Gardens, Kew. The administrative and advisory side became part of the Board of Agriculture and, as one of its functions, it assumed responsibility for the production and revision of advisory leaflets on crop pests and diseases. J. C. F. (later Sir John) Fryer, who had been Entomologist to the Board since 1913, became the first Director of the Laboratory, which moved to Harpenden in 1920.

Fryer obviously had a guiding hand in the production of the pest leaflets and he wrote the introductions to the collected editions published in the 1920s and 1930s. In view of recent criticism of the large amount of biological information given in the leaflets, it is interesting to note that Fryer's introductions to the bound sectional volumes (1921–30) contained a spirited defence of their technical content. 'It is sometimes complained', he wrote, 'that . . . (they) contain too much "technical matter" and that all that is required is "how to recognise the pest" and "how to control it". Now recognition and control are exactly the objects at which each leaflet aims, but the necessary information cannot be given without the introduction of some so-called "technical matter", particularly in relation to the habits and life of the pest. . . . Insects, though very small, are creatures with their own habits and ways, living in a perfectly wild state just like birds, fish, or any wild animal. In order to fish successfully, whether for minnows or sharks, the first essential is to understand the habits of the fish. This applies also to shooting, no matter whether the game is a tiger or a rabbit. . . . The same considerations apply to insect pests — to hunt or control them successfully it is essential to know as much as possible of their habits. . . . A useful leaflet, therefore, must contain some technical matter.'

Responsibility for the crop pest leaflets remained at the Plant Pathology Laboratory (later the Harpenden Laboratory) even after the formation of NAAS. The task of technical editing was undertaken by various members of the Entomology Department, but for many years this work was fitted in with other duties. However, with the rapid changes in pesticide usage in the 1950s and early 1960s and the increasing complexity of pest control, the time required for leaflet revision increased. This led to the creation of a new post with the prime function of keeping the existing pest leaflets up to date and editing new ones. This post was filled in 1965 when the present Editor joined the Ministry. In the following 20 years, 22 new leaflets and 500 revisions were edited, some leaflets passing through eight editions during that time.

Final editions In 1985, with the development of a more commercially orientated advisory service, ADAS policy on publications changed. In the interest of economy it was decided to cease production of encyclopaedic or educational material whenever possible and to publish only concise information that was important in commercial practice. The so-called permanent leaflet series was deemed to fall within the former category, so publication ceased with the 1985 editions — exactly one hundred years after the publication of Charles Whitehead's first monographs.

COMMON AND SCIENTIFIC NAMES OF INVERTEBRATES

All common and scientific names of invertebrates have been updated and conform to the fourth edition (1989) of *Invertebrates of Economic Importance in Britain — Common and Scientific Names*, produced by the Harpenden Laboratory of the Ministry of Agriculture, Fisheries and Food. The authorities for the scientific names were abbreviated in the leaflets, but to comply with current practice they are given here in full with the exception of Linnaeus, which is abbreviated to L.

PESTICIDES

This book mentions insecticides, acaricides, nematicides, molluscicides, fungicides, herbicides/weedkillers, and soil sterilants, all of which are now included under the term 'pesticides'.

The use of pesticides (together with their advertisement, sale, supply and storage) is controlled by the Control of Pesticides Regulations 1986, which implement Part III of the Food and Environment Protection Act 1985. It is an offence to use a pesticide that has not been approved, or to use an approved pesticide on crops or in situations other than those for which an approval has been granted or in ways which are not included in the conditions of approval relating to use.

Throughout this book a 'recommended' pesticide is one that is *approved* under the Control of Pesticides Regulations 1986 and *recommended* by a marketing company for use on the crop concerned against that particular pest. The marketing company's recommendations are printed on the product label or in a leaflet accompanying it.

An annual list of pesticides approved for agricultural, horticultural and other use is produced jointly by the Ministry of Agriculture, Fisheries and Food and the Health and Safety Executive. The list is subdivided into products approved for professional use and those approved for amateur use. New approvals and amendments to approvals are listed monthly in *The Pesticides Register*.

APHIDS AND OTHER PLANT BUGS

Hemiptera

1
Apple and pear suckers

Fig.1.1 Nymph of apple sucker on underside of apple leaf (× 12). Note waxy threads being produced

Apple sucker (*Psylla mali* (Schmidberger)) used to be one of the more serious pests of apple, but nowadays it is so well controlled by routine sprays that infestations are rarely seen except in gardens or unsprayed cider orchards.

Pear sucker (*Psylla pyricola* Förster), however, has increased in importance in recent years and large populations have become established in some orchards in the pear-growing areas.

Apple sucker

Nature of damage

In spring the immature apple suckers (nymphs) feed by sucking the sap, mainly in the blossom trusses although leaf buds may also be attacked. Large populations can cause brown discoloration of petals on partly opened blossom buds or even death of these buds. The discoloration resembles, and is sometimes mistaken for, frost damage.

Description and life history

Apple sucker overwinters in the egg stage and has only one generation per year. It lives only on apple. The eggs are oval and pale straw-coloured, barely visible without magnification but easily seen with the aid of a hand lens. They are laid mainly on the fruit spurs, generally along leaf scars, and to a lesser extent at the base of the leaf buds or scattered on the twigs (Fig. 1.2).

In south-east England, hatching begins when varieties such as Cox's Orange Pippin or Bramley's Seedling are at the bud-burst stage and is complete by green cluster. Thus in a normal season eggs hatch during April in Kent, but slightly later in more northerly areas. The newly hatched nymphs have a distinctive flattened body and are orange-brown with red eyes. Drops of sticky fluid (honeydew) and very conspicuous white or iridescent waxy threads (see Fig. 1.1) are produced by the nymphs and are usually seen easily on infested trusses. No sooty mould is associated with honeydew on apples (*cf.* pears). After the second moult the nymphs become bright green and rudiments of wings can be seen (Fig. 1.1).

Four to six weeks after hatching, the first adults appear. They are small insects, 3 mm long, with bodies that are greenish-yellow at first but darken later; the wings are transparent with green veins and are folded over the body when at rest (as in Fig. 1.4). The adults live and feed on the apple tree throughout the summer and early autumn. Egg laying begins towards the end of August and continues for about one month; the adults then die.

Fig.1.2 Eggs of apple sucker on apple spur (× 16)

Pear sucker

Nature of damage

Pear sucker nymphs feed on developing leaf and blossom trusses in the spring and on leaves in the summer. Blossoms may be killed when infestations are large. Sooty moulds grow on the honeydew excreted by the nymphs, so that blackened shoots and foliage (Fig. 1.3) are an indication of pear sucker infestations; these moulds persist on the wood into the dormant season. The moulds may also cover the fruits, spoiling their appearance.

Pear suckers can also cause damage later in the season when the young of the third generation feed on the developing fruit buds. This may result in death or a reduction in vigour of fruit buds, thus causing a decrease in the following season's crop. Very large infestations in late summer may lead to premature defoliation in September or October.

Description and life history

Unlike apple sucker, pear sucker overwinters in the adult stage (Fig 1.4). The adults are similar in appearance to those of apple sucker but are slightly smaller (about 2 mm long) and dark brownish-black. In winter they can be found resting on pear trees, but some disperse to trees and other vegetation outside the

Fig.1.3 Sooty mould on pear leaves

Fig.1.4 Adult pear sucker on leaf (× 20)

orchard. In spite of the cold they often make short flights on bright sunny days.

Eggs, which are smaller than those of apple sucker, are laid on pear spurs and shoots between late February and full blossom. At first they are straw-coloured but gradually they become orange. Hatching usually begins at bud burst and continues until the end of the flowering period. The nymphs (Fig. 1.6) have red eyes and orange-yellow bodies which darken with age. They live on the blossom and leaves of the trusses and mature to form the first summer generation of adults. Eggs of the summer generations are laid on the leaves, mainly along the mid-rib (Fig. 1.5). During the summer the nymphs live in drops of honeydew on the underside of leaves.

There are usually three generations per year with peak adult populations occurring about early June, late July and mid-October. There is some overlap of generations during the summer. In the absence of control measures, even small numbers of adults surviving the winter can give rise to very large populations by the end of the summer.

During the growing season the presence of pear suckers may be detected on a dry day by the droplets of honeydew on the underside of the leaves. The honeydew attracts flies, wasps and bees, which feed on it; the presence of these insects on pear trees is usually a sign of infestation by pear sucker.

Natural enemies

Predators, particularly anthocorid bugs, feed on the eggs and young of both apple and pear suckers.

Several parasitic wasps have been recorded from pear sucker nymphs of the first generation.

Control in orchards

Winter washes

Formerly, a winter wash was the standard method of controlling apple sucker, but this treatment is expensive and is effective only if all the bark is thoroughly wetted. It is less effective against pear sucker because only

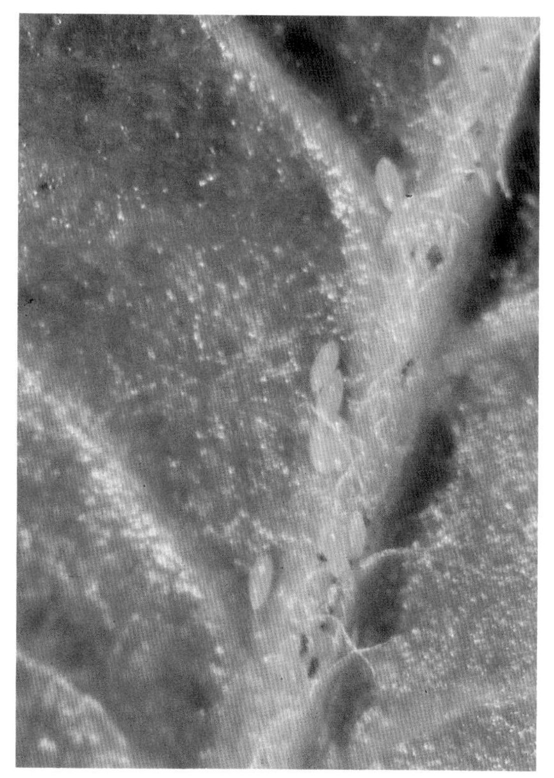

Fig.1.5 Eggs of pear sucker on leaf (× 12)

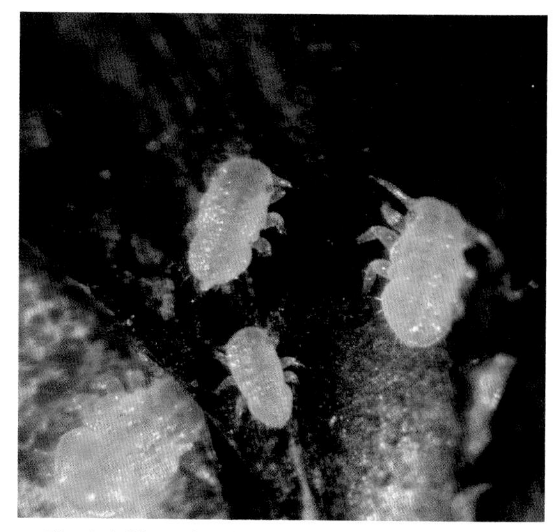

Fig.1.6 Nymphs of pear sucker on twig (× 18)

those adults overwintering on the trees will be killed.

Spring and summer sprays

Apple sucker is seldom seen in regularly sprayed commercial orchards as it is controlled by the insecticides currently applied at green cluster against aphids and caterpillars (see pages 19 and 156). For large infestations, as in cider orchards, one of the insecticides that is most effective against apple sucker should be used.

Pear sucker In many years pear sucker populations are kept below pest level by the effects of weather and predators. In years when chemical control is necessary it is important to use insecticides that are relatively harmless to predators, particularly anthocorid bugs. Broad-spectrum insecticides should only be used as a last resort.

Pear sucker populations have become resistant to many insecticides, but in cool, wet years some of these chemicals may still control the pest. Insecticides belonging to several chemical groups are available for pear sucker control and most populations are susceptible to at least one of them. Experience of resistance in pear sucker in Europe and North America has shown that it is essential to use the effective insecticides carefully to prolong their useful life and to delay the development of resistance to them for as long as possible. Spray programmes should comprise the minimum number of applications.

Trees should be inspected regularly; light infestations of pear sucker can be tolerated with little resultant economic loss. If adult suckers are very numerous in early spring, a recommended insecticide spray should be applied either on a sunny day in early March to kill the adults or at white bud to kill the hatching nymphs. If adults are not abundant, treatment should be delayed until after the blossom period; then, if branches and fruit become sticky with honeydew, an effective insecticide should be applied, preferably using one that does not harm predators.

An early spray of certain insecticides will give some control of aphids and caterpillars pre-blossom. Where codling moth (see page 108) may cause economic damage to pears, pheromone traps should be used to monitor codling moth activity and sprays applied for this pest only when they are essential, thus minimizing the harmful effects on predators.

If this approach to the control of pear sucker is adopted, the need to spray against this pest should be relatively infrequent and the useful life of the effective insecticides should be prolonged.

Control in gardens

Apple sucker Some winter washes will give excellent control of apple sucker and will also depress populations of pear sucker. They will also control overwintering stages of aphids and kill moss and lichens growing on the branches. Care must be taken to prevent drift contaminating vegetables or scorching the foliage of evergreens and the tender growth of ornamentals.

As an alternative, a spray of a recommended insecticide at green cluster will also control aphids and may control caterpillars.

Pear sucker Infestations of pear sucker are uncommon in gardens and are unlikely to develop before the end of June. Any such infestation can be controlled by spraying with a recommended insecticide.

2
Apple aphids

Fig.2.1 Apple fruit and foliage damaged by rosy apple aphid, showing down-curled and distorted leaves and absence of red patches

Eight species of aphid (greenfly) may be found on apple in Britain and five of them are common pests capable of causing serious damage. Four of these are described here: the rosy apple aphid (*Dysaphis plantaginea* (Passerini)), the rosy leaf-curling aphid (*Dysaphis devecta* (Walker)), the green apple aphid (*Aphis pomi* Degeer) and the apple–grass aphid (*Rhopalosiphum insertum* (Walker)). The fifth species, the woolly aphid, is described on page 78.

Recognition and damage

Apple growers will readily recognize aphids as a group, but they should also be able to identify the more injurious species and know their different habits, so that they can apply the most effective control measures.

Apple aphids are most numerous in the orchard in April, May and June, and the worst damage is caused during this period. With the exception of the green apple aphid, they are either absent from apple, or present in very small numbers, from the end of June until the autumn. The green apple aphid sometimes reappears in large numbers after midsummer.

The main distinguishing features of the four species are given in Table 2.1, together with their occurrence, damage potential and a summary of the damage caused. The damage is described below in greater detail.

The rosy apple aphid (Fig. 2.3) can cause severe injury. Infested leaves become severely down-curled and distorted (Fig. 2.1), and may turn yellow but never red; severely damaged leaves turn brown and drop off. Infested young shoots remain short and twisted, and the fruits on infested trusses are small or distorted, and ripen prematurely (Fig. 2.1).

The rosy leaf-curling aphid infests older trees with rough bark and usually appears on the same trees year after year. The damage is conspicuous and resembles that done to foliage by the rosy apple aphid except that bright red areas develop on the upper side of infested leaves, as well as yellowing and pronounced distortion (Fig. 2.2). This species does *not* cause small distorted fruit.

Fig.2.2 Apple shoot damaged by rosy leaf-curling aphid, showing down-curled and distorted leaves with red patches. (In foreground note cluster of yellow eggs laid near leaf tip by a ladybird)

The green apple aphid lives mainly on the young extension growth, causing leaf-curl and reduction in growth. This species is not usually common in commercial orchards in spring, but it may become numerous in June or July as a result of immigration from other host plants nearby. It seriously affects the growth of young trees and nursery stock. Established trees are not harmed by light to moderate infestations, which are sometimes controlled by natural enemies.

The apple–grass aphid is usually found on therosette leaves of blossom trusses. Unless very abundant, the aphids cause little apparent harm apart from slight leaf-curl.

Life history

All four species overwinter in the egg stage on apple or related trees. The small shining black

Table 2.1 Four aphid pests of apple: recognition, occurrence and damage

Aphid	Occurrence	Damage potential	Damage
Rosy apple aphid (greyish or pinkish, waxy)	Common in some years	Very great (the only species causing direct injury to fruits)	Severe leaf distortion; yellowing – never red. Small, distorted fruit.
Rosy leaf-curling aphid (bluish or pinkish, powdery)	Very local (no dispersal phase)	Great	Severe leaf distortion; yellowing; red patches.
Green apple aphid (uniformly green)	Common in some years as summer immigrants to orchards	Moderate (Great in nurseries and on young trees)	Leaf-curling. Reduced extension growth.
Apple–grass aphid (yellowish-green with two broad, darker stripes along back)	Very common	Small	Slight leaf-curling.

eggs (Fig. 2.4) hatch over a period of two to three weeks, beginning between mid-March and early April according to the temperature.

The main hatching period coincides with the opening of the fruit buds of apple. With the rosy apple and apple–grass aphids, hatching normally begins at bud break of Bramley's Seedling or Cox's Orange Pippin, reaches a peak at bud burst and ends at about the mouse-ear stage; it is rather later for the rosy leaf-curling aphid and the green apple aphid.

The newly hatched aphids crawl up the twigs and collect on the opening buds, where they feed on the sap through their fine stylets. While thus exposed the aphids are extremely vulnerable to insecticides but, as the buds begin to open, the aphids move down into the crevices where they are well protected from insecticides.

All these aphids are female and on reaching maturity they produce living young. Successive generations do likewise until the last one, when males as well as females appear for the only time in the life cycle. After mating, the females lay the overwintering eggs.

Infestations on apple generally reach a peak in late May and early June. During this period winged aphids develop and fly off to start new colonies on other plants.

The habits of the different species are described below.

Rosy apple aphid Eggs are laid in the crevices of rough bark on the spurs and smaller branches. In June and July winged aphids are produced and migrate to plantains – the herbaceous hosts. However, colonies may persist on apple and breeding will continue into August if new growth is available on the trees. The return migration from plantains to apple occurs in early autumn.

Apple–grass aphid Eggs are laid on the roughened areas of the spurs and small branches. In the south-east of England, the migration to grasses begins in May and few aphids remain after petal fall (late May to early June). The return migration to apple is in the autumn. In Kent, above-average numbers of migrants and eggs have been observed to follow wet summers favouring the growth of grass.

The green apple aphid lives only on trees, common hosts being apple, pear, quince, hawthorn and rowan. The eggs are laid in autumn on the young wood (Fig. 2.4 shows a heavy infestation). Winged aphids produced in June

Fig.2.3 Colony of rosy apple aphid on underside of apple leaf (× 10)

and July infest the growing tips of young shoots of the same kinds of tree.

Rosy leaf-curling aphid Eggs are laid from mid-June to mid-July and are inserted under flakes of bark or in deep crevices on the main branches and trunk; they are thus well protected. This species occurs only on cultivated and ornamental apples. The life cycle is completed unusually early, the colonies dwindling to nothing in July.

Control in orchards and nurseries

It is most important to obtain control before breeding colonies become established and the leaves or young fruits are injured. This is normally done by spraying an aphicide in spring, usually before or, if infestations are light, just after blossom when hatching is complete. Further sprays may be required during the growing season. As an alternative to spraying in the spring, winter washes can be applied during the dormant period to kill eggs, but they are expensive, less effective than spring aphicides and more difficult to apply. They are rarely used nowadays in commercial orchards.

Spring sprays

Control of apple aphids is usually considered as part of the general strategy of pest control in the orchard rather than as a separate operation because apple sucker (see page 11) and caterpillars of winter and tortrix moths (see pages 153 and 136, respectively) need to be controlled at the same time as aphids. The best time to spray is at late green cluster, by which time the eggs of all these pests have hatched and the immature stages are feeding on the rosette leaves or among the clusters of flower buds.

Fig.2.4 Heavy infestation of eggs of green apple aphid on apple twig (× 2)

Control measures are not necessary for the apple–grass aphid unless at least half the trusses are infested with five or more aphids. A spray should be applied at late green cluster if rosy apple aphid or rosy leaf-curling aphid is present; a second spray after the blossom period is sometimes necessary.

Many insecticides are effective against aphids and other pre-blossom pests, but some are highly selective against aphids. These specific aphicides are useful because they are not toxic to beneficial insects or mites.

Contact insecticides do not give such consistent control of aphids as do systemic insecticides, especially when populations are large. A systemic for aphids can be combined with a contact insecticide to control caterpillars, or a mixed product can be used.

Summer sprays

Infestations sometimes occur after the blossom period, usually because of the omission or failure of earlier sprays or because of reinfestation by winged forms of the green apple aphid. Although this species usually causes most trouble at this time, especially in the nursery and on young trees, the rosy apple aphid and the rosy leaf-curling aphid continue to infest young shoots. Post-blossom sprays against the apple–grass aphid are unnecessary.

Summer infestations, which usually cause much leaf-curling, are effectively controlled by one of the recommended systemic insecticides. The time of application and the required number of post-blossom sprays vary with the season and the degree of infestation. In the nursery it may be necessary to spray at three- to four-week intervals from early June until August. On young orchard trees one spray in late June or early July may be sufficient.

Winter washes

Winter washes can be applied, but success depends on thoroughly wetting the bark on all parts of the tree. Thus a high-volume spray is essential and satisfactory control is usually obtained only by hand spraying. However, winter washes fail to control the rosy leaf-curling aphid because its eggs are laid in protected situations under the bark.

Interplanted strawberries and vegetables are injured by winter sprays and therefore need protection, which can usually be provided on only a limited scale.

Some winter washes kill the eggs of aphids and of apple sucker and also help to clean the trees of mosses, lichens and loose bark. They must be applied while the trees are fully dormant, i.e. from leaf fall until mid-February.

Other winter washes, which are applied slightly later, kill aphid eggs, give partial control of apple sucker and reduce the incidence of apple mildew. Some also give partial control of winter moth (see page 156). These sprays should be applied between early March and the time when the buds are breaking.

Control in gardens

Spring and summer sprays Both the rosy apple and rosy leaf-curling aphids are common on garden trees. If rosy apple aphid is present, it is essential to spray to avoid spoiling the fruit. This, and the other aphid species, can be controlled by a thorough spraying at green cluster with a specific aphicide or with a recommended systemic insecticide. Any colonies appearing later in the year can be 'spot-treated' with one of these insecticides.

Winter washes Alternatively, a winter wash may be applied in the dormant season, but it is expensive and will not control the rosy leaf-curling aphid. Before application, all vegetables and evergreen herbaceous plants should be covered with newspaper or polythene etc. to prevent damage from drift. All parts of the apple tree must be thoroughly wetted by the spray.

3
Black bean aphid

Fig.3.1 Field bean plant infested with black bean aphids

The black bean aphid (*Aphis fabae* Scopoli) or blackfly is a common and widespread pest of field and broad beans, dwarf French and runner beans, sugar beet, mangel, fodder beet, spinach beet and spinach. It also occurs on fat-hen, poppy and dock. Very similar black aphids occur on other cultivated plants and weeds including rhubarb, dahlia, nasturtium, dock and thistle.

Description

The bodies of both the winged and wingless forms of the black bean aphid (Fig. 3.2) are black or brownish-black, sometimes with an olive-green tinge. The legs are much paler and small white patches of wax may be seen on the backs of some of the wingless individuals. The winged founders of the colonies, and the incipient colonies themselves, are rarely noticed unless one searches carefully in the folded young leaves and flower buds among which

Fig.3.2 Adult (winged and wingless) and immature black bean aphids on underside of bean leaf (× 4)

they feed. It is usually several weeks later, when the offspring of the winged founders have multiplied, that growers notice the familiar black 'blight' on the plants (Fig. 3.1).

Life history

The black bean aphid overwinters as an egg on spindle (*Euonymus europaeus*), and on the snowball tree (*Viburnum opulus roseum*) grown in gardens. The eggs hatch from late February to early April. The young aphids feed on the developing buds and on the leaves as they unfold. Attacked leaves become curled, so that heavily infested trees are easily recognized in early May. The aphids that hatch from winter eggs are all wingless females, which reproduce without mating. The offspring are wingless, but the next generation consists almost entirely of winged females which, in late May and early June, fly away to the herbaceous summer hosts. At this time of year, field and broad beans, beet seed crops, sugar beet and mangels are the most important crops colonized. The first aphids to arrive tend to colonize the plants on the headlands to a greater extent than those within the crop.

The rate at which the colonies develop, and therefore the extent of the infestation, depends on several factors of which weather is the most important. In warm, sunny weather development is rapid and the young may become mature in eight days but, if it is cold and wet, maturity may not be reached for 18 days.

Wingless aphids predominate in the early summer but, as the plants become crowded with aphids and the crop matures, more winged aphids are produced. The time at which this occurs varies in different parts of the country and with the degree of infestation. In the earliest districts, peak infestations occur at about the end of June in years when black bean aphid attacks are severe, but about a month later when attacks are slight. The winged aphids from the early colonies migrate within the crop and invade later crops of dwarf French and runner beans, root crops (particularly sugar beet and mangels) and other crops. Peak infestations on these crops normally occur in mid-July in eastern and southern England and up to a month later elsewhere.

In the autumn, winged females fly back to the winter hosts where they give birth to egg-laying females; winged males mate with these

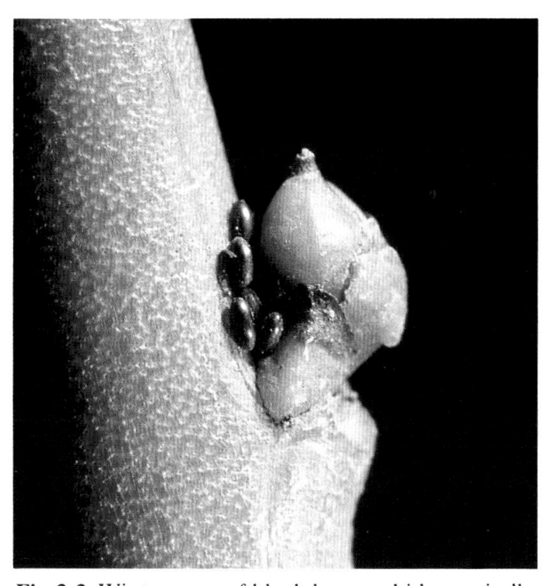

Fig.3.3 Winter eggs of black bean aphid on spindle

females on the winter hosts. Eggs are laid in the axils of buds, in cracks in the bark and, when aphids are numerous, all over the twigs. The eggs (Fig. 3.3) are small and dark green at first but soon turn shiny black.

In mild winters wingless aphids may survive on various hosts including evergreen *Euonymus*, autumn-sown or volunteer beans, and weeds. This occurs more frequently in south and south-west England and south Wales. Winged aphids developing from these colonies may migrate to the summer hosts up to three weeks earlier than those from spindle or *Viburnum*.

Nature of damage

Direct damage

Damage is caused mainly by the loss of sap and by injury to the plant tissues during feeding; large numbers of aphids soon reduce the vigour of the plants. Broad beans are slightly more susceptible to damage than field beans, while spring-sown field beans, particularly those sown late, are much more susceptible than autumn-sown crops. The flowers of beans are damaged and pod development may be retarded or prevented. On beet seed crops and root crops the feeding of the aphids results in leaf-curling, stunted plants and yield reduction.

There is a tendency for serious infestations of black bean aphid to occur in alternate years, so that a severe attack can be expected in a year following little or no infestation.

Virus spread

The black bean aphid is a vector of one of the viruses causing yellows and the virus causing mosaic of beet. The aphid may transmit these viruses from the overwintered beet seed crop to the root crop and the resulting diseases will reduce the yield further.

A more important virus vector — the peach–potato aphid (*Myzus persicae* (Sulzer)) — also occurs on the beet and mangel seed and root crops. This and other green aphids, e.g. vetch aphid (*Megoura viciae* Buckton) and pea aphid (*Acyrthosiphon pisum* (Harris)), move on to beans from clover and other overwintering legumes in May or June, bringing other harmful viruses; the most common of these cause bean leaf roll, bean yellow mosaic and its pea mosaic variant. Control measures used against the black bean aphid will also prevent infestations of green aphids, unless insecticide resistance has developed (see **Root crops,** page 25).

Forecasting attacks

Since 1970, area forecasts of the likelihood of serious infestations on field beans in southern England, East Anglia and the Midlands have been available. Growers should consult their local adviser for up-to-date information.

If the forecast indicates that serious attacks are *probable*, a preventive treatment should be applied to field beans just before flowering. If attacks are forecast as *possible* or *unlikely*, a preventive treatment should be applied only to those crops with more than five per cent of the plants on the south-west headland infested at the beginning of flowering.

Untreated crops should be examined at intervals from the start of flowering (early June) until pod formation (mid-July): favourable conditions may allow aphid numbers to build up to the extent that eradicant treatments become necessary.

Natural enemies

The black bean aphid is attacked by the larvae of certain small parasitic wasps. In addition, ladybird beetles and their larvae, and the larvae of some species of hoverfly, feed on aphids and destroy considerable numbers. However, the reproductive powers of aphids are so great in warm seasons that in certain years these natural enemies only delay and do not prevent the rapid increase of the aphids. Most complete control is given by measures which assist the natural enemies to keep aphid numbers at a low level. Complete eradication of predators by chemicals applied to kill the aphids may allow reinfestation of late-sown crops. In addition, if predators are killed by the insecticide, any surviving aphids will be able to multiply rapidly.

In warm, humid weather the aphids may be attacked by a fungus which causes a great decrease in numbers. The winged aphids, which fly from beans and seed crops to the root crops, can carry the fungus with them and pass the infection to their young before themselves succumbing. The spraying of crops when the aphids have largely succumbed to parasites or fungus is unnecessary and unprofitable.

Control measures

Field beans

Pollinating insects While several insecticides are able to control aphids efficiently, special attention is necessary to ensure that the method of control does not harm pollinating insects, especially bees. Attention to area forecasts and examination of crops at the correct time to ensure that only those at risk are treated will do much to reduce hazards to pollinating insects. If crops do require treatment, correct timing of an appropriate insecticide will prevent or reduce the infestation without harming beneficial insects. Following this procedure can, to a great extent, even out the large annual variations in yield traditional to this crop.

Preventive treatment For a successful preventive treatment it is necessary to use an aphicide that is relatively persistent. Granular insecticides applied to the foliage are more persistent than sprays and are recommended for use just before flowering when aphid attacks are likely.

Sometimes in Midland counties, where aphids arrive later, only the headland becomes sufficiently infested to require treatment. Treating a 9-m swathe around the edge of the field should prevent damage to the crop, but it should not be assumed that this treatment will prevent an infestation in the rest of the field.

Where ground equipment is used, the earlier the treatment the less the wheel damage to the crop, so a preventive treatment is advantageous in this respect. Another advantage of early treatment is that insecticide application before flowering safeguards pollinating insects.

Eradicant treatment Where no preventive treatment has been applied, an eradicant treatment may become necessary if aphid colonies develop. A late treatment will coincide with the flowering period of the crop, so that danger to pollinating insects should be one of the main considerations when deciding which of the effective aphicides to use at this time.

Several insecticides are recommended for use as sprays to control black bean aphid. Some of these can be used without undue risk to bees, but with most insecticides this depends on applying them only in late evening and, if possible, in dull weather. Application should be made in at least 450 litres of water per hectare.

Granular insecticides may be used as an eradicant treatment and are much less dangerous to pollinating insects than are sprays. In very dry conditions granules may act more slowly.

An eradicant treatment should be applied by high-clearance equipment with narrow wheels or by aircraft.

Broad beans and beans for seed

These are more valuable crops and a preventive treatment should be applied unless the area forecast is *unlikely*.

Winter beans Treatment of winter beans is not generally worth while because they are relatively unattractive to aphids flying into the crop and are not, therefore, heavily colonized.

Pollinating insects The crop needs maximum pollination by wild bees or honey bees if it is to yield to the full. Lack of pollination will make the crop taller (as pod set will be postponed), later to harvest and more prone to lodge in summer storms. Bruised haulm in wind-damaged crops is more susceptible to attack by chocolate spot disease.

Wild bees can be sufficiently numerous to pollinate small fields (up to four hectares), but it is desirable to import honey-bee colonies to larger fields. However, beekeepers are reluctant to move hives to a crop if there is a serious risk of bees being poisoned by sprays. Treatment before bloom safeguards both wild and honey bees.

Runner and dwarf French beans

These are generally attacked later than field beans. The granular insecticides are too persistent for use on runner and dwarf French beans after sowing but can be used at sowing in a band 25 mm below the seed. This treatment is not, however, recommended on very light soils as leaf scorch may occur.

The overlapping of podding with the prolonged flowering period restricts the choice of sprays to the less persistent ones. All sprays should be applied in the late evening to safeguard pollinating insects: the visits of these insects increase yields and advance the picking of runner beans by several days. The optimum time for spraying is when infestations are beginning to become established or just before flowering, whichever is the earlier.

Beet, mangel and spinach seed crops

These crops need protection from aphids, especially the peach–potato aphid, in late summer and autumn. Therefore, seed-furrow treatments of a granular insecticide are usually applied to crops sown in late July and in August; such treatments control black aphids also.

In the year of seed production it is necessary to protect the crops from damage by black aphids, as well as to control the spread of yellows to beet and mangel root crops by the peach–potato aphid. Foliage-applied insecticide granules are the most effective, but recommended sprays will also give good control. Although these crops are less attractive to bees than are beans, hazards occur when bees either work the crop or fly over it. Therefore, sprays are best applied in the evening and in dull weather.

Treatment should be applied early, before damage occurs.It is wise to inspect crops from mid-May onwards. Only about 4.5 m all round the crop need be examined, but the examination at the growing tips of the plants should be thorough. Treatment of the field is likely to be worth while when aphids are on any headland plants in late May or on half the plants in mid-June. If the infestation is very patchy it may only be necessary to treat the patches, allowing insect predators to deal with scattered infestations elsewhere.

Contractors using high-clearance machines spray tall seed-crops efficiently. Aerial application can be effective, but over-run on to nearby flowering plants, such as clover, can cause hazards to bees.

Root crops

In most seasons control of black bean aphid has to be considered with that of the peach–potato aphid and virus yellows. However, in some seasons, especially in the northern and western beet-growing areas, control is aimed solely at black bean aphid. Spray warnings, based on aphid populations and the stage of crop growth in the area, are issued and should be heeded.

Insecticides are usually applied to control both aphid species. In most areas peach–potato aphid is at least partially resistant to organophosphorus sprays (see page 64). Therefore, sprays of certain organophosphorus insecticides cannot be recommended against black bean aphid as such treatment is likely to give only partial control of peach–potato aphid.

In gardens and allotments

Many gardeners sow their garden beans early, so that plants are well grown by the time the black bean aphids arrive. Normally, infestations start at the top of the plant and spread downwards. If the infested tips are pinched out early and destroyed, the spread of the aphids will be checked and the bean clusters below will develop better. This is a useful preventive measure but is rarely successful in keeping the crop entirely free from attack throughout the season.

Several insecticides which will give good control of aphids are available for garden use.

4
Cabbage aphid

Fig.4.1 Cabbage aphids on Brussels sprout plant

Cabbage aphid (*Brevicoryne brassicae* (L.)) is a common and destructive pest which in some seasons causes serious losses of broccoli, Brussels sprouts, cabbage, cauliflower, oilseed rape and swede. By the time a crop has become heavily infested serious injury will already have occurred, so attacks must be checked early.

This aphid can also transmit the virus that causes cauliflower mosaic disease.

Various wild plants, including charlock, shepherd's purse and wild radish, are also hosts of the cabbage aphid.

Recognition and damage

The wingless aphids are bluish-grey with a mealy covering and are usually found in colonies on the underside of the leaves (Figs 4.1 and 4.3).

The first signs of attack are small, bleached areas on the leaves. Infested leaves become yellow, curled and more or less crumpled. In severe attacks they are almost destroyed, but even lightly infested leaves never recover completely. The effects of infestation are worst on young plants, which may be stunted and, in unfavourable weather, may die.

Infestations should not be allowed to spread to the hearts of broccoli or cabbage, or the buttons of Brussels sprout near harvest time, because the presence of even dead aphids is enough to reduce the market value of the produce.

Life history

Small, elongate, shiny black eggs are laid in the autumn on the stems and leaves of broccoli, Brussels sprout, cabbage and kale. The winter is normally spent in the egg stage (Fig. 4.2). The adult aphids live on well into the winter and, in mild districts or mild seasons, small colonies persist until the spring. In most years in the south and south-west of England, few eggs are laid and the aphids overwinter mainly as active stages in the more sheltered parts of brassica plants. However, after severe winters most of the surviving population will be derived from overwintering eggs. Egg hatch is completed between the end of February and the end of April. The young aphids feed on the leaves and shoots, but by May they have largely migrated to the flower buds and stems, few remaining on the leaves. The aphids continue to live on plants left to flower for the production of seed or from negligence. The stems of Brussels sprouts etc. that have survived ploughing-in or heaping-up may also harbour the pest.

From the end of May to July, the colonies on old Brussels sprouts, cabbage plants, oilseed rape, etc. produce winged aphids. These fly away and infest newly planted Brussels sprouts and other brassicas. They are at first inconspicuous on the new crop, but in favourable weather they multiply rapidly from August to October.

Aphid numbers

Aphid numbers vary widely from year to year. This variation may be partly due to attacks by various insect enemies, but the weather may have a greater influence: warm, dry weather in the second half of summer favours a rapid build-up, while cool, wet weather hinders it.

Prevention and control on commercial crops

As the cabbage aphid is a carrier of cauliflower mosaic virus, the control of the aphid will help to reduce the spread of the disease. Another carrier — the peach–potato aphid (see pages 52 and 60) — which is often numerous on brassica plants, will also be checked by these control measures.

Sources of infestation

The most effective control measure is to eliminate the sources of infestation before the pest can spread to the new crops. All brassica crops should be destroyed before the beginning of May by rotavation or discing, followed by ploughing to bury the plant debris. Alternatively the plants can be heaped and burnt.

Examination of plants

Cabbage aphid infestations build up rapidly but sporadically and locally. It is important to examine plants regularly. When examining brassica plants, small bleached areas and crumpled leaves should be looked for and a search made for pale bluish, powdery aphids on the underside of the leaves and especially in the heart of young plants.

General control measures

The programme should be planned to give effective control without the need for emergency clean-up procedures, which are seldom satisfactory. The importance of regular examination of plants in seedbeds and in the field to ensure the optimum timing of control measures cannot be over-emphasized.

Systemic insecticides, which are absorbed and translocated within the plant, are particularly effective for cabbage aphid control; these are available either as granules or as liquid concentrates for application as sprays. The choice of chemical depends on the period of protection required, the interval before harvest, the equipment available for application, the cost and, for granules, the maximum quantity allowed on the crop in one season. If very late applications are necessary near harvest, a spray of short persistence or a chemical with fumigant action should be used.

Granules For maximum effectiveness, granules should not be used before mid-May. Some granules should be applied only at transplanting, whereas others are also suitable for application to direct-drilled crops. Some can be incorporated in the peat used for peat blocks. Others can be applied also as a foliar treatment later in the season.

Granules have the advantage that only very small amounts are applied per unit area, but equipment capable of accurate metering of these quantities is necessary.

Certain granules applied against cabbage aphid will also give some control of cabbage

Fig.4.2 Eggs of cabbage aphid on Brussels sprout stem

Fig.4.3 Colony of cabbage aphid on underside of cabbage leaf (× 8)

root fly (see page 242). In addition, there are mixed granular formulations that are intended primarily for the control of cabbage root fly but will give useful early control of cabbage aphid.

Sprays The application of sprays must be really efficient. Enough spray should be applied just to wet the foliage: on established crops 500–1000 litres/ha are necessary depending on crop size and density. Additional wetting agent may be required to ensure wetting of the waxy aphids and leaves.

Period of protection Granules give the longest period of protection, usually about six weeks. Sprays persist for about 10 days or less.

The effectiveness and persistence of all systemic insecticides may, however, be reduced in drought conditions because systemic action is lost. Where the plants cannot be kept growing by irrigation, control by sprays can be improved by increasing the volume used per unit area, adding a wetting agent and spraying in the evening. Better control in drought conditions can also be obtained by using an aphicide which has a powerful fumigant action; for maximum effect such a chemical should be applied when the air is still.

Brussels sprouts

This crop is at risk for an extended period. Special care should be taken to ensure maximum efficiency of control measures to prevent infestation of the buttons.

Occasionally, heavy infestations build up on early spring-planted crops in early May, but the main danger period is June and early July, when winged aphids are most abundant and are infesting new crops. Therefore, crops sown or planted without soil application of granules should be examined frequently from mid-May to mid-September, and crops treated with granules at sowing or planting should be checked at frequent intervals from five or six weeks after sowing or planting until mid-September.

Treatment should be applied if a substantial proportion of plants are infested, even if colonies are very small. Alternative programmes are as follows.

(i) For planting or direct drilling after mid-May, granules can be applied to the soil, followed by foliar sprays as necessary and/or a foliar application of certain granules as late as possible but before the buttons become infested and not later than six weeks before harvest.

(ii) If the soil has not been treated with granules, sprays should be applied as often as necessary. The whole programme can be based on sprays alone, or on sprays with a foliar application of certain granules as a final treatment.

Programme (i) is likely to give the best results.

For eradication of infestations in August and September, pendant lances are essential to improve spray coverage of the lower leaves and buttons, as it is impossible to reach these with overhead applications.

Cabbage and cauliflower

For crops transplanted or direct drilled after mid-May, soil treatments can be used as for Brussels sprouts (see above). Alternatively, sprays or foliar applications of certain granules may be preferred. Again, the aim should be to treat during the early stages of aphid colony formation before plant damage and distortion occur.

Brassica seed crops

These should be examined from mid-April to May and sprayed promptly if potentially damaging infestations are found.

Oilseed rape (see also page 182). Control measures are not usually necessary, but in exceptionally warm, dry summers heavy infestations can develop on flowering heads, causing shrivelling of pods and serious loss of yield. Spring rape is more liable to suffer economic damage than the earlier-maturing winter varieties. Aphid attack while the crop is still flowering is more important than later infestations.

Protection of bees Flowering brassica seed

crops, including oilseed rape, are extremely attractive to bees. If sprays are necessary, every effort should be made to minimize bee losses: an insecticide of low toxicity to bees should be used, sprays should be applied in the early morning or late evening, and local bee-keepers should be given early warning that spraying will occur. For early infestations in kale seed crops, a granular application of a recommended insecticide can give good control without endangering bees.

Brassica seedbeds

These should be sited well away from seed crops to prevent them from becoming infested by aphids dispersing from the crops. Infested beds should be sprayed with an aphicide as necessary so that plants are clean when transferred to the field. Some insecticides applied for cabbage root fly control also give protection against early aphids.

Turnip and swede

Turnips are not attacked by cabbage aphid, but heavy infestations may build up on swedes. Infestations on swedes can cause considerable loss of yield and may be worth controlling, even on those grown for fodder.

Prevention and control in gardens and allotments

All old broccoli, Brussels sprouts,cabbage and other brassica plants should be pulled up and burnt or otherwise destroyed before mid-May.

Before plants are set out, they should be examined carefully. If there is any sign of aphids, the plants should be dipped in the spray fluid of a recommended insecticide. Good quality rubber or PVC gloves must be worn when dipping or handling treated plants.

When aphids are found on growing plants, a recommended insecticide spray should be applied *immediately*. Particular attention should be paid to the under surface of the leaves.

5
Capsid bugs on fruit

Fig.5.1 Apples damaged by common green capsid

Several species of capsid bug occur on fruit. The majority are beneficial and prey on pests such as spider mites, apple sucker and leafhoppers, but the two species described here are plant feeders and cause injury to many kinds of fruit and ornamental plants. The apple capsid (*Plesiocoris rugicollis* (Fallén)), formerly a major pest of apple in England, was eradicated from commercial orchards soon after DDT came into general use in the late 1940s; it is now only an occasional pest of this crop in gardens and unsprayed orchards. The common green capsid (*Lygocoris pabulinus* (L.)) is a much more common species and attacks cane and bush fruits, occasionally strawberry, a wide range of shrubs and herbaceous plants and, to a lesser extent, apple, pear and other tree fruits.

Recognition and damage

Both species are bright green, resembling the colour of the foliage on which they live. They are very similar in appearance and can be

Fig.5.2 Adult apple capsid (× 12)

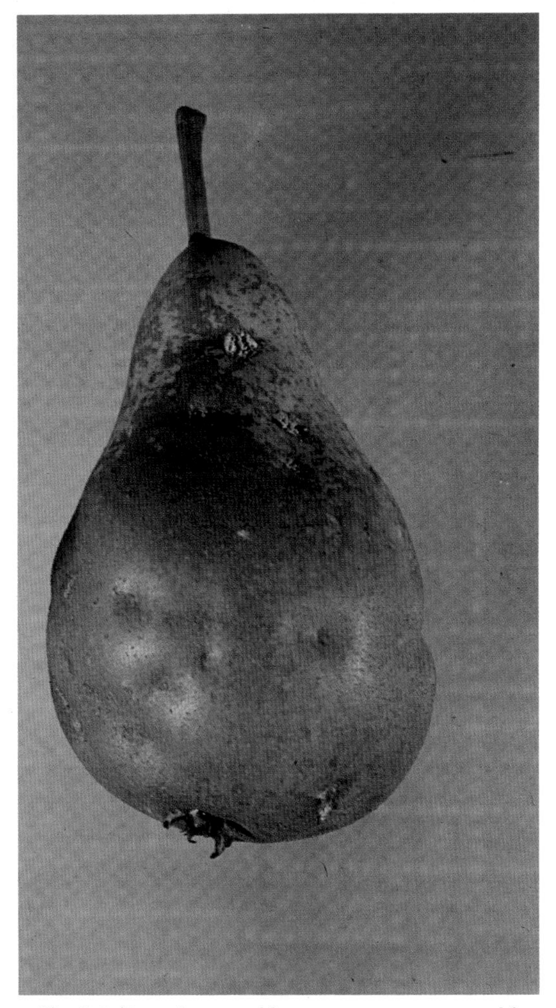

Fig.5.3 Pear damaged by common green capsid

distinguished only by small differences in structural characters. The immature stages (nymphs), apart from their smaller size and absence of wings, are similar in shape and colour to the adults. The mature insect (Fig. 5.2) is about 5 mm long, with pale green wings which are translucent near the tip and are held horizontally on the body when at rest. All stages display great activity when disturbed, running quickly over a leaf to hide on the other side or, in the case of adults, taking to flight.

Capsid bugs possess hollow stylet-like mouthparts with which they pierce plant tissue and suck the sap from the cells. Their favourite feeding sites are those where young growth is present, such as unfolding leaves, shoot tips, flower buds and fruitlets. The cells immediately surrounding each feeding puncture are killed by a toxic substance injected from the salivary glands. The first sign of attack is the appearance of small brown spots on the young leaves, each marking a feeding puncture. As the leaf grows, only the surrounding healthy cells expand and a hole develops at the feeding site, often with some deformation in leaf shape (Fig. 5.5). Young shoots are also attacked, leading to a check in growth and malformation, or the terminal bud may be killed, causing the growth of an excessive number of side shoots and resulting in a badly shaped, stunted tree or bush.

Fruits are also attacked. At first, small brown spots appear on the fruitlets. On apples

Fig.5.4 Black currant foliage severely damaged by common green capsid

these develop into irregular corky areas or discrete corky pimples (Fig. 5.1), depending on the variety, and the fruits become malformed. On pears each feeding puncture often remains separate, appearing on mature fruits as a small depression or eruption of the skin with a rough corky surface (Fig. 5.3); in severe attacks the fruits may become badly malformed with depressions replacing the skin eruptions. Blemished fruits are of little or no market value. On strawberry, attack by the first generation of the common green capsid causes malformed fruits which resemble those resulting from poor pollination.

Because both species of capsid bug cause the same type of injury to leaves, shoots and fruit it is not possible to identify the species from the symptoms. As a general rule injury to apple may be caused by either species, whereas the common green capsid is usually responsible for damage to all other kinds of plant (e.g. Fig. 5.4).

Life history

Common green capsid

This species has two generations each year. It overwinters as eggs laid in crevices in the bark, or under the bark, of various woody hosts. Young tender shoots are the preferred egg-laying sites; rootstock suckers are especially prone to attack. Hatching begins just after mid-April and continues until mid-May. In relation to bud development of apple (varieties Bramley's Seedling and Cox's Orange Pippin), the first eggs hatch shortly after green cluster, about 50% have hatched by the beginning of the blossom period, and hatching is complete by petal fall. On the black currant variety Baldwin, hatching coincides with the blossom period.

The immature stages feed on leaves at the tip of young shoots of cane and bush fruits; on apple and pear the rosette leaves of the blossom trusses are first attacked and later the young shoots and fruitlets.

In general the infestation of fruit trees and bushes ends when adults of the first generation appear during June. Some of these adults lay eggs in the current season's canes of blackberry, loganberry and raspberry; the eggs hatch after about two weeks, giving rise to a second generation on these plants. Other adults leave the plants on which they matured and migrate to a wide range of annual and herbaceous perennial species such as potato, dock, fat-hen, nettle and thistle. From eggs

Fig.5.5 Apple foliage damaged by apple capsid

laid in the stems and petioles a second generation develops on these plants and gives rise to adults from mid-August onwards. These adults return to the woody plants, including fruit trees and bushes, to lay eggs which form the overwintering stage.

Apple capsid

Unlike the common green capsid, the apple capsid has only one generation each year. Eggs, laid in crevices in the bark of shoots or branches, have begun to hatch by the green cluster stage, about mid-April, and hatching is completed by early May. Thus hatching is slightly earlier than for the common green capsid.

The young capsid bugs first feed on the rosette leaves surrounding the blossom trusses, where the brown feeding marks can be clearly seen. Later they feed on the developing fruitlets and the new foliage arising from the spurs and the shoots. Adults appear between late May and mid-June and live for about four to six weeks; active life ends with the laying of eggs during late June and July.

The original host plants of the apple capsid were various species of willow and bog myrtle, on which it now occurs sparsely. Among fruit trees and bushes this species is found mainly on apple but will also attack red and black currants and gooseberry.

Control in orchards and plantations

Common green capsid

This species causes sporadic damage in many orchards in southern England. Where capsids move into the orchard from neighbouring shelter (nettlebeds, gardens, etc.), barrier spraying of 20–30 m along the nearest headland may make the spraying of the whole orchard unnecessary.

Control measures must be directed against the young capsids. These hatch during a three- to four-week period which coincides, at least in part, with the blossom periods of bush and tree fruits. The blossom period is a time when insecticides must not be applied because of the risk of killing pollinating insects and thus reducing fruit set.

On apple and pear, it is essential to eliminate all young capsid bugs by the end of the hatching period so as to prevent injury to the fruitlets; this injury is liable to occur immediately after the blossom period. A recommended insecticide spray should, therefore, be applied without delay at petal fall in orchards where there is a risk of this pest occurring. Honey bees sometimes continue working the flowers of apple after the petals have dropped; if this is occurring the spray should be applied on a dull day.

Currants and gooseberry should be sprayed at the end of the blossom period.

On strawberry, the timing of treatment is critical as malformed fruits can develop from blossom buds or flowers which have been attacked. A spray should be applied immediately before the flowers open. A second spray may be needed on this crop as soon as flowering has ended, but damage to strawberry is infrequent nowadays.

On raspberry and loganberry, a spray should be applied just before the flowers open; this coincides with the end of the hatching period.

Apple capsid

A recommended insecticide should be sprayed at the green cluster stage. This spray will also control caterpillars of winter moth and fruit tree tortrix moths (see pages 156 and 140).

Control in gardens

Some insecticides recommended for control of capsid bugs are suitable for use in gardens. They should be applied at the times given above.

6
Cereal aphids

Fig.6.1 Grain aphids on wheat ear

Five species of aphid (greenfly) occur commonly on cereals in Britain between April and September: the bird-cherry aphid (*Rhopalosiphum padi* (L.)), the rose–grain aphid (*Metopolophium dirhodum* (Walker)), the grain aphid (*Sitobion avenae* (Fabricius)), the blackberry–cereal aphid (*Sitobion fragariae* (Walker)) and the fescue (or grass) aphid (*Metopolophium festucae* (Theobald)). The cereal leaf aphid (*Rhopalosiphum maidis* (Fitch)) has been recorded occasionally on barley leaves. The apple–grass aphid (*Rhopalosiphum insertum* (Walker)) forms colonies on cereal roots and may assume importance in drought conditions. Vagrant winged forms of other aphids may also settle on cereals for short periods but do not reproduce there.

Description and life history

The bird-cherry aphid (Fig. 6.2) is a small, brown to brownish-olive green aphid with rusty-red patches at the hind end of the body. Like many aphid species, it overwinters in the egg stage on a woody host, in this case the bird-cherry (*Prunus padus*); this is a common wild plant in the north of England but is only found as a cultivated plant in parks and gardens in southern England. In mild winters this aphid may survive in the 'summer' (adult) form on grasses and on autumn-sown cereals, where it is often found on the shoot just below the soil surface. In late May and June the aphid is usually found on the lower leaves of cereals, particularly oats and barley, and on grass seed crops (see page 195). When very numerous it may spread all over the aerial parts of the plants. It shows a preference for sheltered areas. On oats, it is often found inside the leaf sheath and, on maize, inside the leaves sheathing the cobs.

The rose–grain aphid (Fig. 6.3) is a medium-sized, ovate, light green or, rarely, pink aphid; green specimens have a darker, bright green stripe down the centre of the back. Eggs are laid on wild and garden roses in October and November; winged forms produced in the spring migrate to cereals and grasses. This aphid is usually seen on the lower leaves of cereals, particularly wheat, at about the time the cereals are coming into ear, but individuals also collect on the top leaves in the later growth stages of the plant. The species can also overwinter in the adult stage on grasses. When living on grasses it prefers moist situations.

The grain aphid (Figs. 6.1 and 6.4) is a large, long-legged, yellowish-green, green or reddish-brown to nearly black aphid. It has no alternate woody winter host. All stages of the life cycle, including the winter eggs, are spent on grasses and cereals. It can overwinter in the adult stage (as wingless or winged females). It is rarely numerous on cereals until late June but can then cause farmers much concern when large numbers are seen on the upper leaves and emerging ears.

The blackberry–cereal aphid (Fig. 6.5) is very like the grain aphid in size and shape, but its colour varies little from dull green. It overwinters in the egg stage on bramble (*Rubus* species), migrating in June and July to cereals and grasses where it is generally seen on the ears. It is considered of much less importance than the grain aphid.

The fescue aphid (Fig. 6.6) resembles the rose–grain aphid but does not have a darker stripe along the back and is more yellowish. The entire life cycle, including the overwintering egg, is passed on grasses and cereals. During mild winters the aphid may survive in the adult form on grasses.

Aphid numbers

Aphid build-up is known to differ on the various cereals, indeed on different varieties of a cereal. Numbers vary considerably from year to year and some of this variation can be explained by the weather in May and early June. Heavy rain or very high temperatures can cause a sudden collapse of an infestation. Overwintering behaviour, parasites, predators and fungus disease can have a marked effect on aphid numbers.

Nature of damage

Aphids colonize cereals in May and June and may become sufficiently numerous to cause direct feeding damage. They also colonize newly emerged seedlings in September and October and then it is their role as virus vectors which is important, i.e. they infect seedlings with the virus and also transmit it from infected to healthy plants.

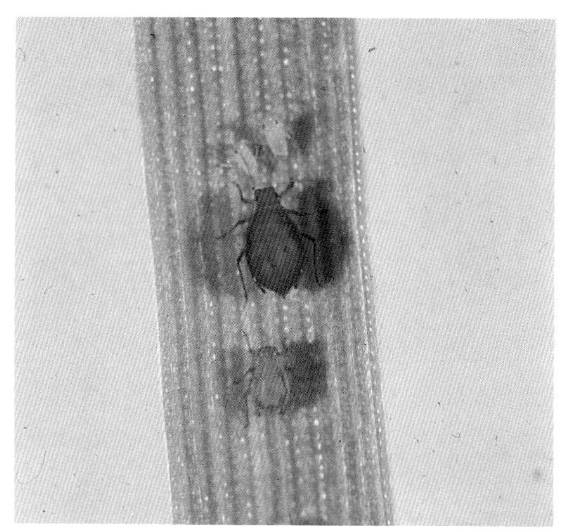

Fig.6.2 Bird-cherry aphids: adult and immature stages (× 7)

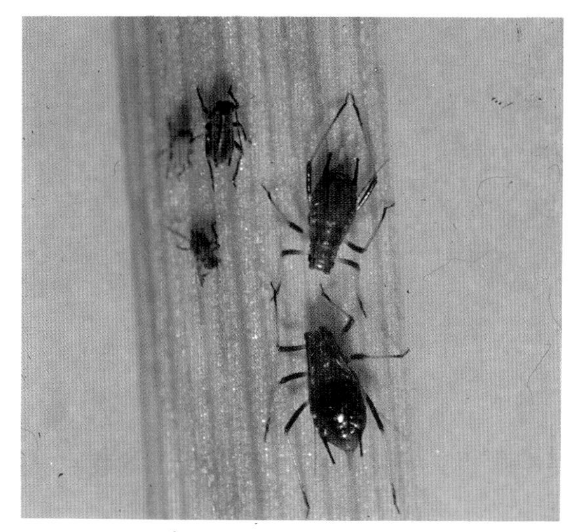

Fig.6.3 Rose-grain aphids: adults (× 7)

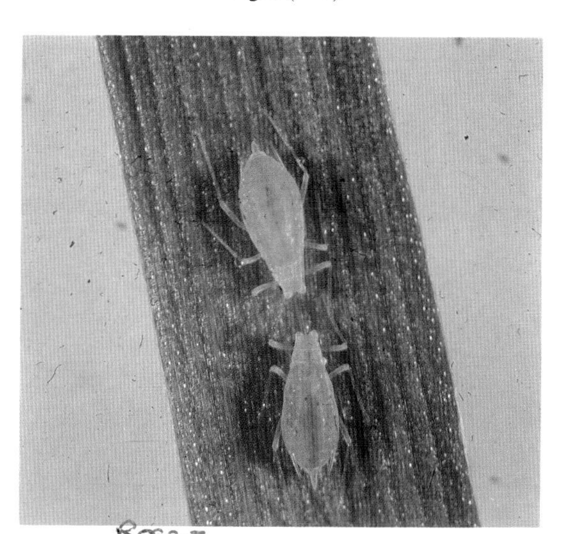

Fig.6.4 Grain aphids: adult and immature stages (× 7)

Fig.6.5 Blackberry–cereal aphids: adult (winged and wingless) and immature stages (× 7)

Feeding damage

Direct injury to cereals is caused during feeding either by the aphids sucking sap from the tissues, so removing water and nutrients, or by the injection of toxic substances present in aphid saliva. In addition, during feeding, aphids produce honeydew, which makes the ears sticky and creates a suitable medium for the development of sooty moulds. Honeydew is also a sugar source for many flies, including pest species such as wheat bulb fly.

Although spending much of its life on grass hosts, the grain aphid causes concern when it is seen on the leaves, stems and ears of cereals. Winged females move to cereals in midsummer and produce colonies of small, red-brown or green, wingless aphids which feed on the leaves, stems and developing ears. The number of aphids present on the ear in relation to the stage of ripening determines the extent of damage. Early attacks, or attacks on backward or late-maturing crops before grain formation, cause blindness or shrunken kernels

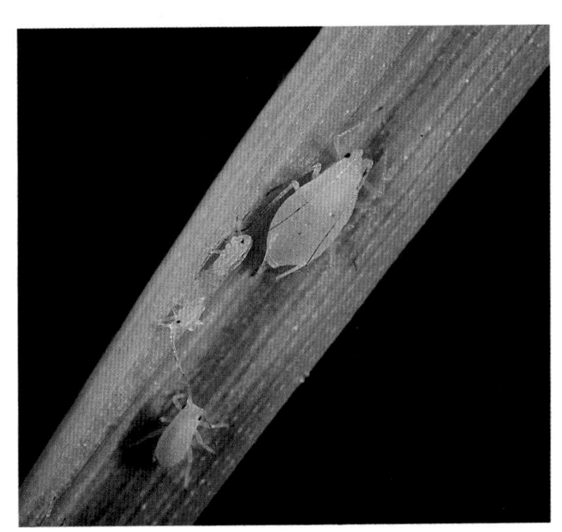

Fig.6.6 Fescue aphids: adult and immature stages (× 7)

(decreasing the number of grains or yield per ear, respectively). An attack of grain aphid when the grain is nearly ripe is of little importance.

The rose–grain aphid normally feeds on the lower leaves of cereals and comparatively little is known about grain losses caused by this aphid. However, when severe outbreaks occur, as in 1979, very large numbers are found on all leaves and their presence from ear emergence onwards can reduce yield.

Virus spread

In the autumn there is a movement of cereal aphids back to their winter hosts. These migrations often start before the winter cereals have emerged. Winged bird-cherry aphids, however, are often found in September and October and infest young cereal crops. Damage can be severe and may spread until cold weather stops aphid movement. Winged grain and fescue aphids may also move from grasses to winter cereals when making short flights in November and December. Survival of the aphids is good in the west and south, particularly on forward dense crops, but is poor in the north and east.

Barley yellow dwarf virus (BYDV) is the most important virus affecting cereals and is widespread in grasses. It occurs in oats, wheat, barley, maize and most grass species. Severe strains stunt plants, inhibit root formation, delay or prevent heading, and reduce yield. These strains are most efficiently transmitted by the bird-cherry aphid. Mild strains, which are most efficiently spread by the grain and rose–grain aphids, also cause yield loss, but effects are usually less than those of severe strains. Grasses are an important reservoir of the virus.

The incidence of BYDV varies from season to season (as do cereal aphid numbers), although September-sown crops in many parts of Britain may be at risk in most years. In some areas of intensive cereal-growing, barley, wheat and oats sown before mid-October are particularly at risk. These areas include the coastal parts of southern England and most of south-west England and south Wales.

Autumn infections of BYDV usually show up in the spring as discrete patches 2-3 m across, mostly in cereals in arable rotations; patches are often much more extensive in cereals grown after ploughed-out grass. The presence of large numbers of volunteers, especially of winter barley, and grassy stubbles may also increase the risk of virus infection. Summer infections are generally confined to scattered plants.

Most strains induce the following colour changes:

Oats: reddish-purple leaf colouring (Fig. 6.7), sometimes with yellowing, stiffening and ear blasting.

Wheat: leaf reddening and yellowing, especially toward the tips, with associated necrosis. Affected leaves have an upright, stiff appearance.

Barley: chrome-yellow discolorations starting at the leaf tips and margins, but soon affecting the whole leaf. Leaves may have alternating pale and dark green stripes along their length. In some winter barley varieties the awns of affected plants may be coloured purple.

All cereal varieties are susceptible to infection but some show a degree of tolerance, i.e. although they may become infected by the

Fig.6.7 Winter oats (right) severely affected by bird-cherry aphid and yellow dwarf virus compared with healthy plant (left)

virus, their growth and yield are less seriously affected. Wheat is more tolerant to BYDV than are oats or barley. BYDV symptoms in barley are similar to those produced by nitrogen deficiency, acidity or waterlogging. Virus-infected plants, however, remain stunted even under optimum growing conditions.

Not all cereal aphids transmit all virus strains. Many species can transmit BYDV, but in Britain only three species are thought to be important. The bird-cherry aphid is most important in southern England, while the grain aphid is more important in northern England. The rose–grain aphid is not an efficient vector in small numbers but will cause within-crop spread in some seasons.

Normally, 16–24 hours' feeding is needed for the aphid to acquire the virus, although 15 minutes have sufficed experimentally. A virus-carrying aphid usually takes 4–8 hours to infect a healthy plant. Once infected, aphids carry the virus for life, but they do not pass it on to their offspring.

Mild winters favour survival of adult bird-cherry and grain aphids, so that when winged stages are produced on grasses or winter cereals in mid-May (bird-cherry aphid) or later (grain aphid) a relatively high proportion may already be carrying the virus. It is from this source that infection is carried into spring cereals. Cereal aphids may spread virus from ploughed-up grassland to succeeding cereal crops; if the grass has not been destroyed well before sowing, the aphids may survive on it until the seedling plants have emerged.

The earlier that crops are sown in the autumn the more likely they are to be infected with BYDV, irrespective of where they are grown. Crops sown after mid-October are rarely infected in the autumn.

Control measures

Autumn

Where autumn infection by BYDV is a problem, i.e. on early-sown crops or in virus-prone areas, spraying with a recommended persistent aphicide when the aphid migration has ended will reduce the spread of virus within the crop. The optimum timing for such treatment is the last week in October or the first week in November, except on crops drilled very early in September when a spray applied in mid-October gives better results.

In areas where ploughed grassland is an important source of infection, special precautions are needed to ensure that aphids do not transfer to the new crop. Suitable treatments include applying a desiccant herbicide spray to the grass before ploughing or an insecticide spray to the cereal crop at emergence.

Summer

The main problem occurs when the grain aphid colonizes the ears of winter wheat during flowering. Experimental work suggests that, if the weather favours an increase in the aphid population, treatment is justified when two-thirds of the total number of ears in the field have one or more aphids on them at the beginning of flowering. Treatment at this stage can prevent a potential loss of up to 20% of yield, but sprays applied at later growth stages generally result in smaller, insignificant yield responses. Since predators and parasites help to regulate aphid populations, an insecticide which has the least harmful effect on beneficial insects should be chosen whenever possible.

7
Currant and gooseberry aphids

Fig.7.1 Black currant shoot damaged by currant–sowthistle aphid

Several species of aphid (greenfly) infest the foliage and young shoots of currants and gooseberry in Britain and two species live on the roots. The most important species frequently cause much crumpling of the leaves and stunting of the young shoots during spring and early summer. Severe attacks may result in premature leaf-fall and small fruit. Both foliage and fruit may become sticky with honeydew excreted by the aphids and dirty with sooty moulds that grow on the honeydew.

Cucumber mosaic virus in black currant and red currant is transmitted by the first six species of aphid mentioned below and also by other aphid species that only visit currants casually. Gooseberry vein-banding virus is transmitted only by currant and gooseberry aphids.

Life history, recognition and damage

The currant–sowthistle aphid (*Hyperomyzus lactucae* (L.)), the currant–lettuce aphid (*Nasonovia ribisnigri* (Mosley)), the black currant aphid (*Cryptomyzus galeopsidis* (Kaltenbach)), the red currant blister aphid (*Cryptomyzus ribis* (L.)), the gooseberry aphid (*Aphis grossulariae* Kaltenbach) and the permanent currant aphid (*Aphis schneideri* (Börner)) hibernate in the egg stage on currant or gooseberry bushes. The eggs are laid either at the bases of the buds or along the shoots where loose rind or cracks are present; after a few days the eggs turn shiny black.

Hatching begins when the buds are breaking and is complete before any blossoms open. The first two generations usually consist only of wingless aphids. Winged forms occur in subsequent generations. Winged aphids of some of the species migrate to other host plants during the summer. After several generations on these plants, aphids fly back to currant or gooseberry bushes in early autumn; the sexual forms mate and overwintering eggs are laid. The gooseberry aphid, two forms of the black currant aphid, and the permanent currant aphid are present on fruit bushes throughout the year.

The currant root aphid (*Eriosoma ulmi* (L.)) and the gooseberry root aphid (*Eriosoma grossulariae* (Schüle)) hibernate in the egg stage on elm trees and live on the roots of currants and gooseberry respectively during the summer.

The currant–sowthistle aphid is the most common species on black currant. It is also found occasionally on red currant. Both wingless and winged forms are green, the latter having blackish markings.

In March and early April aphids hatch from eggs and crawl to the nearest fruit buds, where they feed on the young growth protruding beyond the tips of the bud scales. Later they move into the compact clusters of blossom buds; there, many are protected by the bud scales until these are shed at the late grape stage. Subsequent generations mostly colonize the tips of young shoots, causing a yellow mottling and down-curling of the leaves (Fig. 7.1) and often stunting the shoots. The third generation is winged and migrates in late May and June to sowthistles, where breeding continues throughout the summer. In autumn, winged aphids return to currants and produce a generation of wingless, egg-laying females. These mate with winged males which fly back from sowthistles, and overwintering eggs are laid.

The currant–lettuce aphid resembles the currant–sowthistle aphid except that it is darker green and shiny. It occurs on gooseberry, and occasionally red currant, where it infests the tips of the young shoots in spring, causing slight leaf-curl and shortening of the internodes but not mottling of the leaves. The summer generations are serious pests of lettuce (see page 52) on which crop some aphids overwinter instead of migrating back to gooseberry.

The black currant aphid is fragile in appearance and whitish or cream-coloured. There are three forms: one migrates from black currant to hemp nettle in early summer, another lives on black currant throughout the year, and the third lives entirely on red and white currants. These aphids live on the underside of mature leaves without causing any deformation, but occasionally populations become large in June with the result that fruit and foliage are contaminated by honeydew.

The red currant blister aphid is similar to the black currant aphid in appearance and size, but it is easily distinguished by the blisters that appear on infested leaves (Fig. 7.2). These are red to purple on red and white currants and yellowish-green on black currant. Winged aphids migrate to hedge woundwort in June and July, where breeding continues until autumn when the return migration to currant occurs.

The gooseberry aphid is green, or greyish-green, with a waxy bloom. It lives on currants and gooseberry throughout the growing season and also on plants of the willowherb family in summer. The aphids hatch at the same time as the currant–sowthistle aphid. From May onwards their presence can be detected by slight leaf-curling; dense colonies of aphids can be found on the terminal part of the shoots.

The permanent currant aphid is blue-green with a waxy bloom. It is locally common on currants, where it lives in dense colonies on the shoots, causing more severe distortion of the foliage than the gooseberry aphid: each leaf is bent down at its junction with the petiole and there is a tight mass of foliage near the shoot tip.

The currant root aphid attacks the roots of currants. It is light red to brownish-red and covered with a whitish wool-like secretion. Large infestations of this species may retard the growth of young bushes. In autumn, winged aphids migrate to elm where eggs are laid. These hatch in the spring and colonies develop on leaves, which become blistered and rolled. Winged aphids reinfest currants in late spring.

The gooseberry root aphid is similar in appearance, size and habit to the currant root aphid except that it attacks the roots of gooseberry and flowering currant (*Ribes sanguineum*) and is light reddish to yellowish-white.

Fig.7.2 Red currant leaf damaged by red currant blister aphid

Other aphids In addition to the aphids already described, four other species are occasionally found on fruit bushes.

The gooseberry–sowthistle aphid (*Hyperomyzus pallidus* Hille Ris Lambers) causes a characteristic yellow vein-banding on gooseberry leaves and migrates in summer to *Sonchus arvensis* (the field or perennial sowthistle).

The currant–yellowrattle aphid (*Hyperomyzus rhinanthi* (Schouteden)), a blackish-green aphid found in northern England and in Scotland, causes leaf-curl of currants and migrates to yellow rattle.

The currant stem aphid (*Rhopalosiphoninus ribesinus* (van der Goot)), a dark brown aphid, colonizes the old wood of currants growing in shady situations.

The red currant–arrowgrass aphid (*Aphis triglochinis* Theobald), which is usually brownish-green, causes down-curling of the leaves of red and black currants.

Natural enemies

Currant and gooseberry aphids are attacked by ladybirds, the larvae of lacewings and hover flies, and various minute parasitic wasps. They are sometimes destroyed by parasitic fungi. The winter eggs are eaten by tits and other small birds.

Control in plantations

In commercial crops of currants and gooseberry it may be necessary to apply an insecticide spray against aphids in the spring, preferably just before flowering. Since the most important aphids overwinter on the crops in the egg stage and have hatched soon after bud burst, excellent control is usually obtained. Further foliar sprays may be necessary during the growing season, especially if the tips of young shoots become infested by winged forms of the gooseberry aphid or permanent currant aphid. However, the safety interval between application and harvest must be observed. Further restrictions may apply to crops grown for processing. Flowering bush fruits are attractive to bees and should not be sprayed with an insecticide.

Aphid attacks may also be prevented by killing the overwintering eggs with a winter wash during the dormant period.

Growing-season sprays

In established plantations where there is a full canopy of foliage, aphids, which normally live and feed on the underside of leaves, are a difficult target to spray. For this reason systemic aphicides are generally most effective. However, in well-managed plantations where there are few aphids, non-systemic insecticides are usually adequate and may be more appropriate if other pests are a problem.

Other pests frequently occurring on bush fruits include common green capsid (see page 32) and black currant leaf midge (*Dasineura tetensi* (Rübsaamen)) (on currants only), both of which emerge during or shortly after the blossom period. Post-blossom sprays against these pests may give adequate control of aphids.

Black currant Routine applications of an acaricide are normally made to commercial crops to control black currant gall mite, the vector of reversion virus (see page 354), and these will give some control of aphids, capsid and leaf midge. Alternatively, a recommended systemic insecticide may be used; some systemic insecticides will also control capsid.

Red currant and gooseberry Gall mite is not a significant pest of these crops, so it is only necessary to consider control of aphids and capsid. Both can be controlled by spraying with a recommended insecticide at the end of the blossom period.

Dormant-season sprays

A winter wash will give excellent control if all the shoots and branches are wetted. The buds must be dormant when the spray is applied; this generally means spraying no later than the end of January, although February applications are usually safe on black currant. Sprays should be applied only on dry days and any nearby vegetables should be protected from damage by drift.

Control in gardens

Like the commercial grower, the gardener has the choice of spraying with a winter wash to kill the overwintering eggs (see **Dormant-season sprays**) or spraying during the growing season. For the latter, a systemic insecticide is usually more effective: various garden products are available. Success with non-systemic insecticides depends on wetting the underside of the leaves, where most aphids are feeding. Sprays are most effective if applied just before the blossom period.

8
Glasshouse whitefly

Fig.8.1 Tomatoes damaged by sooty mould following an attack of glasshouse whitefly

The glasshouse whitefly (*Trialeurodes vaporariorum* (Westwood)) is well known to most growers of glasshouse plants. It is not a native British insect but has been established in Britain for many years. It can survive the winter on weeds outdoors, at least until such plants are killed off by a succession of hard frosts. After mild winters infestations in glasshouses often originate from these overwintering sources, but the pest can be brought on to the nursery on plants from other sources. A closely related species — the cabbage whitefly (*Aleyrodes proletella* (L.)) — lives outdoors but only on brassicas and related plants.

Nature of damage

Many species of plants grown under glass are liable to attack by whitefly. Those subject to severe attacks are aubergine, cucumber, dwarf French bean, pepper, tomato and a large number of ornamentals, including fuchsia, *Gerbera*, *Pelargonium*, poinsettia, primula, *Solanum* and occasionally chrysanthemum.

Both the adults and the young suck sap from the foliage and, when present in large numbers, cause a general weakening of growth. In addition, the insects produce a sticky excretion known as honeydew, which covers the leaves and prevents them from functioning normally. This honeydew frequently becomes dark brown because of fungal growth (sooty moulds). If an infestation is allowed to increase unchecked, much of the foliage may be killed and the crop seriously decreased (Fig. 8.1) Tomatoes and cucumbers may have to be cleaned before sale.

Fig.8.2 Glasshouse whiteflies on underside of dwarf French bean leaf

Description and life history

Adult whiteflies (Figs. 8.2 and 8.3) are about 1 mm long and are snowy white owing to a covering of white, mealy wax. They usually settle on the underside of the upper leaves of a host plant but fly fairly readily if disturbed.

The eggs are laid erect in circular groups on the underside of smooth leaves but are more scattered on hairy leaves. Each female lays about 200–250 eggs at a rate of about eight per day during a life span of 3–6 weeks. The eggs are yellowish when laid but become black within 2–3 days; the incubation period is nine days at 21 °C.

Pale green, flattish larvae hatch from the eggs. They are active for a few days but then settle down and remain motionless until they mature. These immobile stages (Fig. 8.4) are called 'scales'. Like aphids and scale insects, they feed by means of stylets which are inserted in the tissues of the plant to take up sap. Before becoming adult, a whitefly passes through four stages, the last of which corresponds to the pupal stage of other insects. The duration of the immature stages varies with temperature; at 21 °C it is about 18 days. Thus the total period from egg to adult at this temperature is about 27 days.

Fig.8.3 Glasshouse whitefly: 'scale' (left) and adult (right) on underside of leaf (× 12)

Preventive measures

Before plants are taken into a glasshouse, it is important to ensure that none of them is already infested. Because of the risk of introducing an insecticide-resistant strain of whitefly, this is just as important when the glasshouse is already infested with whiteflies as when it is free from infestation. If necessary, a suitable insecticide should be applied as a routine precaution. The pest may enter in spite of this precaution, but the risk of an outbreak will be reduced.

Weeds within houses can harbour and carry over whitefly infestations from one crop to the next; this carry-over in empty houses in mild winters is particularly important. Weeds inside houses should be eradicated at all times. Weeds in the immediate vicinity of houses should also be kept down as a general routine.

Yellow sticky traps placed in empty houses are efficient at removing adult whiteflies.

Biological control

Two types of biological control are in use: one depends on a parasite, the other on a fungal pathogen.

Use of parasite

Whiteflies are parasitized by certain minute chalcid wasps, one of which — *Encarsia formosa* Gahan — has been used successfully for many years to control whitefly infestations on a range of glasshouse plants.

Each female parasite can lay 50–200 eggs during its life span of 10–14 days. Each egg is inserted into an advanced scale stage of the whitefly, which is subsequently killed by the developing parasite larva as it feeds on the scale contents. Scales containing parasites are easily recognized because they are black (Fig. 8.4).

Use of this parasite has shown that successful control can be obtained if the parasite is established on the plants when natural infestations of whitefly are small.

On cucumber The method of parasite use preferred by many growers on nurseries where whitefly is an habitual pest is to introduce parasitized scales weekly at the rate of 12 500/ha, starting as soon as plants are available. This weekly introduction should continue as long as whitefly adults can be seen. If the number of whitefly adults exceeds one per 10 plants, the rate of introduction of parasitized scales should be increased for at least four of the weekly introductions: advice on the actual rate should be sought.

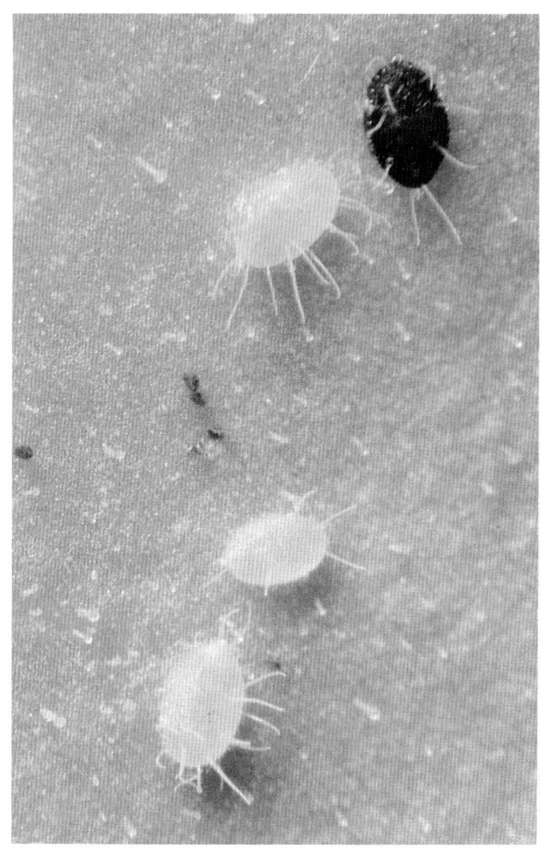

Fig.8.4 'Scales' of glasshouse whitefly (× 35). The black one is parasitized by *Encarsia formosa*

On tomato Most growers introduce parasitized scales weekly at the rate of 5000/ha, starting as soon as plants are available. This low rate of introduction is so cheap that growers usually continue weekly introductions into July or August whether whiteflies are seen or not. If more than one adult whitefly per 10 plants is seen, introductions of parasitized scales should be increased for at least four weeks; the rate of introduction is normally two parasitized scales for every whitefly seen, but advice on the actual rate should be sought.

On ornamentals The parasite controls whitefly equally well on ornamental plants, but no precise rates of use have been established. It is generally agreed that 5–10 parasites/m^2/week should be used when the density of whiteflies is no more than one per two square metres. To ensure even distribution, the introduction sites should be as closely spaced as is commercially practicable. With young rooted cuttings it is important to introduce parasites early, and with certain crops, e.g. fuchsia, *Pelargonium* and poinsettia, control should be initiated when the stock beds are planted.

In some ornamental crops, another aid to control is the use of other plants that are good hosts of the whitefly. These are called 'banker' plants. They attract whiteflies away from the crop and allow the parasite to breed in larger numbers than is possible on the crop. Plants such as tomato, dwarf French bean and tobacco make ideal 'banker' plants.

After parasite establishment Once the parasite is established on the crop plants, it is essential to avoid trimming off, or removing from the glasshouse, leaves with black scales from which parasites have yet to emerge. Empty scales have an emergence hole which is clearly seen when the leaf is held up to the light.

Occasionally, despite what appears to be good initial establishment of parasites on a crop, whitefly numbers become excessive. This situation can usually be corrected on tomatoes and cucumbers by spraying the tops of the plants with a suitable insecticide. Most other plants may be similarly treated, but precautions against injury of the crop should be taken by first testing the insecticide on a few plants. Soft soap is useful as an insecticide in these circumstances as it does not harm the parasites. In addition, yellow sticky traps can be hung about 15–30 cm above the crop canopy.

Small groups of heavily infested plants may be found in a crop where the parasite is otherwise exerting satisfactory control. These infestations can be controlled by selective spraying in the patches only.

Supply of parasites Parasite material supplied by commercial producers consists of leaves or cards populated with black, parasitized whitefly scales. Adult parasites usually emerge within a few days after receipt but, in the event of delay in transit, it is advisable to open the container inside the glasshouse to be treated.

Use of fungal pathogen

Verticillium lecanii, a fungal pathogen that attacks whiteflies and thrips, is proving very useful against glasshouse whitefly in situations where the crop is grown in high humidities, i.e. on propagation benches or under screening material used to regulate day length. The pathogen can be sprayed in any crop when external factors cause short periods of high humidity. The disease takes about 1–2 weeks to develop on the insects and attacks the young as well as the adults.

Chemical control

Many recommended insecticides give some control, but strains of whitefly resistant to one or more of them have become established. The selection of the most suitable chemical therefore depends on whether resistant strains are present, as well as on the species and stage of development of the host plant and on safety to the operator. With edible crops, safety to the consumer must also be considered.

With the increasing use of biological control techniques, it is necessary to choose the chemicals and methods of application which will be least harmful to natural enemies. Different situations demand different treatments. Thus, while most soil-applied systemic chemicals are ideally suited to integration with biological control systems, some beneficial species also survive high-volume sprays through the inadequacy of spray cover or, in the case of *Encarsia* parasites, by application of sprays during the black-scale stage. However, the predatory mite *Phytoseiulus* (see page 363), leaf miner parasites (see page 271) and the adult *Encarsia* parasite are highly susceptible to many pesticides; in particular, the use of treatments with a fumigant action should be avoided.

Most recommended insecticides are only effective against whitefly adults though some also kill eggs and young scales. Few, if any, are effective against the larger, older scales, so that repeated treatments at 3–5 day intervals (depending on the temperature) are essential for several weeks before control can be achieved.

Spraying at high volume to wet both leaf surfaces thoroughly is normally the most effective method of controlling whitefly. Where there is a choice of formulation, liquid concentrates are generally preferable to dispersible powders, which leave a white deposit on the plants.

Smokes are convenient to use in small, reasonably airtight glasshouses. They should be applied in the late afternoon or evening in still weather and when a temperature of 18–21 °C can be maintained for 4–6 hours. Glasshouses treated with smokes must be securely locked and labelled during treatment and then ventilated for an hour before being entered.

A recent development in the control of this pest is the use of a new type of insecticide — an insect growth regulator — which is effective against whitefly but does not harm the parasite.

Disease control Many fungicides can be used without harming *Encarsia formosa*, but intensive use of some products can be harmful. Advice should be sought on the most suitable products to use.

9
Lettuce aphids

Fig.9.1 Lettuce plant (right) collapsed following a heavy attack of lettuce root aphid compared with normal plant (left)

Several species of aphid (greenfly) live on the leaves of lettuce, both under glass and outdoors. When they are heavily infested the leaves may appear curled and blistered, and the plants become so stunted that they fail to heart (Fig. 9.4(c)). The honeydew excreted by the aphids forms a sticky layer over the leaves to which dust and cast aphid skins adhere; this alone can seriously reduce the value of the crop.

As well as causing direct damage to the plants by their feeding, some of these aphids can spread viruses that cause severe stunting and distortion of the leaves (Fig. 9.4(b)).

The two species of aphid that are most commonly harmful to lettuce are the currant–lettuce aphid (also known as the lettuce aphid) (*Nasonovia ribisnigri* (Mosley)) and the peach–potato aphid (*Myzus persicae* (Sulzer)). The lettuce root aphid (*Pemphigus bursarius* (L.)), which feeds on the roots of lettuce, can also cause severe loss of plants, especially in dry summers.

Currant–lettuce aphid

This is probably the chief aphid pest of lettuce, under glass and outdoors, in all seasons. Glasshouse, tunnel, frame and cloche lettuce may suffer severe attacks.

Both winged and wingless forms of the aphid are green with black markings (Fig. 9.2). Eggs are laid in late autumn on the twigs of currants and gooseberry. The aphids hatch in early spring and start to form colonies on the young leaves, causing some leaf-curl. About May winged aphids appear and leave the currants and gooseberries to fly to lettuce, hawkweeds and other plants of the same family, on which successive generations are produced until September or October. There is then a return migration of winged aphids to the winter host plants where the overwintering eggs are laid.

Although this life cycle is the normal one in Britain, and probably the only one in the north, in most years in the south the breeding of 'summer aphids' continues on successive crops of outdoor lettuce throughout the winter.

Peach–potato aphid

This is an abundant and widespread species (described on page 60). It is a very frequent visitor to lettuce but usually colonizes only the outside leaves and does little direct damage. However, it is the chief vector of lettuce mosaic virus. It overwinters in the egg stage on peach or nectarine trees, or as wingless females on brassicas, particularly savoys, in mangel or fodder beet clamps and on lettuce under glass or polythene.

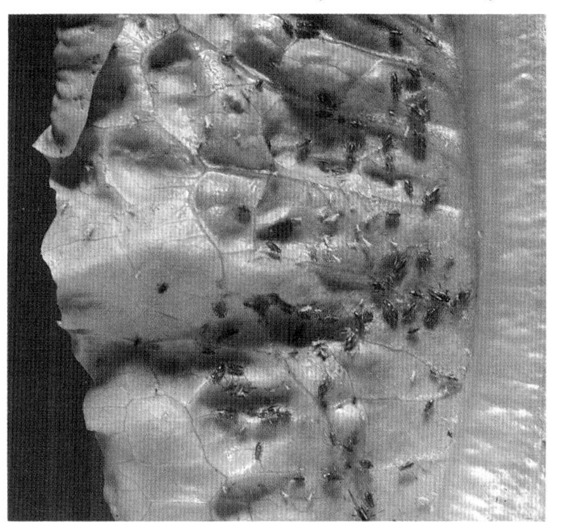

Fig.9.2 Colony of currant–lettuce aphid on lettuce leaf (about life size)

Other leaf-feeding aphids

Of the three other species of aphid that are most commonly found on lettuce leaves, the potato aphid (*Macrosiphum euphorbiae* (Thomas)) can be a serious pest under glass and polythene and also on outdoor crops. The glasshouse and potato aphid (*Aulacorthum solani* (Kaltenbach)) and the shallot aphid (*Myzus ascalonicus* Doncaster) can both be serious pests of glasshouse crops and the former species can be troublesome on outdoor crops. Small numbers of these aphids are liable to move on to lettuce from weeds and from other glasshouse crops. They can breed very quickly, even during the winter, if the lettuce is being forced. (For descriptions of the potato aphid and the glasshouse and potato aphid see page 61.)

Lettuce root aphid

This aphid is harmful mainly to outdoor lettuce in summer, especially during a drought (Fig. 9.1). It, too, is a migratory species. Eggs are laid on Lombardy poplar and black poplar where, in spring, the newly hatched insects live inside characteristic flask-shaped galls on the leaf stalks. During June winged aphids develop in the galls. When the galls dry and open, these aphids escape and migrate to lettuce and certain weeds including sowthistle, where they give rise to the first of several generations of wingless aphids on the roots. These wingless aphids are yellowish-white and made conspicuous by the white, waxy secretion which covers them (Fig. 9.3).

In late summer and autumn, winged aphids

Fig.9.3 Roots of lettuce plant attacked by lettuce root aphid

Fig.9.4 Winter lettuce: (a) treated before transplanting; (b) affected by mosaic and free from aphids; (c) untreated and heavily infested with aphids

appear and make the return migration to poplar. A small residue of root aphids may survive the winter in the soil — even in the absence of host roots — and colonize lettuce planted in the same ground during the following spring.

Natural enemies

Aphids infesting outdoor lettuce in the summer may be attacked by small rove beetles, hover fly larvae, ladybirds and their larvae, and by minute parasitic wasps. These insect predators and parasites are much less active in the winter, even in heated glasshouses where aphids are numerous. Lettuce root aphids may also be attacked in their galls in the winter host by anthocorid bugs. In addition, lettuce aphids may be killed by parasitic fungi.

Spread of virus

Several species of aphid are able to transmit viruses that cause diseases of lettuce. The peach–potato aphid is an efficient vector of lettuce mosaic virus, cucumber mosaic virus and beet western yellows virus but does not carry dandelion yellow mosaic. This last virus is comparatively rare but causes severe mosaic and necrosis; it is spread by several aphid species, including the glasshouse and potato aphid and the shallot aphid. The currant–lettuce aphid can transmit cucumber mosaic virus.

Lettuce mosaic virus is seed-borne and the risk of spread can be reduced by the use of virus-tested seed stocks, which should contain less than 0.1% infection. Spread can be further reduced by maximum isolation of crops, especially seedlings, and by efficient aphid control.

Control of leaf-feeding aphids

On glasshouse crops

To avoid planting clean seedlings in an infested house or other structure, all weeds and crop debris should be removed as soon as possible after harvesting the previous crop, and the house kept clean and empty for at least 7–10 days. If this is not possible, or if other crops are present in the structure, a recommended aphicidal aerosol or fumigant should be used to kill any aphids a day or two before planting. These formulations are only effective if the structure is reasonably leak-proof and properly closed down for the duration of the treatment. If possible, periods of bright sunshine and strong wind should be avoided and a temperature of 18–24 °C maintained during treatment as this increases the efficiency of the aphicides.

Seedlings Infestation of seedlings can usually be avoided by raising them in weed-free isolation from other crops. They should be examined regularly for signs of aphid infestation and, if any is found, promptly sprayed with a recommended aphicide. One or two days before transplanting they should be sprayed to ensure that the transplants are free from aphids. Where seedlings are raised in soil blocks the same care is needed: if any aphids are found, the seedlings should be sprayed one or two days before they are set out in the glasshouse or polythene tunnel.

After planting out, crops should be looked over at regular intervals. When aphids are found, treatment should be applied immediately as the market value can be adversely affected by the presence of dead aphids following treatment that has been delayed until the crop is nearly mature. In glasshouses and large polythene structures treatment is simple, but crops in frames and cloches are awkward to treat, even with sprays. This emphasizes the need to plant clean seedlings in clean land when using low structures; sometimes it is possible to treat the plants either just before they are covered or when the covers are removed.

On outdoor crops

Winter lettuce All rubbish and residues from old lettuce crops near a proposed winter lettuce seedbed should be ploughed in or otherwise destroyed before the seed is sown, and the ground kept free from weeds. Seedbeds should be treated in the same way as beds for glasshouse crops to ensure that aphid-free seedlings are transplanted. If thoroughly treated, the crop will remain clean until cutting time because reinfestation by winged aphids does not take place on a large scale before May.

Summer lettuce The control of aphids in the outdoor summer crop is more costly because winged migrant aphids may reinfest the crop at almost any time from May to September. To ensure that clean lettuce is marketed the crop must be examined regularly for aphids. If aphid colonies are found at any time up to three weeks before cutting, the crop should be sprayed with one of the more persistent recommended systemic insecticides aimed at the heart of the plants. Examination of the plants should continue to within a week of cutting. If aphids are found, a spray of a recommended insecticide with a short harvest interval can be applied, but dead aphids may remain on the crop and reduce its market value.

If aphids are carrying mosaic virus, infection of the crop usually occurs before the aphids can be killed with insecticide. Aphid control after migration is not, therefore, an effective means of controlling lettuce or cucumber mosaic. Hence all crops must be kept clean throughout their life and weed control must be thorough, even in mature crops. If crop residues cannot be ploughed in immediately after harvest, both crop and weeds should be sprayed with an insecticide to kill aphids before they migrate to other crops. Crop residues must be completely buried otherwise aphids will migrate from the exposed parts.

Insecticide resistance

Resistance to insecticides in the peach–potato aphid is now widespread. In addition, the more highly resistant strains are increasing in frequency within populations. It is important that wherever possible, particularly where resistant aphids are suspected, high-volume sprays are used to ensure adequate cover of the plants and thus maximum effectiveness.

Control of lettuce root aphid

Cultural measures

As Lombardy poplar is a winter host of the lettuce root aphid, lettuce growers should not plant these trees as a windbreak. Growers should also discourage the local planting of poplars but should not seek to remove existing stands of these trees.

Immediately after harvesting the autumn lettuce crop the ground should be cleared, so that no debris, particularly cut root stumps, is left in the ground over the winter. This measure will greatly reduce the risk of aphids remaining in the soil and infesting lettuce planted during the following spring. Also, if possible, some rotation should be practised to avoid spring-sown crops going into land that carried infested lettuce the previous year.

Many varieties of lettuce commonly grown in Britain are very susceptible to root aphid attack, but resistant varieties are available and offer a worthwhile means of combating this pest.

The effects of lettuce root aphid infestation are worst when the plants are dry at the roots, so irrigation is highly beneficial. Therefore, where the pest is known to occur and where the soil is becoming dry, irrigation is recommended on crops showing signs of damage.

Chemical measures

Chemical treatment cannot give control once aphids are established on the roots, so preventive treatment is important where resistant varieties are not being grown. The existing insecticides recommended for control of lettuce root aphid are only partially effective. Depending on the insecticide used, it should either be applied to the soil before drilling or planting, or, if the crop is to be transplanted, incorporated in the compost before making blocks/modules. However, such treatment is only worth while where trouble is expected, i.e. where the aphid is common, and on lettuce sown or transplanted between mid-April and the end of June.

Unless the land is treated with a suitable insecticide, wingless aphids may remain in the soil and infest lettuce planted in such soil in the following spring.

10
Plum aphids

Fig.10.1 Plum shoot damaged by leaf-curling plum aphid

Three species of aphid are commonly found on plum, damson and blackthorn: these are the leaf-curling plum aphid (*Brachycaudus helichrysi* (Kaltenbach)), the damson–hop aphid (*Phorodon humuli* (Schrank)) and the mealy plum aphid (*Hyalopterus pruni* (Geoffroy)). Successful pest control on plum depends to a large extent on control of these aphids, which are the most important pests of this crop. In addition to the direct damage done to the tree, these aphids are vectors of the virus which causes plum pox (sharka) — a serious disease of plum, damson, peach and ornamental *Prunus*.

Recognition and damage

Leaf-curling plum aphid The presence of the leaf-curling plum aphid is easily recognized by the tightly curled leaves which occur on infested branches (Fig. 10.1). The small, rather rounded, shining green aphids live on the underside of the curled leaves. This species is probably the most common and injurious plum aphid. On infested branches shoot growth is reduced, fruit is undersized and the curled leaves usually turn yellow and fall prematurely.

Damson–hop aphid and mealy plum aphid Infestation by the damson–hop aphid causes only slight leaf-curl, whereas that by the mealy plum aphid causes none at all. Because of the absence of obvious symptoms, the early stages of infestation by these aphids may be overlooked and their presence is often not noticed until the honeydew which they secrete is seen glistening on the upper surface of the leaves. Sooty moulds (produced by *Cladosporium* spp.) develop on the honeydew, resulting in a dull, blackish film on the contaminated surfaces.

In the absence of characteristic leaf symptoms, aphids of these two species can easily be recognized by their colour: the damson–hop aphid is yellowish-green with darker green stripes along the middle and sides of the upper surface; the mealy plum aphid (Fig. 10.2) is green with a bluish-grey tinge and has a white powdery covering over the entire body.

As shoots become heavily infested, extension growth ceases and the debilitating effect of sap extraction by large numbers of aphids can reduce fruit size. In addition, both fruit quality and the photosynthetic efficiency of the leaves can be further reduced by sooty moulds growing on the film of honeydew.

Life history

Leaf-curling plum aphid Eggs are laid in the autumn near the base of fruit buds and hatch about four to six weeks later; hatching is almost complete by the end of December. The young aphids feed on the base of fruit buds; nymphs of the second generation are often present before leaves appear in the spring. During this period the aphids are purplish-brown and thus harmonize well with the bud scales and bark. As the buds open, the aphids move on to the flower buds and expanding leaves and the first sign of leaf-curl is usually visible by the end of the blossom period. The change in feeding site by the aphids — from bud to leaf — is also accompanied by a change in colour: those which develop on leaves assume the typical shining yellow-green colour.

Fig.10.2 Colony of mealy plum aphid on underside of plum leaf (× 10)

Only wingless aphids are present until about mid-May. Thereafter, winged forms occur in increasing numbers and migrate to various annual or herbaceous plants such as aster, chrysanthemum and clover; all aphids have left plum by late June or early July. The progeny of the winged migrants often cause stunting and distortion of young bedding plants such as asters, and later infestation of clover flowers can reduce the yield of seed. Both winged and wingless aphids are produced during the summer months on these host plants.

Winged aphids return to plum in September; these produce a generation of wingless egg-laying females, thus concluding the active life cycle.

Damson–hop aphid In contrast to the previous species, eggs laid on plum, damson or blackthorn in the autumn do not hatch until April. The first two generations consist only of wingless aphids; some of the third and subsequent generations are winged and migrate to hop, where they cause severe reduction in growth and yield if not controlled by insecticides. The first migrants usually appear in the second half of May and migration to hops occurs from late May until the latter half of July or early

August. Successive generations of wingless aphids occur on hops from June to September, when winged aphids are produced which return to plum, damson or blackthorn to produce the egg-laying generation.

The wingless aphids produced on plum etc. in the spring and early summer are essentially colonizers of actively growing shoot tips. As infested shoots become crowded with aphids, many of the newly formed adults disperse in search of other shoots which are in turn colonized. Thus, in the nursery, where hedges, shoots from layer beds, or closely planted rootstocks form a continuous or almost continuous canopy, the progeny from a single aphid can lead to infestation extending along many metres of row. In the absence of effective chemical control, this movement along a row will continue until all accessible shoots have been colonized, or until natural enemies have drastically reduced the population, which does not usually occur until well into August.

Mealy plum aphid Eggs are laid in autumn on twigs and at the base of buds on plum, damson and blackthorn, but they do not hatch until April. The early generations consist only of wingless aphids; subsequent generations produce both winged and wingless forms. Winged aphids first appear about mid-June, rather later than for the two previous species, and migrate to waterside grasses and reeds. This migration continues until early or mid-August. During September winged aphids, which develop on the summer host plants, return to plum and produce the egg-laying generation.

In contrast to the other species, which colonize the tips of growing shoots, this aphid lives mainly on fully expanded leaves. Infestation is seldom noticed before June, but populations on plum continue to increase until July. The entire lower surface of the leaves can become covered with aphids, but the most obvious effect is severe contamination of fruit and foliage by honeydew and sooty moulds.

A closely related species — the mealy peach aphid (*Hyalopterus amygdali* (Blanchard)) — lives on peach and has the same summer host plants as the mealy plum aphid.

Natural enemies

Plum aphids are eaten by many predators such as ladybirds and their larvae, the larvae of hover flies and lacewings, and anthocorid bugs. They are also parasitized by the larvae of minute wasps. These natural enemies seldom become numerous on fruit trees until aphids have already caused some damage, but they are often responsible for reducing the intensity of attack and ending an infestation earlier than would occur otherwise. In addition, the progeny of the predatory species feed on aphids during the summer and by reducing populations in August and September they can influence the level of infestations in the following year.

Transmission of plum pox

Aphids are the only known insect vectors of plum pox virus, which causes a serious disease of plum and peach in eastern Europe and was first found in England in 1965. Of the species mentioned here, the leaf-curling plum aphid and the damson–hop aphid are known to transmit the virus, while there is some evidence that the mealy plum aphid may also do so. In addition, the virus can be transmitted by the peach–potato aphid (see page 60).

Natural transmission by wingless aphids is most likely in situations providing easy access to neighbouring plants, such as in the nursery or in orchards where the canopies of adjacent trees interlock. This results in a linear or patchy pattern of infected trees. Winged migrants of the damson–hop aphid and the peach–potato aphid, which do not normally breed on plum, will acquire and transmit the virus while probing in search of a suitable host plant. Transmission by winged aphids produces a more random pattern of infected trees.

Adherence to a strict programme of aphid control, coupled with frequent inspection from late April until early August, can practically eliminate the risk of transmission by wingless aphids. This is particularly important on nursery stock. Even if eradication cannot be achieved, populations should be kept at a very low level to minimize the movement of wingless aphids to neighbouring plants. Such measures will also greatly decrease the number of winged aphids produced locally, although others will arise from unsprayed cultivated and wild hosts in the neighbourhood.

Control in orchards and nurseries

Plum aphids can be controlled either by winter washes or by spring or summer sprays. Winter washes kill the eggs and any aphids that have hatched, whereas spring and summer sprays kill the aphids after all the eggs have hatched.

The damson–hop aphid has developed resistance to a wide range of insecticides, including organophosphorus, carbamate and pyrethroid compounds. The forms of this aphid which occur on plum and damson are as resistant as those on hops. Like other plum aphids, damson–hop aphid is a vector of plum pox virus, so special care should be taken to select an insecticide which is effective against this species and will ensure that trees are kept as free from aphids as possible throughout the season. As winter washes are toxic to both resistant and non-resistant aphids, these chemicals have been used more widely against this pest in recent years.

If leaf-curling plum aphid is present, the first symptoms are visible just after the blossom period and spraying should not be delayed more than 7–10 days or severe leaf-curling may develop. In some seasons a white bud spray may be necessary to prevent leaf damage during the blossom period.

Winter washes

Plum aphids can be controlled by one application of a winter wash during the dormant season, but for good results the trees must be thoroughly wetted all over so a high-volume spray is necessary. Spray drift will damage nearby herbaceous plants in leaf.

Spring and summer sprays

Spring applications of insecticides are cheaper than winter washes, but their effectiveness is limited by the development of resistance. Many insecticides are recommended for use as spring sprays; the normal time of the first application is at white bud. Where damson–hop aphid is resistant to one or more chemical groups, insecticides belonging to other chemical groups should be used. It should be noted, however, that use of aphicides from the same group on both plum and hop is likely to hasten the onset of resistance to that group.

Winter moth caterpillars will also be controlled by spraying at the white bud stage with certain insecticides recommended for aphids. If plum sawfly (*Hoplocampa flava* (L.)) is a problem, the spray against aphids can be delayed until the cot-split stage (7–10 days after petal fall) as some recommended aphicides are effective against plum sawfly as well as aphids.

Mealy plum aphid infestations are often overlooked and, as they can persist until August, may require an additional spray of insecticide in May or June, if not properly controlled in the spring.

Control in gardens

Leaf-curling plum aphid and mealy plum aphid — two common species in gardens — can be controlled by a thorough application of a winter wash in late December or the first half of January. It is essential to protect nearby herbaceous plants in leaf from spray drift, using newspaper or plastic sheeting etc.

Alternatively, a recommended systemic insecticide can be applied at the end of the blossom period if leaf-curling plum aphid is present, or delayed until infestation by other aphid species is observed. Trees should be inspected for infestations of mealy plum aphid on the underside of the leaves in May and June (remembering that this aphid causes no leaf-curl).

11
Potato aphids

Fig.11.1 Colony of the potato aphid on potato foliage showing 'false top roll'

Four species of aphid (greenfly) occur commonly on the leaves of potato: the peach–potato aphid (*Myzus persicae* (Sulzer)), the buckthorn–potato aphid (*Aphis nasturtii* Kaltenbach), the potato aphid (*Macrosiphum euphorbiae* (Thomas)) and the glasshouse and potato aphid (*Aulacorthum solani* (Kaltenbach)); a fifth, the violet aphid (*Myzus ornatus* Laing), may occur in large numbers on senescing and blight-damaged foliage in late August and September. A sixth species, the bulb and potato aphid (*Rhopalosiphoninus latysiphon* (Davidson)), occasionally infests the stolons (underground shoots). Several of these species occur on seed potatoes in store.

Description and life history

The peach–potato aphid is a medium-sized, pale green or pink species with a rather barrel-shaped body (Fig. 11.2). It overwinters as adults or young stages on a wide range of herbaceous and brassica crops and weeds, and on chitting potatoes, mangels and beet in store, or less commonly as eggs on peach. The eggs hatch in early spring. Winged aphids develop in the third generation and migrate in May or June to a wide range of summer hosts including the potato. Aphids may multiply extremely rapidly on the summer host, reaching a peak in early to mid-July. The winged forms produced then migrate as the potato haulm matures. The disappearance of the remaining aphids is often hastened by parasites and predators, although there is occasionally another small increase in numbers in late August or September. In late summer sexual forms may be produced which migrate back to the winter host, or asexual forms may survive the winter in sheltered situations. Potatoes are frequently colonized by the peach–potato aphid, although other aphid species may outnumber it on a particular crop.

The buckthorn–potato aphid is a small, bright yellow or yellow-green species that overwinters as eggs on buckthorn. These eggs hatch in April. Migration to potato takes place in mid- or late June and the return migration to buckthorn in early September. Numbers on potatoes may be extremely large, but they vary considerably from year to year.

The potato aphid is a relatively large, green or pink aphid with a long body and a dark stripe along the middle of the back. The young forms are covered with a powdery coating of wax. This species may overwinter as eggs but more usually hibernates as wingless aphids on herbaceous hosts or chitting potatoes; it migrates to potato in May and June. In years of heavy infestation there is often a dispersal migration in late July, but there is rarely a noticeable autumn migration. Potato aphid infestations are widespread but only reach substantial numbers in occasional years. Unlike the other leaf aphids, there is a marked tendency for this species to multiply on the flowers and near the tips of the shoots, where, if very numerous, it can cause 'false top roll' (see page 63 and Fig. 11.1).

The glasshouse and potato aphid is a medium-sized, shiny, yellow-green or brownish aphid which occurs on a wide variety of hosts in glasshouses throughout the year; it can also overwinter outdoors. It is often found on potatoes in the field but is rarely numerous. It also lives on chitting potatoes during the winter (Fig. 11.3).

Fig.11.2 Peach–potato aphids: winged and wingless adults and immature winged form (× 16)

The violet aphid is a small, yellow-green or whitish-green aphid with black spots. It overwinters as adults or young on glasshouse crops, or on weeds in sheltered places. Migration to potato occurs in small numbers from May to July; large infestations develop only on senescing or damaged foliage in late August and September. Unlike other species, it often occurs on the upper surface of leaves. It produces few winged forms on potato.

The bulb and potato aphid is a shiny, dark green aphid which lives on the stolons of potatoes on silt soils in the Fens near Peterborough and occasionally elsewhere. It is also found on bulbs and on chitting potatoes during the winter.

Nature of damage

Aphids feed by sucking sap from the plant, causing direct damage by injection of toxic substances in the saliva, and by removal of water and nutrients. In addition they acquire viruses from infected plants and pass them on, thus acting as virus vectors. Yield losses due to virus diseases are far more serious than those caused directly by the feeding of the aphids, although even these losses may be substantial.

Virus spread and seed potato production

Leaf roll virus and potato virus Y (the cause of severe mosaic) are the most damaging viruses transmitted by aphids. Plants infected with these viruses yield poorly, if infection is early, and produce infected tubers. Plants grown from these tubers are virus infected and yield even more poorly. Such plants also provide sources for further spread of the viruses through the rest of the crop. The effects of aphid-borne virus may be cumulative in potato stocks.

Seed production in the British Isles is concentrated in the relatively wet and cool or exposed northern areas, where aphids on potato are much less numerous and usually less active. In these areas aphid migration to potato crops normally occurs late in the season, making it possible to raise stocks of the basic grades substantially free from virus.

Fig.11.3 Glasshouse and potato aphids on sprout of chitting potato

The Seed Potato Classification Scheme resulted in a great improvement in the health of ware potato crops, so that for many years virus diseases caused little loss of yield. However, mild winters in the early 1970s allowed more aphids and infected groundkeepers to survive from year to year, and virus levels in ware- and some seed-producing areas increased to a level at which they caused appreciable losses. The return to more severe winters and cooler summers in 1978 and 1979 restored the situation. However, it continues to be most important to prevent virus infection of crops from which any grade of seed is to be saved. Leaf roll remains the predominant virus in seed-producing areas, but virus Y is widespread and sufficiently damaging in ware areas to have contributed to a decrease in cultivation of the more susceptible varieties, such as King Edward, in favour of those with greater resistance.

Although migrating aphids of many species alight and feed on potato, only a few are virus vectors and most are relatively inefficient in this role. The peach–potato aphid is responsible for most virus spread in the field; the other species are of less importance in this respect.

There are important differences in the way the viruses are spread by aphids. An aphid must feed on a plant infected with leaf roll for at least six hours to acquire the virus and subsequently feed on a healthy plant for at least two hours to pass it on. However, once it has acquired the virus it retains it for life. Thus the spread of leaf roll can largely be prevented by the use of insecticides. By contrast, virus Y can be quickly transmitted to a healthy plant by an aphid which has fed on an infected plant for a few minutes only. Although the spread of virus Y has sometimes been reduced by the use of aphicides, there is evidence to show that their use can actually increase the spread. This is possibly because the aphids are not affected quickly enough and the sprayed plants are distasteful to them, causing them to move about more, probing as they go. Virus Y is unlikely to be spread over long distances by aphids because the ability to pass it on is lost after about 12 hours.

To raise healthy stocks of seed potatoes, some knowledge of aphid behaviour is desirable. A migrating aphid is largely wind-borne and may travel many miles. Its main flight is followed by a series of very short flights in search of a suitable host; most virus spread occurs as a result of these short flights. Virus can also be carried for short distances in the crop by the wingless forms. Thus, spread occurs mainly within crops or between adjacent crops. It is essential to use uninfected seed for crops from which seed is to be saved and to isolate these from any less healthy crop. The main migratory flight of the peach–potato aphid takes place in May or June and there is a second migration in late July or August. As these late migrants may carry virus from ware crops, the haulm of seed crops should be burnt off as soon as there is an acceptable yield. Haulm destruction by specified times is now required for certain crops of basic seed potatoes produced in England and Wales.

Growing seed in ware-producing areas

With the help of insecticides and timely haulm destruction it has been possible to raise seed of a reasonable standard in many districts outside the accepted seed-producing areas. This is usually cheaper than buying in seed, the grower has control over its management from production to planting the following year, and it avoids the risk of chilling or mechanical damage in transit which increases susceptibility to infection by fungi and bacteria. An additional advantage with early varieties is that home-grown seed may produce an earlier marketable crop than classified seed from upland areas.

Local advisers should be consulted on the suitability of any particular district for raising home-grown seed, because where aphid activity is great the risk of serious virus infection outweighs the possible advantages. When raising seed in ware areas, classified seed of at least AA standard should be planted to ensure a low percentage of initial virus infection. Isolation of seed crops from crops produced from home-grown seed is essential to reduce carry-over of virus, but if seed is to be saved from only part of a crop it should be taken from the centre because most virus is introduced near the edges of a crop.

In ware areas, as in seed-producing areas, it is important to plant as early as practicable to allow early burning-off and so reduce virus spread by the summer migrants. Crops should be protected by insecticides from emergence to burning-off (see **Control measures**).

Growers can ensure that their home-grown seed is suitable for planting by submitting tuber samples for virus testing.

The Seed Potatoes Regulations 1991 require all seed potatoes for sale to have been classified and also specify conditions for their production. In addition, The Plant Health (Great Britain) Order 1987 requires that all land growing seed potatoes for marketing be soil-sampled for potato cyst nematodes (see page 425) before the crop is planted.

Direct damage caused by aphids

Very heavy infestations of aphids can seriously damage the haulm and reduce the yield of maincrop potatoes; even moderate numbers can cause an appreciable loss of yield. Much of the damage appears to be caused when the tubers are bulking in late July or August; in years when aphids reach peak numbers before mid-July there is probably much less effect on yield. Aphid damage also makes the leaves more susceptible to damage caused by some fungicides.

False top roll

Heavy infestations of the potato aphid may cause 'false top roll' (Fig. 11.1). This rolled condition of the upper leaves may be distinguished from leaf roll virus symptoms because it usually occurs earlier in the season (late June or early July), patches of plants are affected and there are normally live aphids, white cast skins or parasitized aphids present. It probably occurs to some extent in most varieties, particularly second earlies and early maincrops; it is seen frequently in Désirée and King Edward. False top roll has occurred in most of the main potato-growing areas and can make inspection of crops entered for seed classification very difficult, particularly when the aphid damage is aggravated by drought. Seed crops with heavy aphid infestations are likely to be downgraded or refused a certificate.

Control measures

Seed crops or ware crops from which seed is to be saved should be protected against virus, from plant emergence until burning off, by recommended insecticide granules and/or sprays. Incoming virus infection in ware crops is less significant than the potential losses caused by direct feeding of aphids.

Prevention of virus spread

A granular insecticide applied at planting gives protection until at least late June, except in very dry conditions. The granules can be applied in the planting furrow with an applicator attached to the planter. Certain granules used for control of potato cyst nematodes will suppress early aphids on potatoes.

To protect the crop until burning-off, granular treatments should be followed by aphicide sprays. When appropriate, a fungicide should be combined with the aphicide sprays in order to avoid a separate operation, so reducing wheeling damage, which can decrease yields by an average of 1.25 tonnes/ha. When sprays are applied from the air the additional cost will be offset by the absence of wheel damage. Where granules have not been used at planting, sprays may be required from 80% emergence.

The development of resistant strains of the peach–potato aphid (see below) has restricted the choice of effective aphicide sprays in some areas. A persistent systemic insecticide should be used wherever possible to ensure prolonged and effective control. An insecticide with a selective action against aphids is available but lacks persistence. It does, however, allow beneficial insects to survive, so reducing the risk of a late infestation.

Prevention of direct damage

Routine treatments should be avoided wherever possible because they accelerate the development of insecticide resistance (see below). In most years it is cheaper and effective either to combine an insecticide spray (see above) with a blight spray or to apply an insecticide alone if aphids arrive before treatment of blight is required. For maximum effectiveness the insecticide should be applied when the number of aphids is approaching an average of three to five per leaf (i.e. a true leaf comprising a number of leaflets) on a sample consisting of equal numbers of upper, middle and lower leaves. This criterion does not apply to crops from which seed is to be taken: such crops should be sprayed as soon as aphids are seen. Where a spray has been used, the aphid population will rarely build up to a damaging level a second time.

Insecticide resistance

Resistant peach–potato aphids now occur in most parts of Britain. Some strains of this aphid cannot be effectively controlled by any aphicide. A high level of resistance should be suspected if (i) one of the more potent insecticides, when correctly applied, fails to kill the smaller aphids on the lower leaves within a week, or (ii) infestations develop soon after emergence of crops treated with insecticide granules, assuming soil conditions have been suitable for uptake, i.e. not too dry. It should be noted that the other aphid species described on page 61 are currently controlled by any of the aphicides recommended for use on potato; destruction of these aphids may not prevent virus spread if insecticide-resistant peach–potato aphids have survived, often undetected, on the lower leaves. Where highly resistant aphids occur, moderate control can be achieved by applying recommended aphicide sprays that contain either a carbamate or a proprietary pyrethroid/organophosphorus mixture. Potato crops known to be at risk can be protected by carbamate-type granules applied at planting.

Seed storage

The peach–potato aphid, potato aphid, glasshouse and potato aphid, and bulb and potato aphid all breed on the sprouts of seed potatoes in store and weaken the shoots. The peach–potato aphid and the glasshouse and potato aphid also spread virus to healthy tubers, which are very susceptible to infection at this stage. Aphids may infest the tubers before they are brought into store or when stores are left open during the autumn. Aphids may also breed on weeds in glass chitting-houses. If bought-in and home-grown seed must be kept in the same store, they should be kept as far apart as possible to prevent cross-infection. When migration has finished (about mid-November), potato stores should be fumigated with a recommended insecticide smoke generator. It is often difficult to achieve good distribution of smoke in a tightly filled store; trays of seed should be spaced as evenly as possible. The use of several small smoke generators, rather than one larger generator, and also one or more heaters will improve the distribution of the smoke. If possible the temperature of the store should be raised to 15.5 °C during and just after fumigation. It is also difficult to achieve good distribution of the smoke in very large stores and a careful watch should be kept for a later build-up of aphids. A second fumigation of all stores is advisable early in the new year.

12
Scale insects on fruit trees

Fig.12.1 Mussel scales on bark of apple tree (× 1½)

Scale insects are widely distributed in Britain, living on the bark of many kinds of woody plants such as fruit trees and bushes, ornamental trees and shrubs, and forest trees. Most are small, similar in colour to the bark of the host plant and not easily noticed unless they occur in abundance. Others are quite conspicuous, either because their colour contrasts with that of the bark or because a white woolly mass is produced by the female when eggs are laid.

Following the introduction of regular spray programmes, scale insects practically disappeared from commercial orchards. However, recent changes in orchard pest control have led to scale insects, particularly the mussel scale, becoming more important again in some localities, especially on older trees.

This description includes all scale insects that are likely to occur on fruit trees and bushes in Britain. These are the mussel scale (*Lepidosaphes ulmi* (L.)), the oystershell scales — oyster scale (*Quadraspidiotus ostreaeformis* (Curtis)) and pear scale (*Quadraspidiotus pyri* (Lichtenstein)) — the brown scale (*Parthenolecanium corni* (Bouché)), the woolly currant scale (*Pulvinaria ribesiae* Signoret) and the woolly vine scale (*Pulvinaria vitis* (L.)). Fruit under glass may also be attacked by several tropical scale species, in particular the hemispherical scale (*Saissetia coffeae* (Walker)) and the brown soft scale (*Coccus hesperidum* L.). These two species appear similar to brown scale.

Reference is also made to the San José scale (*Comstockaspis perniciosa* (Comstock)), which occurs on fruit in nearly all the warmer regions of the world and has spread through southern Europe in the past 50 years. It is rated as a serious pest and, consequently, many countries have introduced legislation in an attempt to restrict its further spread on nursery material and on fruits.

Nature of damage

Heavy infestations, in which the bark is encrusted with scales, have a debilitating effect on the host. A secondary effect, more noticeable with some species than others, is the contamination of foliage by the honeydew excreted by these insects. At first the upper surface of the leaves assumes a glistening, sticky appearance, but later it becomes black and unsightly owing to the growth of sooty mould fungi on the honeydew.

Host plants

Mussel scale In Britain the host plants are apple, bilberry, blackthorn, cotoneaster, hawthorn, heath, heather, *Sorbus* and many others. Populations on hawthorn, heather and other wild plants are believed to be the main sources of infestation of orchards.

Oyster scale In Britain this scale has been recorded on apple, apricot, birch, cherry, currant, nectarine, peach, pear and plum. Birch is considered to be the original wild host of this species.

Pear scale In Britain this insect has been recorded from several localities in southern England on apple, peach, pear and poplar. In Europe it also occurs on ash (*Fraxinus* spp.), which is considered to be the original wild host of this species. In the past, pear scale has frequently been confused with oyster scale.

Brown scale This species is found in Britain on acacia, blackberry, broom, cob-nut, cotoneaster, currant, gooseberry, ivy, nectarine, peach, plum, raspberry, rose, vine, etc. It has been recorded from over 300 host plants.

Woolly currant scale In Britain this scale frequently occurs on black, red and white currants, gooseberry and also on rowan.

Woolly vine scale In Britain this species has been recorded on apricot, peach and vine and on some ornamental plants grown under glass or in sheltered positions outdoors.

San José scale This scale attacks over 250 different host plants and is a destructive and widespread pest of fruit trees, but it has not yet been recorded in Britain.

Description

Scale insects belong to the group of sucking insects (Homoptera) that includes greenflies, whiteflies, suckers and mealybugs. Most scale insects are adapted to a sedentary existence and, as a result, the females are wingless and have mere traces of legs and antennae. Apart from the adult males, the only mobile stage in the life cycle of most species is the newly hatched young. This is called a 'crawler'; it is this stage which disperses, being carried by air currents or on insects or birds, and starts new infestations on other trees. The crawlers then settle down to feed on plant sap taken up through extremely fine mouth stylets which also serve to anchor the insect. The brown scale is exceptional in that all the immature stages retain a limited degree of mobility, but the adult female once settled usually remains fixed on the same spot.

The mussel and oystershell scales secrete a waxy, shield-like covering which is supplemented each time the insect moults by additional wax and the cast skin. Thus, immature stages and adult females are similar in appearance, differing only in the size of the waxy covering. This covering protects the soft body of the female and her deposited eggs. In the brown and woolly (cushion) scales, the back of the female hardens but not until near or at maturity. This hardening protects the eggs of the brown scale, which are laid under the body. However, the eggs of the woolly scales are laid in a white waxy egg sac in the form of a woolly pad or cushion produced under and behind the hind end of the body. The waxy covering of mussel and oystershell scales and the hardened skin of the brown scale often remain on the bark for some years after the females have died, thus giving an exaggerated impression of the population of living insects.

The adult male is like a normal insect but with only one pair of wings and no mouthparts. It is seldom seen as it is minute and delicate and has a very short life.

Most species live on the bark of trunks or branches, but the immature stages of the brown scale live on the underside of leaves in the summer and move to twigs and small branches before leaf fall in the autumn. Crawlers of the mussel and pear scales may settle on fruits, causing reddish-purple spots to appear at the feeding sites; these spots (Fig.12.2) are indistinguishable from those caused by San José scale on fruit in other countries.

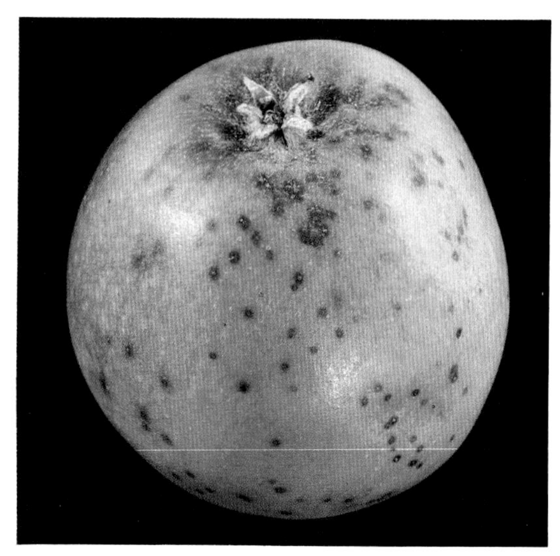

Fig.12.2 Pear scales on apple (variety Cox's Orange Pippin)

Individual species can be identified only by specialists. For general purposes the shape, size and colour of the female scale are the only means of recognition. These details are given below.

The **mussel scale** (Fig. 12.1) is elongate, like a minute mussel shell, 3.5 mm long and greyish-brown.

Both **oystershell scales** (e.g. Fig. 12.3) and the **San José scale** are circular, rather flat, about 2 mm in diameter and grey.

The **brown scale** (Fig. 12.4) is oval, very convex, 6 mm long and shiny brown.

The **woolly currant** (Fig. 12.5) and **woolly vine scales** are oval, wrinkled and rather flat. The former is 4 mm long and dark brown, while the latter is 5–8 mm long and chestnut-brown. When the egg sac is present a large white woolly mass protrudes and raises the hind part of the scale off the bark.

Life history

The oystershell and woolly scales are bisexual. The mussel scale and the brown scale have parthenogenetic and bisexual races, the

Fig.12.3 Pear scales on bark of apple tree (× 6)

Fig.12.4 Dead female and live second-stage nymphs of brown scale on loganberry stem (× 9)

distribution of which depends on the locality and on the host plant. Both forms occur in Britain, but the bisexual form of the mussel

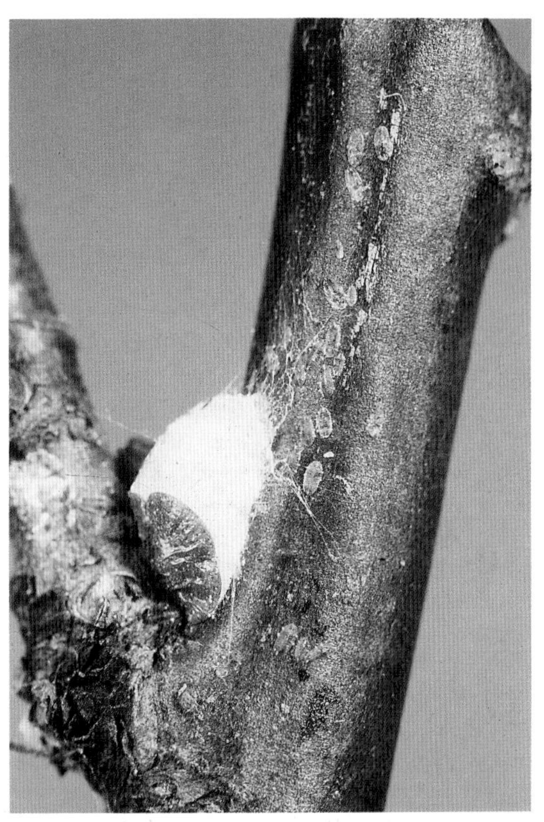

Fig.12.5 Female and young of woolly currant scale on red currant stem (× 5)

scale has been found only on bilberry and heather, while the bisexual form of the brown scale occurs on cotoneaster. There is no evidence to suggest that the bisexual strain can spread to adjacent fruit crops. Outdoors in Britain all these species have one generation a year.

Mussel scale

The minute, active crawlers hatch in late May and early June and escape on to the stems on which they wander for three to four days before settling. After settling and inserting its mouthparts in a plant, a crawler soon moults into the second-stage nymph. The adult stage is reached in late July after another moult and, in bisexual populations, the males are on the wing in early August. By the end of August the eggs are laid under the scale and the female then dies. The eggs are white and up to 80 may be laid under one scale. The egg is the hibernating stage in this species.

Occasionally, mussel scales are found on leaves or fruit, but the eggs of such individuals are unlikely to survive the winter.

Oystershell scales

The life histories and habits of the oyster and pear scales are similar. Egg laying begins in late June or early July and continues until the end of July for the oyster scale and as late as September, or even October, for the pear scale. A female oyster scale can lay up to 100 eggs, a female pear scale from 40 to 70.

The eggs hatch within a day or so of being laid. The activity of the crawlers is rather limited and most of them settle near or under the scale of the mother. Consequently, on older branches of a host that has been infested for several years the colonies are very crowded and the new, living scales may be covered by two or more layers of old, dead scales of previous generations; this makes chemical control difficult. Some crawlers migrate to the new growth and gradually spread the infestation.

The prolonged egg-laying period of the pear scale results in considerable overlapping of the developmental stages; some overlap of stages also occurs with the oyster scale. In both species the main hibernating stage is the second-stage nymph, but some pear scales hibernate at other stages. Development is resumed in the spring and the adults of both sexes appear in May or early June.

Brown scale

Hatching begins in mid- or late June and may continue until mid-July. After a few days, the young escape from within the dried parent and move upwards to the vigorously growing green parts of the plant including the leaves, where they feed. In August they moult into the second-stage nymph, which continues to feed and wander around. On deciduous host plants the nymphs move to the twigs and small branches when the foliage begins to die in autumn, but on evergreen host plants they remain on the underside of the leaves.

Both the first- and the young second-stage nymphs are greenish, semi-transparent, flat, oval and no more than 0.5 mm long. The overwintering second-stage nymphs show little activity and little increase in size but gradually change colour, becoming orange or brown to reddish-brown (Fig. 12.4).

In spring the nymphs resume feeding and become active again. In April they moult into adults, which finally settle and remain fixed. In bisexual populations the males also emerge at this time. The females rapidly increase in size and become convex. During May and June each female produces several hundred tiny, white eggs and then dies. Her dead body, the so-called brown scale, protects the eggs and the newly hatched young. These scales may persist on the plant for several years.

Woolly scales

On outdoor plants the life histories of the woolly currant and woolly vine scales are similar, but when the woolly vine scale occurs on plants in glasshouses or in sheltered positions its life history varies according to the conditions.

Woolly currant scale The young crawlers hatch from early June until the beginning of July; they crawl out of the woolly egg sac and move on to young shoots and leaves on which they feed. In the first half of July they moult into the second-stage nymph, continue to feed and gradually migrate to the twigs where they settle. They moult into the third-stage nymph in the second half of August. Adults of both sexes appear about mid-September. After mating, the males die and the females hibernate.

The immature stages and the young females are flat, broadly oval or circular, gradually increasing in size and changing from whitish in the first two nymphal stages to pale greyish-brown in the young female.

Early in spring the females resume their growth, becoming more convex and deeper brown, and in May they begin to spin their egg sac and to lay eggs. Egg laying in each female may last two to three weeks, during which about a thousand eggs are laid.

Natural enemies

Under natural conditions scales are controlled by a complex of natural enemies. The minute chalcid wasp *Aphytis mytilaspidis* (LeBaron) is a common external parasite of mussel, oyster and pear scales. The egg of the parasite, usually one per scale, is laid under the waxy scale close to the body of the insect. The wasp has two generations a year and can feed on the second nymphal stage as well as on the adult female. Several related wasps are also found commonly on oyster scale. Good control of brown scale is frequently achieved by another parasitic wasp — *Coccophagus lycimnia* (Walker). Levels of parasitism by these wasps can be assessed by carefully looking for small circular holes on old scales from which the wasps haved emerged.

The pear scale is preyed on by larvae and adults of the ladybird *Chilocorus renipustulatus* (Scriba) and by the mirid *Pilophorus perplexus* (Douglas & Scott). Predatory mites, anthocorid bugs and other predators are also responsible for destroying large numbers of scales, particularly the vulnerable young nymphs.

The adult woolly scales with their conspicuous white egg sacs are readily preyed on by birds (titmice) and are also attacked by the predaceous larvae of some flies (*Leucopis* sp.) and by egg-parasitic wasps (Mymaridae).

Control measures

Scale insects are not at present a problem on commercial fruit because of the regular spray programmes normally used. In the past, winter washes largely contributed to their eradication, but the continued absence nowadays of mussel and oystershell scales from apple orchards is probably due to the use of broad-spectrum insecticides for control of codling moth and fruit tree tortrix moths (see pages 111 and 140, respectively). These sprays, in late June or in July, are applied soon after the majority of the scale insect eggs have hatched.

In addition to fruiting plantations, it is essential that nursery stock is kept free from scale insects to avoid the risk of distributing infested material.

If fruit trees or bushes require treatment, there is a choice of using a winter wash in the dormant season or a spray in summer. Generally, the winter treatment is preferable as all species can be dealt with at the same time. The present trend towards the use of selective pesticides will encourage outbreaks of scale. The chemicals currently recommended for use against scale insects are broad spectrum and highly disruptive of an integrated pest control programme. For instance, winter washes kill overwintering stages of predatory mites and will result in an increase in the incidence of fruit tree red spider mite (see page 341). Where integrated control is being practised and scale outbreaks are localized, individual treatment of infested trees is a possible solution.

Winter wash

To avoid injury, deciduous trees and bushes should be sprayed with a winter wash when the buds are dormant: the safest period is December and January. Peaches, especially those under glass, must be sprayed before the end of December. This treatment should not be used on Myrobalan plum. Winter washes should not be used on evergreens.

When applying a winter wash, thorough and forceful spraying is necessary to ensure complete wetting of the bark and penetration into crevices in the wood, especially of wall-trained trees.

Summer sprays

Good control of scale insects can also be obtained during the growing season providing sprays are correctly timed and applied so that all the bark and the underside of the leaves are thoroughly wetted. The aim is to spray when the tiny crawlers — the most susceptible stage — are settling down soon after hatching. This is the best time to control infestations on evergreens, but on deciduous woody plants the presence of much foliage and sometimes blossom or ripening fruit makes the task more difficult than in winter.

Two sprays a fortnight apart are often necessary, the first one applied according to species as follows:

Mussel scale—early May under glass or early June outdoors
Oystershell scales—mid-July
Brown scale—early July
Woolly currant scale—mid-June.

13
Willow–carrot aphid

Fig.13.1 Young carrot plants in June showing twisted foliage infested with willow–carrot aphids and shining with honeydew

The willow–carrot aphid (*Cavariella aegopodii* (Scopoli)) is a widespread pest of carrot and often causes considerable loss of yield in early and mid-season crops. Parsnip, celery, parsley and other cultivated umbellifers such as chervil and fennel can also be attacked by this aphid and the willow–parsnip aphid (*Cavariella theobaldi* (Gillette & Bragg)). The willow–carrot aphid overwinters in the egg stage on willow trees, especially the crack willow (*Salix fragilis*) and the white willow (*Salix alba*), and also as an adult or immature stage on host crops that remain in the ground until spring.

Carrot motley dwarf virus complex is transmitted by this aphid.

ON CARROT

Nature of damage

Attacked carrot plants may have twisted and malformed foliage (Fig. 13.1), be stunted and even killed. The foliage may also show reddish or yellow discolorations and, when many aphids are present, appear varnished (Fig. 13.1) from the coating of honeydew secreted by the pest. The cast skins of the aphids adhere to the honeydew so that the plants and the ground below soon look as if they have been sprinkled with cigarette ash. The symptoms are frequently mistaken for those of drought or carrot fly attack (for the latter, see page 244).

Wingless aphids generally begin to arrive on crops in the second or third week of May; thus early-sown maincrop carrots, often at an early stage of growth, can be attacked.

Description and life history

The wingless aphids are small, green and inconspicuous. They settle on the leaves near the main vein where they blend with their background. The winged aphids are darker and more easily seen.

The eggs are laid round the bud axils of willows (Fig. 13.2) and soon become shiny black. They hatch in February or March; if this is before the willow tree breaks bud, the young aphids feed through the bark of young shoots and later move to the foliage and catkins where they start colonies. Winged aphids form in these colonies in May (Fig. 13.3) and migrate to carrots over a period of five to six weeks, usually with a peak in early June. Late seasons may delay migration for two to three weeks.

Typically, the progeny of these winged forms develop to peak numbers about the fourth week of June. The population then declines and usually by the end of July few willow–carrot aphids can be found on carrots. This decline is due mainly to the development of another winged generation, which leaves carrots and disperses chiefly to hedgerow umbellifers and willows.

Many umbellifers die down completely before the winter, but those which do not can support hibernating forms of the aphid capable of producing colonies in the following spring. Winged forms may arise from these colonies in advance of those developing on willows. In this

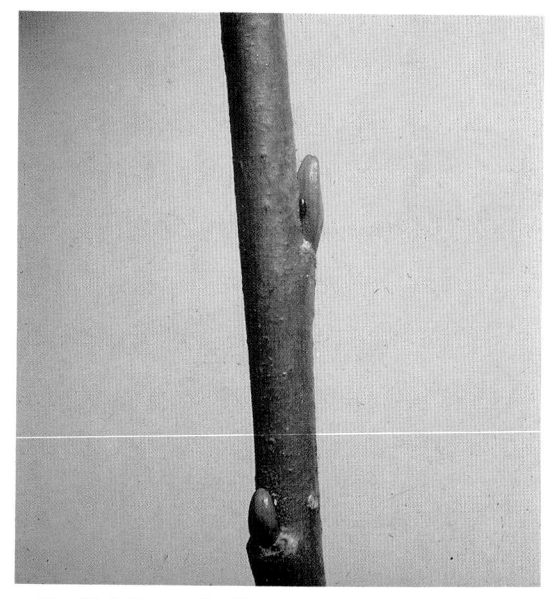

Fig.13.2 Eggs of willow–carrot aphid on willow twig in winter

respect, the most important overwintering sites are carrot crops left in the ground throughout the winter, for the few aphids that may have persisted on carrots in the summer can be supplemented by others returning to carrots in late summer and the autumn. Such overwintering is of particular importance for transmission of carrot motley dwarf virus, because any winged progeny of the hibernating aphids are likely to be carrying the virus when they arrive on new crops.

Aphid numbers

Natural enemies such as ladybirds, hover flies and parasites can reduce populations of the willow–carrot aphid, but the main cause of fluctuations in aphid numbers from year to year is climatic. Temperature and rainfall largely determine how troublesome willow–carrot aphid will be in any particular year. The aphid is unusual in that, during the spring flight period from willows to umbellifers, some winged individuals settle on other willows, where they found colonies of potential winged forms which become adult in little more than a

Fig.13.3 Willow–carrot aphids, including a winged form, on willow shoot in May (× 8)

week. If the weather is warm, these considerably increase the numbers migrating to carrots. Dry sunny weather at the end of May and in early June favours this process and ensures a large-scale overall migration to carrots from, originally, a small number of overwintering eggs. Conversely, in cold rainy weather, very large pre-migration populations on willows may never be able to reach or successfully colonize carrots.

Virus incidence

Willows are not a source of carrot motley dwarf virus complex; high incidence of the disease is closely linked with the hibernation of the aphid on overwintering carrots.

The practice of covering up carrots for late lifting usually enables some aphids to survive on carrot crops even in a severe winter. Late crops and crops that have been damaged by frost, pests or diseases are often left until April, or not adequately destroyed, and can form a major source of virus-carrying aphids in the spring. The rate at which these aphids multiply and migrate to other crops depends on the temperature, but only a few still, warm days are needed for a significant migration. Overwintered carrots should not, therefore, be ignored since they can act as a source of virus, even in a cold spring.

Because hibernation by the aphid occurs on only a few wild umbellifers, these play little part in the spread of virus before the middle of June; even then most of the virus reaching carrots from wild umbellifers is probably transmitted in July, when aphid dispersal from these plants mainly occurs.

For an account of the virus on parsley, see page 76.

Control measures

In carrot-growing areas, willows are too numerous and too large for any practical scheme aimed at reducing this source of the pest. Carrot crops not lifted by the end of March should be sprayed then (preferably with a systemic insecticide of short persistence) to prevent any aphid migration to the new season's crop. This spray treatment should also be applied to any overwintered umbelliferous crops left for seed production.

Crops sown in April or May are most at risk, while those sown in June may often escape attack.

Sprays

Several insecticides are recommended for use as sprays on carrots. They are best applied as a band spray over the foliage in not less than 200 litres of water per hectare. The more persistent chemicals have given the best results. The migration from willows can normally be controlled by two applications to the crop: the first a week after the aphids arrive on the crop and the second 14 days later to coincide with the immigration peak. A third spray application may occasionally be needed if the aphids are particularly troublesome, but there is little to be gained economically by exceeding three applications. Where migrations have occurred from old crops, up to seven sprays of a persistent aphicide have failed to control the aphids and the only hope in such a situation is to find the source of the aphids and deal with them there.

The less persistent chemicals may need to be applied at more frequent intervals and are only suitable for use on carrots near to harvest.

Granules

Certain systemic insecticide granules can be used as a bow-wave treatment or placed 25 mm below the seed to give several weeks' protection. For carrots grown on the bed system, granules should be broadcast over the soil surface and harrowed into the top 25 mm before drilling. Some granular insecticides are recommended for use on mineral soils but not on organic soils.

Granules do not control aphids as effectively as do spray treatments. In a year when aphid attack is intense it will be necessary to supplement granules with sprays; some loss of crop will still occur. Granular treatments also control carrot fly attacking crops lifted in summer or early autumn (see page 249).

Certain granules can also be used on the foliage as a single application at emergence or as a split application, using half of the granules then and half three weeks later. Only pumice-based granules should be used. Foliar applications are less persistent than soil applications.

ON PARSLEY

The seasonal development pattern of the aphid on parsley is similar to that on carrot. Where parsley is grown throughout the year its tightly curled leaves offer concealment and much protection for hibernating forms of the aphid.

Direct damage by the aphid is seldom as serious as that caused by transmission of virus, especially carrot motley dwarf virus which causes a particularly damaging disease in parsley. Until late June or July parsley infected with this virus may seem healthy, but then the green foliage suddenly shows reddish discoloration which soon turns to a bleached yellow. These colour changes are accompanied by stunting of the plants and quickly render the crop unmarketable.

The source of the virus is invariably overwintered crops of carrot or parsley from which virus-carrying willow–carrot aphids fly in spring. These overwintered crops may not show any sign of infection.

Control measures

To prevent parsley crops from becoming affected by carrot motley dwarf virus in summer, it is necessary to tackle the source of the virus. If overwintered umbelliferous crops, especially parsley, must be retained, subsequent parsley crops should be situated as far from them as possible. In addition, these overwintered crops should be sprayed with a recommended aphicide either in November–December, or in March–April before temperatures are high enough to enable winged aphids to emigrate successfully to new parsley crops. Alternatively, overwintered crops should be ploughed in before mid-May.

Spring-sown crops of parsley can be protected against incoming aphids, thus reducing the spread of virus, by application of a recommended granular insecticide. The granules can be applied either at drilling, using the bow-wave method, or as a foliar treatment at the seedling stage provided that the required interval will elapse between application and harvest. Otherwise a recommended aphicide spray should be applied about 20 May and again 14 days later. However, care is needed in selecting the aphicide for the second spray: if less than three weeks remain between its application and the time when the crop is to be cut, one of the less persistent chemicals should be used. Normally, there should be no need for any further aphicide treatment between the first and second crop-cut, or between the second and third if three cuts are to be taken.

14
Woolly aphid

Fig.14.1 Woolly mass secreted by woolly aphid on branch of apple

The woolly aphid (*Eriosoma lanigerum* (Hausmann)) is generally known by the white woolly masses that are so conspicuous on the stems and trunks of infested apple trees during the summer (Fig. 14.1). The 'wool' covers the aphids and is an accumulation of waxy strands which they have secreted.

This aphid originated in the eastern part of North America and is sometimes known as American blight. It was first noticed in England in 1787 and now occurs in all countries where apples are grown, having been distributed on nursery material. The woolly aphid also attacks cotoneaster, pyracantha and ornamental species of *Malus*.

Four other aphid pests of apple are described on page 17.

Description and life history

The woolly aphid is brown or greyish-purple. The entire life cycle is passed on the host tree. The young overwinter in sheltered positions such as in cracks or under loose bark on the tree. They are not covered with 'wool' so are inconspicuous. In March or April they become active and 'wool' is secreted by wax glands on their bodies. Breeding colonies are present by the end of May. A few winged aphids are produced in July and may fly off to infest other trees. Other winged aphids occur in September and produce egg-laying females. Although a single egg may be laid by each female, there is no further development. Thus, the life cycle consists essentially of successive generations of wingless female aphids producing living young. Breeding slows down in the autumn and the adult aphids die during the winter.

During early summer, colonies are found mainly on the spurs and branches, especially at pruning cuts or where the bark is cracked or damaged by implements. Later the infestations may spread to the young growth, particularly the succulent growths arising from main branches or the trunk ('water shoots'). The colonies are seen first in the axils of the leaves and eventually spread along the whole length of the shoot.

In some countries, breeding colonies are found on apple roots below ground level; these provide a source of reinfestation for the aerial parts of the tree. In Britain, root infestation does not occur, although the aphid may occasionally attack the collar of the tree or roots exposed by cultural operations.

Spread to other parts of a tree is mainly due to the wandering first-stage young; these can also be carried to other trees by wind or mechanical means. New infestations arise from the winged aphids that occur in July and from the introduction of infested nursery stock.

Nature of damage

Galls often form at points where aphids have fed; these spoil the appearance of nursery stock and reduce its value. The galls often split open and may then provide a point of entry for disease organisms such as *Gloeosporium* spp.; apple canker may also develop.

Although woolly aphid colonies are so conspicuous, the amount of injury to established trees is probably less than the appearance suggests. At harvest, however, infestations can be a severe nuisance to the pickers.

Resistant rootstocks

Crosses between Northern Spy, which is highly resistant, and various Malling rootstocks have resulted in the Malling-Merton series of apple rootstocks which are highly resistant to attack by woolly aphid. Isolated cases of infestation by local races of woolly aphid have occurred in Australia, South Africa and North America, but these rootstocks should still be useful in countries where root infestation occurs. They do not confer any resistance on the scion varieties worked on them.

Natural enemies

Predators

Several native species of insect prey on the woolly aphid: the most important are earwigs, ladybirds and their young, and the young of hover flies and lacewings. During late June, July and August, these predators are sometimes responsible for a decline in infestations of woolly aphid. Killing earwigs will lead to an increase in the number of woolly aphids.

Parasite

An introduced parasite of the woolly aphid is widely established in south-east England and in some western districts. Although it does not thrive here as well as in warmer, drier climates, in some years it can make a useful contribution to woolly aphid control.

The woolly aphid parasite is a minute wasp-like insect (*Aphelinus mali* (Haldeman)). The female inserts eggs into the bodies of young aphids, and the developing parasite destroys the internal organs of the aphid. When fully fed, the parasite pupates inside the aphid skin, which remains attached to the bark. There are several generations of the parasite during the summer; winter is passed as a mature grub.

Parasitized aphids are easily recognized: first, the amount of 'wool' is reduced, then the body darkens so that eventually the aphid is shiny blue-black and devoid of wax. A circular hole in the back indicates that the adult parasite has emerged.

Survival of the parasite through the winter can be ensured by storing shoots bearing parasitized aphids in a cold store at 4–8 °C from October to May or June. The shoots should then be tied to the branches of infested trees. However, this procedure is pointless where broad-spectrum pesticides are to be applied in the summer for control of codling moth, tortrix moth or red spider mites, because these chemicals kill the parasite.

Chemical control

Infestations on nursery material must be controlled to prevent further spread of the pest. In orchards and private gardens heavy infestations requiring control may develop during the summer.

In nurseries

In the nursery, aphids mainly overwinter on the stools or layers; effective control is possible only if aphids are destroyed before earthing-up begins. A thorough application of a recommended systemic insecticide just before earthing-up will eliminate all aphids. If colonies appear later on the shoots of rootstocks or worked trees, a systemic spray should be applied at the first sign of infestation — to avoid any risk of galling.

After lifting, rootstocks and worked trees should, as a routine, be fumigated, or dipped in a winter wash (except for the roots) and then stood upside down to drain. These treatments will also kill the eggs of other aphids that may be present.

In orchards

As attacks of woolly aphid tend to be sporadic, it is usually unnecessary to apply a routine spray for control of this pest. Most growers watch for signs of infestation and spray only when necessary.

Successful control depends on efficient wetting and good distribution of the spray. High-volume spraying is preferred, as the spray has to penetrate the protective woolly covering of the aphids and reach colonies living in cracks on the underside of branches.

Some control is obtained by sprays applied primarily for control of other pests, e.g. winter washes, and insecticides used at green cluster or petal fall. However, these treatments do not always prevent infestations from developing during the summer.

The most effective insecticides for woolly aphid control are systemic. When applied any time between petal fall and early June a recommended systemic insecticide can maintain satisfactory control throughout the summer. Care should be taken to observe the minimum interval between application and picking the fruit.

Various other pests may also be controlled by this treatment, but fruit tree red spider mite (see page 341) is now resistant to organophosphorus compounds. Systemic insecticides will have a serious effect on beneficial insects.

In gardens

The worst infestations usually develop on the branches of older trees. These are most easily controlled by spot treatment, such as brushing a dilute solution of a recommended insecticide over the infested area.

Infestations on the young shoots, which arise from aphids dispersing from bark colonies, do not usually occur until July or even later and can often be prevented if bark colonies are treated by the end of June. When such infestations occur, a drenching spray of a recommended insecticide should be applied, care being taken to prevent contamination of other edible produce by drift.

THRIPS

Thysanoptera

15
Gladiolus thrips

Fig.15.1 Gladiolus flowers damaged by feeding of gladiolus thrips

Several species of thrips (thunderflies) have been recorded from gladiolus, but the one which breeds extensively on this plant, and the most important, is the gladiolus thrips (*Thrips simplex* (Morison)). This thrips also feeds on several other flower bulb crops including freesia, iris, lily and montbretia. It was first recorded in Britain in 1950 and has since spread to many areas. In the warmer, drier districts, gladiolus crops are likely to become infested at some time during the summer in most years. During cool, wet conditions attacks are less severe, but the thrips population can survive at a low level and build up again under more favourable conditions unless control measures are taken.

Nature of damage

Thrips feed by puncturing the plant cells and sucking the contents; these cells become filled with air and give the characteristic silvery appearance to the damaged leaves (Fig. 15.2) and cause the bleached flecks on the flowers (Fig. 15.1). With severe infestations the flowers may fail to open or the leaves and flower spike may bleach to a golden-brown colour and then wither. Red and purple varieties are considered to be the most susceptible; white and early varieties often escape serious attack.

At the end of the growing season many thrips perish, but survivors move down the plant and invade the corm where their feeding causes dried grey-brown rash-like patches. When infested corms are replanted the new growth is soon invaded.

Fig.15.2 Gladiolus leaf showing speckling caused by feeding of gladiolus thrips

Description

The adult is dark brown, about 1.7 mm long and elongate with a pointed hind end. Both sexes have two pairs of wings; each wing comprises a narrow membrane, which is dark except for a lighter band near its point of attachment to the body, and bristle-like projections forming a fringe. The immature stages are yellowish-orange at first but later darken towards reddish-brown; they lack wings and range in size from about 0.6 mm to 1.7 mm.

Life history

In store, eggs are laid in the flesh of the corm and hatch after one to three weeks, depending on temperature. The young feed for about one to two weeks, after which they pass through a non-feeding, almost immobile, pre-pupal and pupal stage of about one week before emerging as adults. These adults may survive for three or four weeks during which time each female may lay more than 100 eggs. One generation is completed in less than three weeks at 25 °C.

After the corms have been planted, the adults and young invade the developing leaves; they are frequently found in the tightest folds of the leaf sheaths and flower buds, where they continue to breed throughout the growth of the crop. Later in the season they return to the corms and the soil. In Britain they are unlikely to survive freely in the soil from one season to the next. Adults surviving on corms can breed throughout the storage period if temperatures are above 15 °C.

Control measures

In the store

Careful management of stocks is essential if this pest is not to become a serious problem on the holding. Newly acquired corms should be carefully examined for infestation or damage and, if affected, should be rejected to prevent the spread of infestations. The field performance of new stocks should not be assessed without first subjecting them to a full fumigation programme. As an insurance, all corms should be fumigated before being stored. To ensure that all stages, including newly hatched young, are killed, several fumigations are needed. Ideally, after treatment the store temperature should be lowered to about 10 °C to limit further breeding.

During the filling of a store, fumigation should be repeated as new stocks are added and finally, to prevent reinfestation after fumigation, a recommended insecticide smoke may be used. Several small smoke generators of total capacity appropriate to the store volume, and well distributed throughout the store, should be used rather than one or two large generators. If the store is to remain closed, insecticide-impregnated plastic strips at the rate of four per 100 m^3 of store volume can be used instead of a smoke treatment.

In the field

Cultural control When non-infested corms are planted early, the growing crop is less likely to become infested. Maincrop varieties usually suffer most damage. It is, therefore, worth considering planting these varieties on the prevailing windward side of any early varieties to limit the spread of infestation from early to maincrop varieties. Similarly, if the planting of several varieties in adjoining long narrow rows can be avoided, cross-infestation may be reduced. Where the crop is being grown for corm production, removal and destruction of the flowering spike — once the variety has been determined — should reduce the risk of thrips invading the corms at the end of the season.

Examination of plants Early detection and programmed spraying are necessary to keep the thrips in check. Stocks should be examined at the two- to three-leaf stage, immature and adult thrips being sought in the tightest leaf sheaths by splitting these open. In most years air temperatures from about mid-May onwards are suitable for thrips to fly into uninfested crops from elsewhere. The first sign of damage may be silver flecks on the foliage; if this damage is ignored in hot, dry weather, the leaves quickly turn yellow and wither.

Chemical control Recommended insecticide sprays must be applied from late May or early June onwards if thrips have been detected or if conditions are suitable for their activity, i.e. warm (20 °C), settled, sunny weather. Routine spraying at two- to three-week intervals should provide adequate cover on new leaf and shoot growth to prevent damage to the flowers. When the crop is grown for the flowers and therefore only pre-flowering sprays are needed, the first of these should be applied one month before the expected date of flowering, with a second spray a fortnight later. Where large crop areas are to be treated, alleyways should be left to allow access for ground-spraying equipment to treat the crop when fully grown. A spray volume of about 700 litres/ha is necessary if good cover and control are to be achieved with the early treatments.

When the need for later treatment coincides with hot, dry weather, application should be made during the coolest part of the day and the spray applied at high volume until the start of run-off. This is to ensure adequate cover of the crop and to reduce evaporation losses and the risk of scorch.

16
Thrips on peas

Fig.16.1 Pea pod and leaves damaged by pea thrips

The small insects known as thrips, or thunderflies, are abundant in Britain and some of them can cause considerable damage to cultivated plants.Different species breed on grasses and cereals and on fruit, vegetable and flower crops. Two species that attack peas are described here.

The pea thrips (*Kakothrips pisivorus* (Westwood)) is mainly a pest of peas and broad beans in gardens and market gardens, though it is frequently present on peas grown on a field scale. The thrips attack the young pods and young foliage from early June onwards, reducing the quality and yield of pods.

The field thrips (*Thrips angusticeps* Uzel), also known as the cabbage thrips, is a field pest and injures very young pea plants in the spring. Attacks continue for only a few weeks; they usually occur on peas that follow brassicas or are grown close to fields that grew brassicas the previous year.

PEA THRIPS

Nature of damage

Pea thrips is found throughout England and Wales, but damage is usually recorded in districts in south and east England, particularly in hot, dry seasons. The main attacks occur in June and July. The most notable feature of attack is the appearance of silvery mottled patches on the young pods and leaves where the thrips have destroyed the surface cells (Fig. 16.1). On young flat pods such extensive feeding injury results in small bent pods containing few peas. Similar marks appear on the young leaves. Heavy attacks may lead to stunting of the plants; severe early attacks cause blind flowers which produce no pods.

Description

The adult pea thrips (Fig. 16.2, left) cannot be distinguished from other species of thrips without microscopic examination. It is a shiny dark brown insect with some yellow, about 1.7 mm long, elongate, tapering towards the head and pointed at the hind end. It has four wings, usually folded along the back; each wing consists of a narrow membrane surrounded by a broad fringe of hairs. These hairs increase the area of wing sufficiently to enable the insects to fly or drift long distances.

The young pea thrips (Fig. 16.2, right) resembles the adult in shape but has no wings; it is bright yellow or orange with a conspicuous black tip to the abdomen.

Life history

Minute, kidney-shaped eggs are laid in June. They are embedded in the tissues of the stamen sheaths, young pods and occasionally in the petals of a pea or bean flower. Hatching occurs after about nine days and the young feed on the petals, stamens and inner parts of the flower and later the pods. Their method of feeding is characteristic: they pierce the surface tissues of the plant and then suck up the exuding liquid, together with minute fragments from the wound.

After feeding for three or four weeks the young thrips descend into the soil. There they remain inactive until the following spring when they become mature. Winged thrips leave the soil in May or June and search for pea or bean flowers.

Control measures

On field crops control measures against pea thrips can safely be left until signs of attack are seen.Frequent examinations of peas should be made from the appearance of the first pod until the pods are full. The pest can be controlled by spraying with a recommended insecticide.

On smallholdings and in gardens peas are seriously attacked more often. A careful watch should be kept for thrips injury on the young pods and foliage from mid-June until August. If slight but widespread injury is discovered and the weather is warm, a recommended insecticide should be applied without delay.

FIELD THRIPS

Nature of damage

Field thrips is a localized pest on peas in parts of Essex and Cambridgeshire, on the wolds of Lincolnshire and North Humberside and occasionally elsewhere. Damage starts as soon as the seedlings emerge; adults may be found feeding inside the terminal leaflets even before emergence. Damage is usually noticed in April when the crop is 5–10 cm high. Growth is checked, the crop is pale and on close inspection the leaves are seen to be puckered with pale blotches (Fig. 16.3).

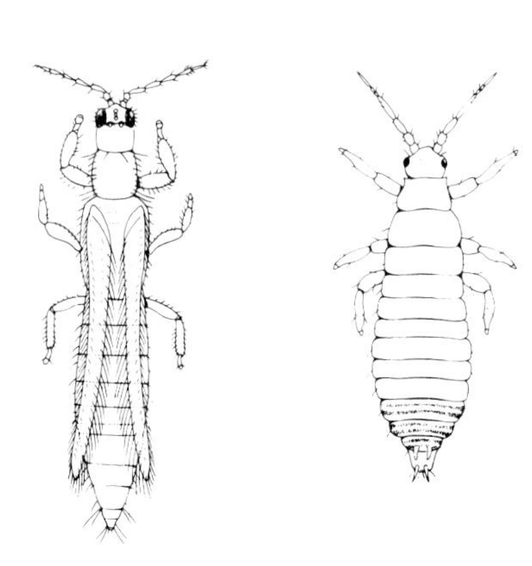

Fig.16.2 Left: adult thrips (× 25). Right: immature thrips (× 30)

Life history

Most crops, particularly brassica seed crops, preceding peas are likely to become infested during the summer months by winged field thrips. The next generation, which is short-winged and flightless, overwinters in the soil. Most individuals emerge early in the following spring and attack peas grown in the same field, but a few may remain in the soil for a further year. The next generation has normal-sized wings and leaves the pea crop from late May onwards.

Control measures

When growth is retarded by cold or dry weather, damage by field thrips is often apparent, but with the onset of better conditions the crop usually grows away. Thus, although sprays applied to dry-harvesting peas at crop emergence have given good control of field thrips and reduced damage, no yield increases or differences in crop maturity have been recorded.

In areas where field thrips is a problem, emerging crops, particularly of vining peas, should be examined closely and sprayed with a recommended insecticide immediately thrips damage is seen. Late-emerging crops usually escape severe damage. As most of the damage occurs early in the life of the crop, sprays applied after mid-May are not worth while.

Fig.16.3 Pea plant showing distortion caused by field thrips

CATERPILLARS

Lepidoptera and Hymenoptera: Symphyta

17
Apple sawfly

Fig.17.1 Apple fruitlets attacked by caterpillars of apple sawfly, showing (left) an entry hole, (centre) an entry hole surrounded by wet frass, and (right) scarring

The apple sawfly (*Hoplocampa testudinea* (Klug)) is a pest of apple in many parts of Britain. Injury is caused by the caterpillars burrowing in the fruit and is sometimes mistaken for that caused by caterpillars of the codling moth (see page 109). All the fruit infested with sawfly falls during June and early July, and some of the remaining fruit is reduced in value by scars caused by young caterpillars that failed to establish themselves (Figs 17.1 and 17.2). Because of its thinning action sawfly attack is less important in years when fruit set is heavy.

Recognition of caterpillar and damage

Caterpillars of the apple sawfly, codling moth and fruitlet mining tortrix (see page 137) are similar in appearance, having a whitish body and black or brown head. They all damage apples by tunnelling in the fruit.

The simplest method of distinguishing caterpillars of apple sawfly from those of codling moth is based on the time of occurrence. In the south of England sawfly caterpillars are fully grown and most have left the fruits by the end of June, whereas very few, if any, codling caterpillars have hatched by this date. At this time an apple attacked by a sawfly caterpillar has a mass of wet, reddish-brown frass (excrement) around a large hole, whereas a codling caterpillar produces a small amount of dry frass which covers a small entry hole.

Caterpillars of the fruitlet mining tortrix occur during June and early July at the centre of a cluster of apples joined together by silken webbing. Damage caused by these tortrix caterpillars may be distinguished by the presence of shallow punctures in the apples and the fact that the deeper tunnels follow a meandering course instead of going straight to the core.

The number and distribution of the fleshy, sucker-like legs (prolegs) on the abdomen of the caterpillars can also be used for identification. In addition to those at the end of the abdomen, sawfly caterpillars have prolegs on segments 2–7 (Fig. 17.3), while they occur only on segments 3–6 of codling and fruitlet mining tortrix caterpillars.

Fig.17.2 Mature apples scarred by caterpillars of apple sawfly early in the season

Description of adult

The adult sawfly (Fig. 17.4) is black above, reddish-yellow below and has reddish-yellow legs. The first adults emerge in the spring at about the time apples are beginning to flower. Peak numbers occur when the trees are in full bloom. The sawflies are active on warm, sunny days, but not in dull, wet weather. They live for about eight or nine days and are attracted only to trees in blossom.

Life history

Eggs are laid singly in open blossoms, usually from below; the female uses her saw-like ovipositor to make a slit in the receptacle just below the base of a sepal. After the ovipositor has been withdrawn, sap oozes from the slit and becomes reddish-brown. This discoloration, which can be seen on the underside of the blossom, provides an easy method for assessing the extent of an infestation. About 30 eggs are laid by each female. Hatching occurs after about 14 days, beginning four or five days after 80% petal fall.

The caterpillar (Fig. 17.3) has a white or cream-coloured body; the head is black at first but light brown later. Soon after emerging, the caterpillar burrows into the fruit, usually starting within the eye (calyx) and moving outwards. On reaching the skin of the fruitlet it turns abruptly and burrows just under the skin, before penetrating towards the core where it eats the seeds. When the seeds have been damaged, the fruit ceases to develop and falls. If the caterpillar fails to reach the core, the fruit may continue to develop but will be marked by the characteristic scars illustrated in Fig. 17.2. The caterpillar migrates to another fruitlet in which it bores straight to the core. Fruitlets attacked in this way can readily be recognized by the large entry hole around which frass accumulates. Each caterpillar normally attacks two fruitlets and may attack a third, but seldom more.

The caterpillar feeds for about four weeks,

then drops from the fruit and forms a cocoon in the soil, usually at a depth of between 8 and 24 cm. It then enters the pre-pupal stage but does not pupate until three or four weeks before it is due to emerge in the spring. Not all the caterpillars pupate in the first spring; some may remain in the pre-pupal stage until the spring of the second year.

Varietal susceptibility

Certain varieties of apple seem to be especially susceptible to sawfly attack; these include Charles Ross, Discovery, Ellison's Orange, James Grieve and Worcester Pearmain. Culinary apples, with the exception of Early Victoria and Edward VII, are rarely attacked by sawfly.

Control measures

Monitoring sawfly activity

Adults of both the apple sawfly and the plum sawfly (*Hoplocampa flava* (L.)) are highly attracted to white, non-ultraviolet reflective surfaces. Sticky traps of this type can be placed in orchards during the blossom period to monitor the activity of these adults, but care must be taken that other common sawfly species (e.g. dock sawfly (*Ametastegia glabrata* (Fallén)) are not mistaken for the pest species.

These traps will also cause a limited reduction in sawfly attack but will not adequately control a serious infestation.

Chemical control

In the past, regular spraying eliminated apple sawfly from many dessert apple orchards but, with the recent reduction in insecticide spray programmes, the pest is now increasing. However, sprays are necessary only where damage occurred in the previous year. On culinary varieties sawfly attack is generally insufficient to warrant control measures.

In orchards Apple sawfly can be controlled very effectively provided the application is correctly timed. A spray of a recommended insecticide should be applied about the time eggs are beginning to hatch, in order to kill the young caterpillars before they have tunnelled under the skin of the fruitlet. Satisfactory control has been obtained by spraying at any time during the seven-day period immediately following 80% petal fall. At this stage, which can be judged by eye, most of the petals have fallen and the majority of those remaining will fall if the tree is gently shaken.

Since eggs are laid throughout the blossom period, many, if not all, will be present at the time of spraying. Irrespective of the insecticide used, these eggs appear to continue development up to their normal time of hatching, but the young caterpillars die shortly after emerging from their eggs. If spraying is delayed beyond the seven-day period after petal fall,

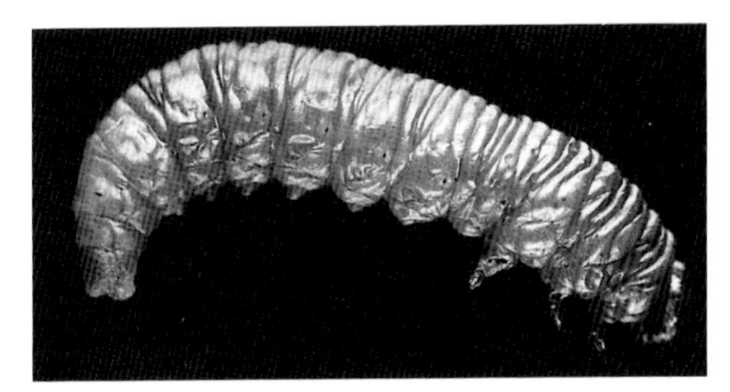

Fig.17.3 Caterpillar of apple sawfly (× 8)

Fig.17.4 Apple sawfly on apple blossom (× 6)

many caterpillars will have hatched and already burrowed some distance beneath the skin of a fruitlet. These caterpillars are killed, but the site of the burrow remains as a corky scar (Fig. 17.2) which reduces the quality of the fruit.

Insecticide treatment at petal fall may also control aphids (see page 19) and tortrix caterpillars (see page 140).

Protection of bees. Where bees have been imported into orchards, or if other bees are visiting the trees, it is desirable not to spray until they have finished working the flowers. Wild flowers, such as dandelions, underneath the trees should be cut.

In gardens, if the trees are small, apple sawfly can be controlled by picking off and destroying infested fruits. It is advisable to start inspecting the trees early in June while caterpillars are still feeding in their first fruit; this can be recognized by the presence of a tunnel just under the skin. All infested fruits should be removed by the end of the third week in June, as caterpillars are fully fed and begin to leave the fruits soon after this time.

Success with an insecticide spray is only possible if equipment is adequate to give a good cover. Care must be taken to prevent contamination, by drift, of produce that will be harvested within a few days of spraying.

18 Cabbage caterpillars

Fig.18.1 Cabbage damaged by cabbage caterpillars

The caterpillars so often found destroying cabbages, cauliflowers and other brassicas are usually those of cabbage white butterflies and cabbage moth (*Mamestra brassicae* (L.)). Caterpillars of the diamond-back moth (*Plutella xylostella* (L.)) sometimes occur in large numbers and can do a great deal of damage. Cabbages and Brussels sprouts may also be attacked by caterpillars of the garden pebble moth (*Evergestis forficalis* (L.)), which in some years are as important a cause of damage as cabbage white caterpillars.

Cabbage white caterpillars attack all vegetables of the cabbage family, flowers such as stocks and, among weeds, charlock and shepherd's purse — in fact, almost any cruciferous plant. They may also be found abundantly on garden nasturtium. They are mainly a pest of allotments and gardens.

The caterpillars of the diamond-back moth and the garden pebble moth are pests of cruciferous crops only, but those of the cabbage moth also attack other plants, such as lettuce, maize, onion, tomato and many ornamentals.

Nature of damage

Small white butterfly

The caterpillars of this species (*Pieris rapae* (L.)) cause the most serious economic damage to commercial crops of brassicas, although the damage is less spectacular than that caused by caterpillars of the large white butterfly. The outer leaves on which the eggs are laid do not suffer much injury from the feeding of the young caterpillars, but these caterpillars soon move to the centre of the plant where they feed on the young leaves and foul them with their excrement.

Small white butterflies fly into large, open fields and, because eggs are laid singly, many plants are attacked; thus commercial crops can be seriously damaged.

Large white butterfly

Since this butterfly (*Pieris brassicae* (L.)) lays its eggs in batches of 20–100 eggs, the caterpillars occur in large numbers on plants and can rapidly reduce them to skeletons. Because only a few plants are attacked and because the butterflies prefer shelter, damage (Fig. 18.1) is more serious in gardens and allotments and on the headlands of sheltered fields. In coastal regions, damage may be particularly severe when there is a large migration of butterflies from the Continent, but immigration has become much less common in recent years.

Green-veined white butterfly

This species (*Pieris napi* (L.)) causes similar damage to that of the small white butterfly, but it is much less common and rarely lays its eggs on cultivated crucifers.

Cabbage moth

Although caterpillars of this moth will feed on a wide range of plants, damage is most serious when it occurs in the heart of brassica plants, particularly cabbages, which the caterpillars foul with their excrement. Attacks are most common in gardens and allotments but severe damage may also occur in field crops.

Diamond-back moth

Caterpillars of this species feed on the lower surface of leaves, often rejecting the upper surface and the main veins. The leaves assume a translucent silvery appearance before becoming brown and shrivelled. In some years diamond-back moth caterpillars feed on developing Brussels sprout buttons, making small holes which enlarge as the sprouts grow. Occasionally, they may attack the heart of brassica plants near to harvest.

These caterpillars cause most harm from midsummer onwards and the damage may be done remarkably quickly. Outbreaks are usually confined to small areas, often near the coast. Such outbreaks generally occur during a hot summer in which the moth is able to complete more than two generations, or when

Fig.18.2 Large white butterflies (× 1). Left: male. Right: female

Fig.18.3 Caterpillar of small white butterfly (× 2)

the British population is reinforced by migrants from the Continent.

Garden pebble moth

Damage caused by caterpillars of this moth has become more serious in recent years, particularly on Brussels sprouts in the Vale of Evesham and in Bedfordshire. The caterpillars feed on the lower surface of leaves and in the heart of plants, initially producing copious webbing.

Other species

The caterpillars of four other moths are occasionally troublesome, feeding usually on the outer leaves of cruciferous plants. These moths are the tomato moth (*Lacanobia oleracea* (L.)), dot moth, (*Melanchra persicariae* (L.)), silver y moth (*Autographa gamma* (L.)) and the flax tortrix moth (*Cnephasia asseclana* (Denis & Schiffermüller)). They rarely justify treatment.

Description and life history

Small white butterfly

The butterflies, which have white wings with black markings, appear from March to April. The eggs are laid singly, mainly on the underside of leaves, and hatch after about a fortnight. The caterpillars are green with a narrow yellow line down the back and have a velvety appearance (Fig. 18.3). When fully grown they are about 34 mm long. Pupation occurs on the plants in June and July. The pupae, which are greenish-brown, are secured to the plant by the hind end and by a silken thread round the middle.

Adults of the more important second generation emerge over a long period from July to September; this makes the timing of control measures difficult. Normally the pupae of the second generation do not develop into butterflies until the following spring.

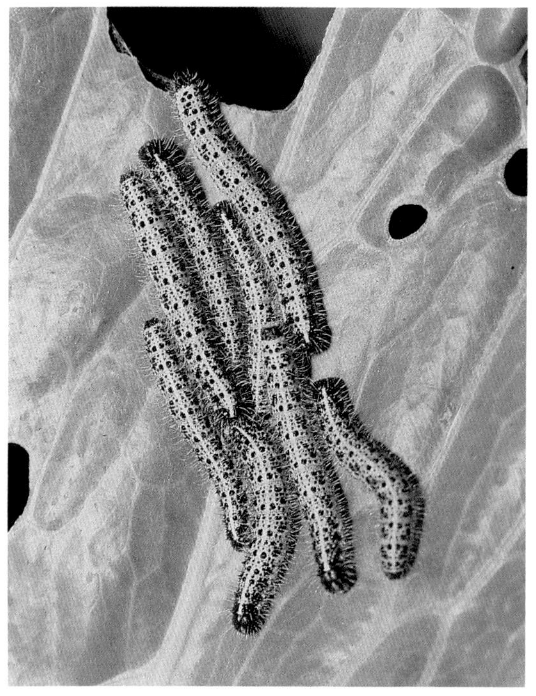

Fig.18.4 Caterpillars of large white butterfly (× 1)

Large white butterfly

This species has a similar life history to that of the small white butterfly, but the adults (Fig. 18.2) emerge a little later in the spring — in April and May — and lay their eggs in batches mainly on the underside of leaves. The eggs hatch after about a fortnight and at first the young caterpillars remain together in a colony. They feed continuously and grow quickly, always tending to keep together, so that many caterpillars may be found on one leaf (Fig.

Fig.18.5 Caterpillar of cabbage moth (× 2)

18.4). The caterpillars are bluish- or yellowish-green, heavily marked with black; they have a yellow line down the back, yellow along the sides and are sparsely covered with hairs. The fully fed caterpillars (about 40 mm long) crawl off the plants in search of sheltered sites where they pupate.

The second-generation butterflies emerge in July and August; caterpillars arising from this generation usually cause the worst damage.

Green-veined white butterfly

The life cycle is the same as that of the small white butterfly. There are minor differences in the appearance of the two species, e.g. the caterpillar of the green-veined white lacks the narrow yellow line down the back.

Cabbage moth

The moths appear towards the end of May and in June but are rarely seen because they fly at night. They are dingy greyish-brown with white markings. Their eggs are globular and laid in batches on the underside of the leaves of any of the numerous food plants. The caterpillars are light green at first but as they become older (Fig. 18.5) their colouring varies greatly. They may remain light green or become brown, dark green or grey on the upper surface with a darker square mark towards the hind end; the legs and under surface vary from green to yellow. The skin is smooth with few hairs. These latter features help to distinguish the caterpillar of this species from the small and green-veined white caterpillars, which have a velvety appearance, and the large white caterpillar, which has conspicuous hairs.

Fig.18.6 Diamond-back moth (× 7)

After feeding for four or five weeks the

Fig.18.7 Grubs of the parasite *Apanteles* sp. beside caterpillar from which they emerged

caterpillars become fully grown (about 42 mm long). They crawl down the plants and burrow into the soil where they form shining brown pupae. During the second half of summer these pupae frequently produce moths which give rise to a further generation of caterpillars. There is much overlapping of the generations, and caterpillars may be found from June to October or even later.

Diamond-back moth

The moths appear in May or June and fly during the day as well as at night. They are small — 6 mm long — brown and grey with conspicuous white marks along the inner margin of each fore wing. These marks meet when the wings are closed and then form the characteristic diamond pattern by which the moth may be recognized (Fig. 18.6).

The eggs are laid on any plants of the cabbage family. The caterpillars are pale green and live in a flimsy web underneath the leaves. After about three weeks they are fully grown (about 14 mm long) and spin cocoons on the leaves and stems of their food plant, turning into pupae from which moths emerge after about a fortnight.

Two or more generations are produced during the summer and moths may be found from spring to autumn. It is believed that the winter is passed in the pupal stage.

Garden pebble moth

The moths appear in May and June and in August and September, but they fly only at dusk and at night. They are small, having a wing-span of about 25 mm. The fore wings are yellowish with brown markings, whereas the hind wings are very pale yellow streaked with grey.

The eggs are laid in small clusters on the leaves of crucifers. At first the caterpillars are yellowish-green with a darker green mid-dorsal line and a yellowish stripe along each side of the body. When fully grown (about 22 mm long) the caterpillar is almost uniformly glossy pale green; it then spins a silken cocoon in the soil below the surface. Caterpillars hatching in June pupate after three weeks, but those hatching in September hibernate in their cocoons and pupate the following spring.

Natural enemies

Cabbage white butterflies are attacked by various insect parasites. Dead or dying caterpillars of the large or small white may be seen with a large number of little yellow cocoons near them. These cocoons are made by the grubs of a small parasitic wasp, *Apanteles glomeratus* (L.), which have fed inside the caterpillar (Fig. 18.7). *A. glomeratus* is a valuable ally in decreasing the numbers of large whites, so the parasite cocoons should not be disturbed nor should insecticides be used unless they are absolutely necessary.

The diamond-back moth is also attacked by insect parasites, which, in most years, apparently prevent its numbers from reaching epidemic proportions. Birds, notably starlings, destroy large numbers of the caterpillars. A fungus disease also kills the caterpillars.

Since the late 1950s, caterpillars of the large white, both in Britain and on the Continent, have been subject to a heavy incidence of a naturally occurring virus disease. Attacked caterpillars, when nearly full-grown, become limp and the body contents degenerate into a brownish liquid which is released on to the foliage and can infect other caterpillars. This virus is probably largely responsible for the decline in immigration of the large white butterfly mentioned on page 98.

Attempts have been made in several countries to control cabbage caterpillars with naturally occurring virus and bacterial diseases; some promising results have been obtained.

Control measures

On farms and market gardens

Caterpillars are very difficult to kill once they have entered the heart of a brassica plant, so these pests are best controlled as soon as possible after they have hatched from the egg.

Control of cabbage caterpillars can be obtained by spraying with a recommended insecticide or a product based on an insect growth regulator which disrupts normal larval development. The recommended rate of some chemicals varies with the species of caterpillar to be controlled. All the insecticides should be applied in 350–1100 litres of water per hectare. As the foliage of brassica crops is difficult to wet, it may be necessary to add more water to some sprays. The addition of extra wetter may also improve spray cover on crops with exceptionally waxy leaves.

Usually only one spray is necessary, but occasionally, when the adults emerge over a long period, a second spray is required. The less persistent chemicals have the advantage that they can be used nearer to harvest. Some insecticides will also decrease numbers of cabbage aphid present at the time of application.

'Drop legs' used to ensure efficient spray cover should be of the flexible type to minimize physical damage to the crop.

In gardens and allotments

In gardens hand-picking should not be despised; in a normal year cabbage white caterpillars can be controlled by hand-picking alone. When the butterflies are on the wing, the crops should be examined at least once a week and any eggs or caterpillars crushed. The sculptured, yellow eggs of these butterflies should be distinguished from the smooth, yellow eggs of ladybirds which sometimes accompany them and which should not be destroyed. The cabbage moth can also be dealt with by hand-picking, but the older caterpillars burrow into the centre of the plant where they cannot be reached.

A few of the less persistent insecticide sprays that are recommended for commercial use are also available to the home gardener. Dusts are often preferred by the private gardener as they can be applied simply. Application should be made when the air is still; it is an advantage if there is dew on the plants.

When diamond-back moth caterpillars are present, the insecticide should be directed at the underside of the leaves.

19
Caterpillars on currants and gooseberry

Fig.19.1 Gooseberry shoot defoliated by caterpillars of common gooseberry sawfly

Caterpillars of several insect species feed on the foliage of currants and gooseberry. Described here are the caterpillars of the magpie moth (*Abraxas grossulariata* (L.)) and three species of sawfly, i.e. the common gooseberry sawfly (*Nematus ribesii* (Scopoli)), the black currant sawfly (*Nematus olfaciens* Benson) and the pale spotted gooseberry sawfly (*Nematus leucotrochus* Hartig).

The caterpillar of the magpie moth is black

and white with yellow markings along the sides, while the sawfly caterpillars are grey-green and more or less speckled with black. The magpie moth caterpillar has only two pairs of abdominal legs, compared with seven in the sawflies, and progresses with a looping action (Fig. 19.3).

Magpie moth caterpillars attack several other shrubs and trees as well as currants and gooseberry, e.g. plum, blackthorn and hawthorn.

Destructive attacks of sawfly caterpillars are liable to occur on currants and gooseberry in gardens and commercial plantations in most parts of the British Isles. The common gooseberry sawfly and the pale spotted gooseberry sawfly prefer gooseberry, but they also attack red and white currants though not black currant. The black currant sawfly infests black, red and white currants.

Nature of damage

Damage is caused by the caterpillars eating the foliage (Fig. 19.1). The overwintered caterpillars of the magpie moth begin feeding as soon as the leaves appear in spring and cease feeding in May or June. Sawfly caterpillars do not hatch until May and damage may occur at intervals throughout the summer. A characteristic feature of sawfly attack is that it usually begins unnoticed in the centre of a bush, spreading more rapidly to the outer foliage as the caterpillars grow. When heavily attacked by magpie moth or sawfly caterpillars, bushes are completely stripped with the result that the season's crop may be lost and the bushes so weakened that only a small crop can be produced the following year.

Description and life history

Magpie moth

The magpie moth (Fig. 19.2) has a black and yellowish-orange body with white wings that have black and yellowish-orange markings; the fore wings are 35–40 mm across. The moths are on the wing at night in July and August. The females lay eggs, singly or in small groups, on the underside of leaves. The eggs are yellow when first laid but turn black later. After about two weeks, minute blackish caterpillars hatch and feed on the foliage. The caterpillars grow slowly until the autumn when, while still small, they hibernate among the dead leaves of their food plant and in sheltered crannies.

As soon as foliage appears in the spring, the caterpillars leave their winter quarters to feed until the end of May or early June. The caterpillars are then black and white with a yellow stripe along each side. When fully fed (Fig. 19.3), each caterpillar turns into a shining black and yellow-banded pupa suspended in a flimsy cocoon among the leaves or twigs of the food plant or on fences, walls, etc. The moth emerges about a month later.

Common gooseberry sawfly

Adults of this sawfly (Fig. 19.4) have clear transparent wings, which are 10–16 mm across, and a body that is black in front and yellowish-orange behind. They appear first in April or May. Eggs, semi-transparent and pearly-white or greenish, are laid on leaves — usually a large number on one leaf in the centre of a bush. They are fixed into shallow slits along the main veins on the underside of the leaves and

Fig.19.2 Magpie moth (× 1½)

Fig.19.3 Fully grown caterpillars of magpie moth (× 1)

Fig.19.4 Female common gooseberry sawfly (× 7)

are fairly conspicuous (Fig. 19.5).

The young caterpillars, which hatch after about a week, have a black head and green body slightly speckled with black. For some days they feed together in a colony, first eating the surface of the leaves and then making holes. As they grow, they disperse over the bush eating the entire leaves. By then the body is coloured sage-green with numerous black spots (Fig. 19.6). When the caterpillars are fully grown, about four weeks after hatching, the black spots disappear and the body becomes pale green with an orange area behind the head and another near the hind end. At this stage the caterpillars leave the bushes and make oval, brown cocoons in the soil; in summer, adults emerge 10–21 days later.

There are usually three generations during the season and the infestation may have increased greatly by the end of the summer. At this time, because of the overlapping of the generations, adults, eggs and caterpillars of all sizes may be found on bushes at the same time. The cocoons of the last generation remain in the soil during the winter and produce adults in the following spring.

Black currant sawfly

The adults of this species are similar in size and coloration to those of the common gooseberry sawfly. They appear first in May and are active in sunny weather, flying and running over the leaves of a host plant.

The eggs are whitish and are laid, singly or in small numbers, in shallow slits along the veins on the underside of the leaves. The caterpillars are green with black markings that are smaller and less conspicuous than those on common gooseberry sawfly caterpillars. The head of the caterpillar is also green with a few black spots. (Note that the head of the common gooseberry sawfly caterpillar is black.) Damage by black currant sawfly is seen first as small holes in the leaves, usually in the centre of the bush. Later the caterpillars eat the entire leaves except the stumps of the main veins.

Fig.19.5 Eggs of common gooseberry sawfly on underside of gooseberry leaf (× 4)

Fig.19.6 Caterpillar of common gooseberry sawfly on gooseberry leaf (× 3)

The black currant sawfly has two or more generations during the summer. The first adults emerge in May and June, whereas the second generation is on the wing between July and mid-September. As with the common gooseberry sawfly, the generations overlap and all stages may be found together towards the end of the season. The insect overwinters in its cocoon in the soil, usually near the surface.

Other sawflies

The pale spotted gooseberry sawfly has only one generation a year. The adults are similar to those of the common gooseberry sawfly except that the body is much blacker. They are on the wing in May and pale yellow eggs are laid on the under surface of leaves. The caterpillars are similar in colour to those of the common gooseberry sawfly; it is not necessary for the grower to distinguish between these two species.

In addition to those described, other kinds of sawfly caterpillars are sometimes found on gooseberry and currants.

Natural enemies

In some years infestations of gooseberry sawfly caterpillars are effectively checked by a parasitic fly (a tachinid). Pearly-white eggs of the parasite may be seen on the caterpillars.

Control measures

It is important to control infestations early in the season before serious damage has occurred.

Magpie moth

As the overwinterd caterpillars of the magpie moth begin feeding early, it is worth examining bushes during the blossom period for signs of injury in gardens in districts where attacks occur. At this stage the caterpillars are almost black. If necessary, a spray of a recommended insecticide can be applied after flowering, but hand-picking of caterpillars is the simplest method of control in gardens.

Sawflies

Sawfly caterpillars begin to appear in May; those of the common gooseberry sawfly may appear at any time after early May and those of the black currant sawfly a few weeks later. They are usually found on leaves in the centre of a bush, where adults tend to lay most eggs. Thorough inspection of bushes is worth while as hidden colonies of caterpillars can soon strip a bush bare of its leaves. Sometimes, only isolated bushes are affected in plantations, necessitating individual control measures.

On commercial plantations Bushes should be sprayed thoroughly with a recommended insecticide. Spraying during the blossom period should be avoided because of the danger to bees. Gooseberries should be sprayed about mid-May when caterpillars appear, then again in early June if necessary. Black currants should be sprayed during the first half of June, while the majority of the caterpillars are still small. A second spray may be necessary 2–3 weeks later.

In gardens Where there are only a few bushes, hand-picking is the simplest method of control, but this must be done before the colonies of young caterpillars have dispersed over the bush. If an attack has been noticed at an early stage, leaves on which eggs have been laid should be picked and destroyed. Otherwise sprays of a recommended insecticide should be applied at the times indicated above.

20 Codling moth

Fig.20.1 Apples damaged by caterpillars of codling moth. Upper left: recent entry hole. Lower left: older entry hole. Upper right: older entry hole at eye. Lower right: damaged apple cut open

The codling moth (*Cydia pomonella* (L.)) is one of the most widespread pests of apple; it occurs in almost every country in which apples are grown. The damage is caused by the caterpillar, which burrows into the fruit (Fig. 20.1).

In addition to apple, codling moth sometimes attacks pear, the fruits of other species of *Malus* and *Pyrus*, quince and walnut. In Britain codling moth is mainly a pest of apple, but occasionally attacks pear, and is more important in the southern half of England than elsewhere. More serious infestations occur in prolonged hot summers. Orchard areas adjacent to apple bulk bins, pallet and box stores, or unsprayed apple trees, are particularly prone to attack. However, as a result of effective sprays, populations are now very small in well-managed commercial orchards.

Description of caterpillar and damage

Two other pests, the apple sawfly (see page 93) and the fruitlet mining tortrix (see page 137), damage apples in a similar way to codling moth. As these other pests should be controlled at a different time, it is important that growers should be able to distinguish them from codling moth. The differences between the caterpillars of these three species and the types of damage caused by them are as follows.

Codling moth The main feeding period is July and August; there is no surface scar on the fruit and the small entry hole is covered with fragments of dry frass (excrement) (Fig. 20.1). The caterpillar is whitish with a black head until almost fully grown; in the last stage it becomes pinkish and the head is brown. It has five pairs of abdominal legs.

Apple sawfly The main feeding period is June; attacked fruits have a conspicuous hole leading straight to the core, with a mass of sticky reddish-brown frass at the entrance. On fruit attacked by young caterpillars there is a rough, ribbon-like scar ending at an actual or attempted point of entry. The caterpillar is dirty-white and has seven pairs of abdominal legs.

Fruitlet mining tortrix The main feeding period is June and early July; fruit damage consists of several shallow, small holes with occasional deeper ones that follow a meandering course. The caterpillar is dirty-white and is smaller than codling or sawfly caterpillars; it has five pairs of abdominal legs.

Description of adult

The codling moth is about 8 mm long and has greyish-brown fore wings with a metallic copper-coloured patch near the tip (Fig. 20.2). The moths are seldom seen because they are active mainly at dusk, resting among the foliage at other times.

Fig.20.2 Codling moth (× 7)

Life history

There is one complete generation of codling moth each year and in warm summers the earliest caterpillars (up to 5% of the total) can give rise to a second generation of adults in August and September. In an average season the first moths emerge in late May or early June. The main emergence occurs between late June and mid-July; numbers then decline until the first generation ends in early August. In warm summers second-generation moths can then be found until early September.

The moths fly and mate on warm, still evenings when the temperature at dusk is 14 °C or more. Eggs are laid singly on the foliage or fruit, generally in the evening. The prevailing temperature affects the number laid, temperatures over 15.5 °C being favourable. The eggs are translucent, flat and round — about 1 mm in diameter — and look like small scales (Fig. 20.3); they are unlikely to be found unless infestations are heavy.

After 10–14 days, depending on temperature, the young caterpillar hatches (Fig. 20.3)

Fig.20.3 Newly hatched caterpillar of codling moth and empty egg-shell (× 9)

Fig.20.4 Cocoons opened to show caterpillars of codling moth inside (× 3)

and immediately searches for a suitable point of entry into a fruit and starts to burrow. In the early part of the season when the fruits are small, at least half the caterpillars enter at the eye (calyx). For the first few days the caterpillar usually feeds in a small cavity just beneath the skin; it then burrows to the core, feeding on the flesh and pips (Fig. 20.1). Sometimes a second fruit is attacked before the caterpillar reaches maturity.

The caterpillar is fully fed after about four weeks. It then leaves the fruit and spins a cocoon under loose bark on the tree, under tree-ties, or in cracks in supporting posts. Second-generation moths arise only from caterpillars that have spun up by late July. Caterpillars leaving fruits that have been picked may spin their cocoons in apple bins, trays or elsewhere. These can lead to new infestations or to the cocoons becoming contaminants in other produce.

The winter is spent as a fully fed caterpillar in a cocoon (Fig. 20.4). The pupal stage occurs in late spring.

Natural enemies

The most important natural mortality is caused by birds, especially tits, feeding on overwintering caterpillars. Predation of the eggs by various insects, such as mirid and anthocorid bugs and earwigs, can also be important. There are several species of parasite which attack eggs, caterpillars and pupae.

Pheromone traps

At dusk female moths emit a scent, or pheromone, which attracts males. The sex pheromone of codling moth has been identified and synthesized. Sticky traps baited with the synthetic pheromone are now used to monitor the flight of the male moths as an aid to the timing of sprays.

An insecticide spray should be applied if five or more moths are caught per trap for two weeks (not necessarily consecutive weeks). Application should generally be made 7–10 days after the second catch of five or more moths. If fewer than five moths are caught per trap per week, it is not necessary to spray. Thus the use of pheromone traps can improve spray timing and sometimes leads to a decrease in the number of sprays applied.

A pheromone trap is more efficient than a light trap in attracting codling moths and is very much easier to operate; only a short time is needed to sort the catch from a pheromone trap because only a few species are attracted to it. Traps baited with codling moth sex pheromone also attract fruitlet mining tortrix.

Control measures

Well-timed sprays of broad-spectrum insecticides are so effective in controlling codling moth that the level of attack in commercial orchards is generally very low. Migration of moths from neighbouring unsprayed apple trees in gardens and abandoned orchards can be an important source of infestation, but female moths do not normally travel more than 100 m. In recent years infestations have generally been more important in the south-west and the West Midlands than in the south-east and east of England.

Routine spraying against codling moth should be avoided. Sprays should be applied according to catches in pheromone traps and assessment of damage at harvest in previous years. It may also be necessary to spray for summer fruit tortrix and fruit tree tortrix (see page 140).

In apple orchards

Where heavy infestation by codling moth and/or tortrix moths occurs, two sprays should be applied: one in the second half of June and another two to three weeks later. The optimum timing for the first spray in an average year is the third week of June plus or minus two weeks, depending on the earliness or lateness of the season. The best date for spraying can also vary considerably within a locality. Spray timing should be related to catches in pheromone traps or to spray warnings.

Where infestation by codling moth is lighter, one spray should be applied as indicated by pheromone traps, usually in the last week of June or the first 10 days of July. If fruit tree tortrix is numerous, a second spray should be applied two weeks later.

There is a wide choice of recommended insecticides. Most of them have a broad spectrum of activity and when applied in the summer may eliminate beneficial insects. Some are also very harmful to organophosphorus-resistant typhlodromid mites, which are important predators of fruit tree red spider mite (see page 344). Where an insect growth regulator is used, application may have to be made 7–10 days earlier than with conventional insecticides.

In pear orchards

In Britain codling moth is usually a minor pest of pear and control is only occasionally necessary. Where control is required, the spray programme recommended above for apple should be adopted. However, the application of certain insecticides against codling moth may lead to a more severe infestation of pear sucker (see page 12) by killing predatory insects. A suitable selective insecticide should be chosen.

In gardens

Sprays will be effective only if timed correctly (see above) and applied thoroughly (which is not usually possible on large trees in a garden).

Insecticide application in midsummer may, by killing beneficial insects, encourage infestation by fruit tree red spider mite and make it necessary to spray against this pest. It will generally be easier for the gardener to take other, less effective, measures against codling moth and to tolerate some damage to his fruit. All loose bark should be scraped off to remove overwintering sites for the caterpillars. Stakes and tree-ties should have as few crevices as possible. Bands of sacking or corrugated cardboard, about 100 mm wide, can be tied round the trunks by mid-July to provide alternative overwintering sites. These bands should be removed after the crop has been picked and either burnt or immersed in a pail of boiling water.

21
Cutworms

Fig.21.1 Lettuce plant showing effect of cutworm attack

The name 'cutworm' is given to the caterpillars of certain moths in the family Noctuidae that feed on plants at about ground level, frequently severing them from their tap roots.

In Britain the most harmful cutworm is the caterpillar of the turnip moth (*Agrotis segetum* (Denis & Schiffermüller)). Caterpillars of the large yellow underwing moth (*Noctua pronuba* (L.)), the garden dart moth (*Euxoa nigricans* (L.)), the white-line dart moth (*Euxoa tritici* (L.)) and occasionally other species also attack some crops.

Most damage is caused to young plants and to those, like lettuce (Fig. 21.1), beet and other row crops, with tender tap roots. Close-planted crops are particularly susceptible, e.g. self-blanching celery, carrots, leeks (Fig. 21.2) and red beet. Potato tubers and large roots such as turnips and swedes are also attacked. Potatoes, particularly in light soils, are liable to suffer considerable damage (Fig. 21.3) when June and early July are warm and dry. Cutworms do not seriously damage grasses or cereals.

Nature of damage

Young cutworms make holes in leaves of plants but cause no serious damage. Older cutworms

Fig.21.2 Leek plant (centre) distorted as a result of cutworm attack compared with normal plants

Fig.21.3 Caterpillar of turnip moth on potato tuber showing typical cutworm damage

generally spend the day in the surface layers of the soil or hidden under leaves, thick grass or stones. They feed at night, sometimes on the leaves of plants, but more often on their stems, both above and below ground level. They may work along a row of plants, cutting them off one after another; it is from this habit that their name is derived.

The signs of attack are poor top growth or wilting (Fig. 21.1), caused by the caterpillars feeding on the stem or roots near ground level; in extreme cases the plant wilts very suddenly. If the root is found to be more or less severed near ground level, the most likely cause is a cutworm, but occasionally the grub of a chafer beetle (see page 163) may be responsible. The culprit will probably be found lying curled up in the soil near the wilted plant (e.g. Fig. 21.6).

Cutworm damage to potatoes (Fig. 21.3) is often confused with slug damage (see page 471). However, cutworm damage is most prevalent in light and medium soils (including peat) during hot, dry summers, while slug damage is most likely on medium or heavy soils after moist summers and/or autumns. A characteristic feature of slugs is that they leave slime trails whereas cutworms do not.

Description

The adults of the four species mentioned above have brownish fore wings, usually with darker markings. The hind wings are pearly-white in the turnip moth (Fig. 21.4) and the white-line dart moth, brownish in the garden dart moth and orange-yellow with a black border in the large yellow underwing moth (Fig. 21.5).

The caterpillars (Figs 21.6 and 21.7) have

Fig.21.4 Turnip moth in normal resting position (× 3)

Fig.21.5 Large yellow underwing moth with wings unfolded to show coloration of hind wings (× 1½)

Fig.21.6 Caterpillar of turnip moth (× 3)

Fig.21.7 Caterpillar of large yellow underwing moth (× 2)

three pairs of true legs and five pairs of 'false' abdominal legs. They are generally dull greyish-brown with faint longitudinal lines along the back and sides, but those of the large yellow underwing may be greenish with distinct black marks along the back (Fig. 21.7). Caterpillars of the large yellow underwing when fully grown are 50 mm long; those of the other three species rarely exceed 35 mm. Because of their colouring all these caterpillars are difficult to see in the soil.

Their dull colouring and the presence of abdominal legs distinguish cutworms from chafer grubs, which are dirty-white, have longer true legs and lack abdominal legs. Chafer grubs may also be recognized by a characteristic bend in the middle of the body and a swollen hind end. Cutworms are sometimes confused with leatherjackets (see page 275), but the latter lack a distinct head, are completely legless and are uniformly grey.

Life history

Turnip moth

The adult moths emerge from late May until the end of June. Eggs (Fig. 21.8) are laid on the leaves or stems of many common weeds and cultivated plants or on soil or debris. Caterpillars hatch after 8–24 days, according to the temperature. For the first week or two they feed on the leaves, but then they descend to the soil and adopt the typical cutworm feeding habit. Most caterpillars feed slowly and are not fully fed until late autumn. They overwinter in the soil and turn into brown pupae in April or May (Fig. 21.9). They do *not* cause damage to crops in the spring.

Young turnip moth caterpillars can survive only where the soil is dry. More of them reach maturity in hot, dry summers than in years when rainfall in June and July is above average. A high survival rate can lead to increased crop damage, especially in light, well drained soils.

Turnip moth caterpillars do most damage in June and July and, when they are very numerous, root and potato crops may be seriously damaged (Fig. 21.3). A small proportion of the caterpillars which hatch in June feed rapidly and pupate at the end of July or in August, producing another generation of moths in September or October. Eggs are laid by these second-generation adults and hatch into caterpillars, but most of these probably die. Those that survive feed through the autumn and into the spring; they then pupate in May, reinforcing the number of moths arising from the first generation. Caterpillars of this second generation are, however, always few in number and do not cause significant damage to crops either in the autumn or the following spring.

Large yellow underwing moth

All stages of this moth are found later in the year than those of the turnip moth. The adults emerge from mid-June to August and caterpillars may be found from July until the following May. These caterpillars may feed during the spring, but they pupate before most spring-sown crops have germinated. However, early brassicas and lettuce may occasionally be at risk.

The caterpillars of this species are less sensitive to wet conditions than those of the turnip moth and infestations sometimes develop in cool, wet seasons. Large yellow underwing moth is occasionally troublesome on protected crops, usually as a result of immature stages being taken under protection with the plants.

Garden dart moth and white-line dart moth

Adult moths emerge in July and August. Eggs are laid in late summer but do not hatch until early the following spring. The caterpillars feed on both leaves and roots until they pupate in June. In some seasons these caterpillars cause damage to root crops, particularly on light soils in the eastern counties of England where late-sown or backward sugar beet crops may be thinned badly.

Control measures

On field crops

Cultural control Cutworms are most common on weedy land as it provides cover attractive to the egg-laying moths and also food for the caterpillars. Crops immediately following heavy weed cover are much more likely to be seriously damaged than those on land that is kept clear of weeds. Because young turnip moth caterpillars die in wet soil, the frequent irrigation of susceptible crops should prevent the development of damaging infestations.

Chemical control Susceptible crops are best treated while the caterpillars are still young and feeding on the foliage or at the soil surface. It is difficult to kill older caterpillars when they are feeding below ground.

Adult turnip moth activity and caterpillar damage are monitored in most regions of England and Wales. Agrometeorologists use temperature and rainfall data to provide an index of turnip moth egg development and caterpillar survival. Spray warnings are issued when weather conditions are suitable for cutworm survival. As routine treatments are likely

Fig.21.8 Eggs of turnip moth (× 10)

Fig.21.9 Pupa of turnip moth (× 2½)

to be unnecessary or wrongly timed, growers of susceptible crops are strongly advised to consult their local adviser from early June for up-to-date information on cutworm survival and spray timing.

Treatments should be applied only during periods of warm weather and when the soil is dry. Young cutworms can be controlled by a high-volume spray of a recommended insecticide. The spray should preferably be applied in at least 1000 litres of water per hectare.

In gardens

In gardens and allotments, hand-picking is often the simplest method of dealing with cutworms. They can be found in the soil near the plants last attacked or collected at night when active on the surface of the soil.

Soil-applied insecticides which will give some control of cutworms are available for garden use.

22
Pea moth

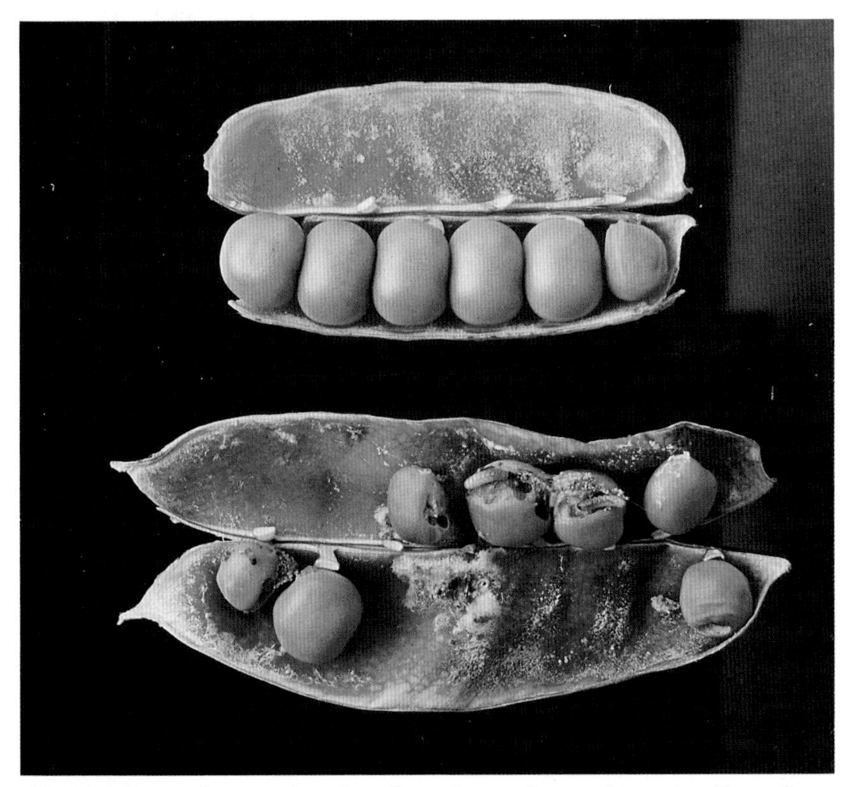

Fig.22.1 Pea pods opened to show (lower) peas damaged by caterpillars of pea moth and (upper) undamaged peas

The pea moth (*Cydia nigricana* (Fabricius)) is one of the most serious pests of field and garden peas. Its caterpillars are found inside pea pods in the summer and are responsible for serious losses in some seasons, especially in the south and east of England. Not only do the caterpillars ruin the peas which they eat (Fig. 22.1), but their presence, even in small numbers, may reduce the economic value of the crop. In crops destined for freezing or canning, only a very small percentage of damaged peas can be tolerated owing to cleaning difficulties.

However, actual damage to peas harvested dry tends to be greater because they are harvested later and exposed to attack for a longer period.

Although varieties of pea differ in their susceptibility to attack, any pea crop in flower or in pod in June or July is liable to attack to a greater or lesser degree.

Certain other leguminous plants, both wild and cultivated, are also known to be attacked. Pea moths have been bred from meadow vetchling (*Lathyrus pratensis*), tufted vetch (*Vicia cracca*) and sweet pea (*Lathyrus odoratus*). Garden lupins and broad beans appear to be immune.

Description and life history

The moths are dull grey-brown with pale and dark markings near the tip of each fore wing (Fig. 22.2) and are about 6 mm long with a wing-span of about 15 mm. In most seasons they begin to emerge from their cocoons in the soil in late May or early June. They may be seen fluttering among pea plants in warm, sunny weather from early June to mid-August and are most abundant in the second and third weeks of July.

The small, flattened eggs (Fig. 22.3) are laid singly or in small groups, mainly on the leaves and stipules. After about a week, the young caterpillars hatch, wander over the plant and eventually enter young pods, where they remain feeding on the peas until fully grown (Fig. 22.1). They then bite through the pods and descend to the soil, where each spends the winter in a tough, silken cocoon. In the spring each caterpillar turns into a pupa from which the moth emerges in early summer. There is only one generation a year.

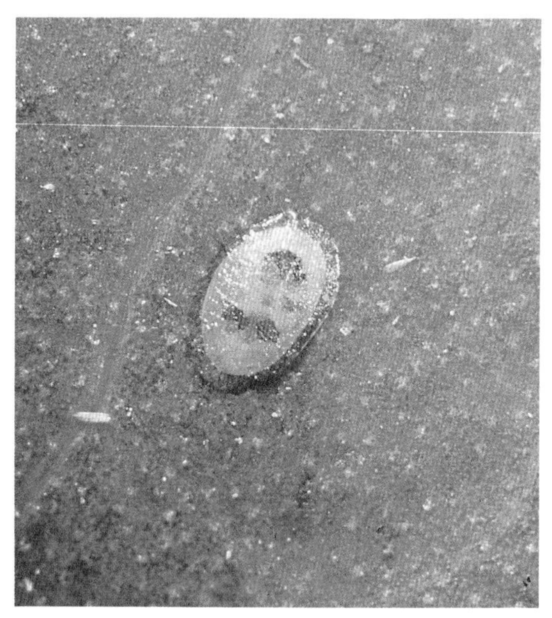

Fig.22.3 Egg of pea moth on leaf (× 40)

Fig.22.2 Pea moth on leaf (× 5)

Natural enemies

Pea moth caterpillars are attacked by four species of hymenopterous parasite and a pathogenic fungus. Although these help to keep the pest in check, insecticides are usually necessary to obtain commercial control.

Cultural control

Peas that come into flower and mature during the flight period of the moth (between early June and mid-August) usually suffer most damage. Those sown either early or very late tend to suffer less damage. Early-maturing varieties should, where possible, be sown for picking before the second week of July. Crops sown after mid-June will mature after the period of moth activity and will, to a large extent, escape attack.

Chemical control

Pea-moth traps

Traps are used to determine the need to spray against pea moth and, where spraying is necessary, the optimum timing of the first spray. The timing of sprays is very important and should be related to the time of moth emergence and the stage of growth of the crop. Spraying before moths and eggs are present, or before the first pods begin to set, is unlikely to be useful.

Traps to catch male pea moths are available commercially. They are supplied as 'systems', each comprising two traps with instructions for use and a calculator to predict the optimum date for spraying. The system is intended for monitoring pea moth populations in dry-harvesting peas. The traps (Fig. 22.4) are baited with a synthetic sex attractant simulating female pheromone. Female moths are on the wing at the same time as the males, so the traps can be used to detect the presence of egg-laying females at very low population densities. These traps therefore enable a grower to predict when control measures should be applied to kill the caterpillars; they also indicate when spraying is not necessary.

In most years spraying should start 10–15 days after the onset of a period of sustained catches of 10 or more moths per two-day period (the threshold). A second spray should be applied about 10 days after the first.

Pea moths do not fly at temperatures below 18 °C; egg development also depends on temperature. Growers who wish to use the egg-development calculator must record accurate maximum and minimum shade temperatures on successive days following capture of a threshold number of moths. These figures can then be used to calculate the expected date of hatch of the eggs laid on the threshold date.

Fig.22.4 Sex-attractant trap catching male pea moths in a field

The timing of an attack on any one farm can vary considerably from year to year. Furthermore, large differences in infestation can occur between fields that are comparatively close together. More than one set of traps may therefore be necessary on farms with several pea crops, especially where altitude, shelter or proximity to previous year's crops vary greatly. In very large fields (more than 50 ha) two sets of traps are recommended. Traps should always be put into the crop by mid-May because after this date large numbers of moths may be caught at the first observation, resulting in a misleading recording of threshold numbers of moths.

Advice on spray timing

In some regions a spray advice service is provided for growers who are using pea-moth traps. A recorded message operates from the end of May and consists of calculated spray dates appropriate to a range of threshold catch dates. This information is based on temperature data from several meteorological stations and is classified separately into geographical areas if temperatures within the region vary sufficiently at that time. The advice is useful only to growers who have traps, and the service should be consulted only after threshold numbers of moths have been caught.

Spray recommendations

Routine spraying against pea moth is *not* recommended. In many instances sprays are not required at all, e.g. on peas grown for animal feed. However, where treatment is necessary, two sprays often give better control than one, but the cost of extra protection may not always be justified economically. This depends on the type of peas grown, i.e. whether for harvesting green or dry, and on the date of harvesting. As a general guide the following spray recommendations are made.

Picking peas Very early drilled peas to be picked near the end of June or beginning of July may not need spraying; those to be picked later will probably need one spray 7–10 days after the beginning of flowering.

Vining peas (for freezing and canning) Growers should be guided by the factory fieldsmen as even very small infestations can lead to rejection of the crop. Spray protection is probably needed from mid-June, although this can vary in different seasons. Some contractors may require growers to spray twice; the second spray should be applied about 10 days after the first. The threshold catch of moths used to calculate the spray timing for dry-harvesting peas is unlikely to be appropriate for vining peas. Therefore, traps are useful in vining peas only to ascertain if moths are present in the crop. If *any* moths are caught in the period from first flower to formation of the first pods, a spray should be applied 7–10 days later.

Dry-harvesting peas When this crop is grown for animal feed, spraying against pea moth is uneconomic. When the crop is for human consumption or for seed the use of pea-moth traps is strongly recommended. As these peas are susceptible to attack for a longer period than other types of pea, i.e. until the pods begin to dry off, two sprays will probably be necessary to give satisfactory control. If pea-moth traps are not used, the first spray should be applied when the peas have reached or just passed the flowering stage, provided moths and eggs are present in the crop (usually after mid-June). A second spray should be applied about 10 days later. Aerial sprays are as effective as ground sprays.

23
Pear and cherry slugworm

Fig.23.1 Slugworm on pear leaf showing damage caused by feeding on upper surface

The pear and cherry slugworm, which is the larva of the pear slug sawfly (*Caliroa cerasi* (L.)),occurs in many parts of Britain, particularly the south. It is sometimes a pest in private gardens but is infrequently recorded in commercial plantations. Pear and cherry trees are most commonly attacked, but apple and plum are also attacked occasionally as well as ornamentals such as almond, flowering cherry, hawthorn and rowan.

Nature of damage

The damage caused by the slugworms feeding on the foliage is very characteristic (Fig. 23.1). The upper layer of the leaf is eaten away, leaving only the veins and the lower surface, which turns brown. Occasionally, larvae of the first generation cause considerable injury to the leaves of young trees in June, but usually damage does not become obvious until the second generation occurs in August.

In addition to the effect on the current season's foliage, damage may be sufficiently serious to cause a check to the following year's growth.

Description and life history

The adult sawfly is a small, black insect about 8 mm long with two pairs of wings. It first appears towards the end of May or in early June. Eggs are laid on the underside of foliage, each egg being inserted in a slit made in the leaf. After about two weeks the eggs hatch and the young larvae make their way to the upper surface of the leaf where they start feeding.

At first the larvae are whitish but soon become dark olive-green or black (Fig. 23.1). They are broad at the head end and taper towards the hind end; they are slimy and slug-like in appearance but possess 10 pairs of very small legs. When fully fed the larvae lose their slug-like appearance and become yellow with a dry, wrinkled skin. This stage is reached in early July, when the larvae descend to the ground and burrow a short distance below the surface. Each larva then spins a cocoon and turns into a pupa inside it.

After about two weeks a second generation of adults emerges and eggs are again laid on the foliage. The majority of the resulting slugworms do not change into pupae when fully fed but remain as larvae within their cocoons until the following spring. A few, however, may pupate in late summer giving rise to a small third generation.

Control measures

Nowadays this species seldom causes injury of economic importance, so control measures are usually necessary only when larvae are numerous. However, on young trees even small numbers of larvae of the first generation can cause considerable foliage injury during periods when shoot growth is slow.

Slugworms are easily controlled by spraying with a recommended insecticide. A spray should be applied in June as soon as the larvae are seen.

24
Plum fruit moths

Fig.24.1 Plum cut open to show caterpillar of plum fruit moth and frass inside

Caterpillars of many species of moth occur on plum trees, but the majority feed only on the foliage and usually occur in small numbers. However, caterpillars of two species feed on the fruits. These are the plum fruit moth (*Cydia funebrana* (Treitschke)) and the fruitlet mining tortrix (*Pammene rhediella* (Clerck)). Both species are widely distributed in Britain; plum fruit moth is the more common pest, whereas fruitlet mining tortrix occurs sporadically and is less often noted attacking plum.

Caterpillars of the plum fruit moth normally feed on the fruits of blackthorn (sloe) but are also locally common pests of cultivated plum and damson. They attack the fruits from June to September and often cause considerable losses. Caterpillars of the fruitlet mining tortrix feed on hawthorn berries and sometimes also attack fruits of plum and apple in late May, June and July. For the description of the fruitlet mining tortrix on apple, see page 137.

PLUM FRUIT MOTH

Nature of damage

The caterpillars of this species, usually one per infested fruit, feed within the flesh, forming a large cavity that becomes filled with brown frass (excrement). If an occupied fruit is cut open, the caterpillar (commonly known as the 'red plum maggot') may be found inside, often lying close to the stone (Fig. 24.1). Infested fruits tend to ripen prematurely and losses can be severe, particularly in years of light fruit set. Where a crop is picked over for the early market, damaged fruits are more likely to be harvested and infestation of the remaining crop is thereby reduced.

Damage by the plum sawfly (*Hoplocampa flava* (L.)) is similar to that caused by the plum fruit moth but occurs much earlier, when fruitlets are smaller; also the hole in sawfly-infested fruitlets is more conspicuous and is marked by masses of wet, black frass.

Description and life history

The adults of the plum fruit moth are dull slate-grey, with indistinct markings on the fore wings, and have a wing-span of about 12–15 mm. They occur from late May or early June to September but are usually most numerous from mid-June to mid-July. The eggs are flat, translucent, more or less circular and about 0.7 mm across; they are usually laid singly on the fruitlets.

The eggs hatch in about 10–14 days and the tiny caterpillars burrow into the fruits to feed in the flesh. A superficial and narrow, brownish mine often marks the path taken by the caterpillar. The young caterpillar is white with a dark head, but the body eventually becomes reddish, reaching a length of about 10 mm. The caterpillars feed for several weeks before becoming fully grown. They then escape from the fruits and form silken cocoons in bark crevices or in similar situations where they overwinter. Pupation usually occurs in the following spring, but under favourable conditions some early-developing caterpillars may pupate during the same summer and produce a partial second generation of adults in August and September.

Pheromone traps

Pheromone traps can be used to monitor the activity of male plum fruit moths. These traps give a useful indication of the likely level of infestation in any particular year and may be used to gauge the optimum time for applying an insecticide. They also give warning of any major second-generation attack against which further control measures might be required.

Control measures

If necessary, a spray of a recommended insecticide should be applied in late June or early July, and again two to three weeks later. With light attacks, a single spray may be adequate, especially if there has been a good fruit set. Care should be taken to observe the minimum permitted intervals between spraying and harvesting.

FRUITLET MINING TORTRIX

Nature of damage

The caterpillars of this species typically web several fruits together. They cause considerable quantities of gum to exude from the damaged areas of plums and damsons; this gum soon accumulates in the surrounding webbing (Fig. 24.2). The caterpillars, usually one per fruit cluster, burrow into the fruits but little or no frass remains within the flesh (cf. plum fruit moth). Some fruits are damaged superficially and these may reach maturity, the damaged areas callousing over, but severely injured fruits will drop prematurely.

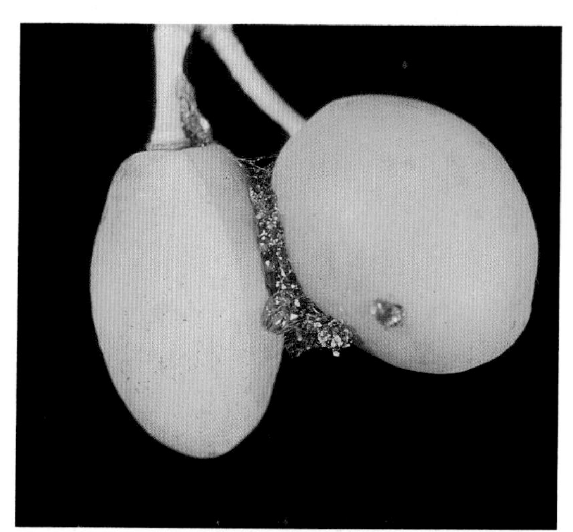

Fig.24.2 Plum fruitlets attacked by caterpillar of fruitlet mining tortrix. Note webbing and gum between fruitlets

Description and life history

Adults are purplish to dark brown, with a metallic orange tip on each fore wing; they have a wing-span of about 9–11 mm. The adults occur in May and June and are active in sunshine. Eggs are deposited singly on foliage close to clusters of young fruitlets and hatch in about two weeks.

The caterpillars, which are whitish and about 5–6 mm long at maturity (cf. plum fruit moth), feed from late May or early June to late June or early July. They then spin overwintering cocoons under loose bark on the host tree or elsewhere. Pupation occurs in the following spring, there being one generation each year.

Control measures

An early insecticide treatment (mid-June) against plum fruit moth (see above) is likely to decrease the numbers of fruitlet mining tortrix, but chemical control of this pest on plum or damson is rarely justified.

25
Stem-boring caterpillars on fruit plants

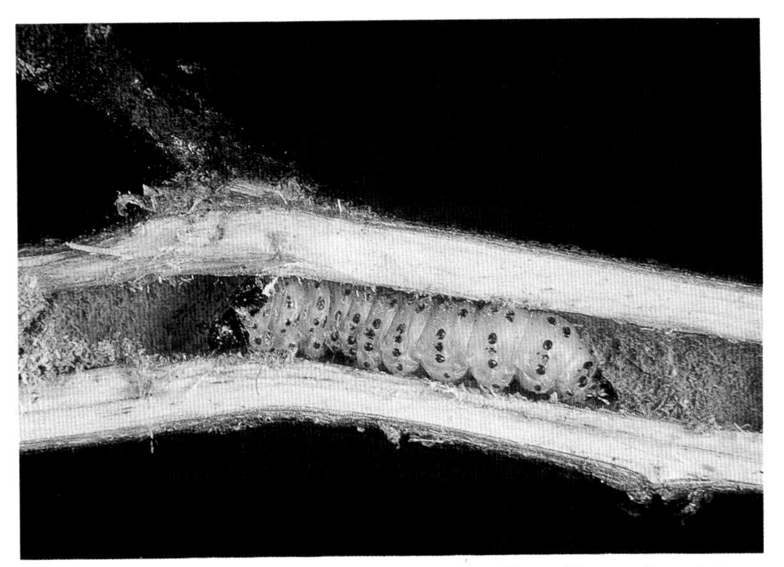

Fig.25.1 Apple stem cut open to show caterpillar of leopard moth in tunnel

Described here are six species of moth whose caterpillars feed in the trunks, branches, stems or canes of fruit trees, shrubs or bushes.

Leopard moth

The leopard moth (*Zeuzera pyrina* (L.)), though generally an uncommon insect, is widespread in the southern half of England, where its wood-boring caterpillars may cause considerable damage to certain trees and shrubs. Those most commonly infested are apple and ornamental *Malus*. Other hosts include ash, birch, cherry, cotoneaster, hawthorn, lilac, maple, oak, pear, rhododendron, *Sorbus* and willows. Damage tends to be more severe than usual following hot, dry summers.

Nature of damage

The young caterpillar bores into a current year's shoot near the tip. When the shoot dies the caterpillar enters older wood farther down the branch and usually a third entry is made in still older wood. On young fruit trees the final entry is in the stem, while on older trees it is usually near the base of a branch.

Symptoms of attack by young caterpillars become obvious in September when dead shoot tips appear or leaves on the apical portions of branches turn colour prematurely. Infested branches usually break on bending. The entry hole (up to 5 mm in diameter) can be recognized by the presence of frass (excrement), which resembles small pellets of sawdust (Fig. 25.3). After 6–9 months of feeding there is often an accumulation of frass outside the entry hole, but later it falls to the ground and may then be the most obvious symptom of infestation (Fig. 25.4). More usually, an infestation is noticed in the year after hatching or even in the spring of the following year. It is unusual to find more than one caterpillar in a tree.

Description and life history

The moths are on the wing from mid-June to early August. They have a grey body and white wings spotted with bluish-black (Fig. 25.2). The female has a wing-span of about 50–60 mm, while that of the male is about 40 mm. Under optimum conditions a female lays several hundred dark yellow eggs, usually deposited singly in crevices in the bark. Eggs are distributed widely but, on a small tree, usually only one is laid; this minimizes competition between caterpillars for food.

Eggs hatch in about two weeks and the young caterpillar, which is pale yellow with small black spots, bores into a young shoot near an axillary bud. Caterpillars then typically tunnel upwards in the pith. Feeding and tunnelling in older wood continue for two or three years (Fig. 25.1); when fully grown, usually in late spring, the caterpillar is about 50 mm long.

The fully fed caterpillar spins a cocoon of silk, mixed with particles of wood, at the base of the feeding tunnel and changes into a bright brown pupa. After emergence of the moth a few weeks later, the empty pupal case can be seen projecting from the tunnel.

Fig.25.2 Female leopard moth (× 1¼)

Goat moth

The goat moth (*Cossus cossus* (L.)), although widespread in Britain, is another uncommon insect. On the rare occasions nowadays when its caterpillars are found, they are more likely to be feeding in mature or old deciduous amenity or forest trees than in fruit trees. The common name of this moth is derived from the goat-like smell of the caterpillar.

Goat moth caterpillars can be distinguished from those of leopard moth by their colour, being reddish-brown above and pale yellow below. In addition, their tunnels extend deeply into the trunk of the infested tree and meander rather than follow a vertical course.

Description and life history

The moth is large, with a wing-span of up to 90 mm, and of drab appearance, having mottled brownish-grey wings and body. Adults emerge in June and July and eggs are laid in crevices in the bark.

The caterpillars feed for three or more years within the trunks of trees. When fully grown they are 70–100 mm long. They pupate in silken cocoons, incorporating chips of wood, usually near the entrance to a feeding tunnel but may sometimes wander away from the tree and pupate elsewhere. Because moths often lay eggs near the entrance holes, trees may contain caterpillars of different ages and an infestation can persist until the tree is killed.

Fig.25.3 Two-year-old apple shoot showing second entry of caterpillar of leopard moth at a node. Note frass outside entry hole

Fig.25.4 Frass on ground near trunk of apple tree, indicating the presence of an older caterpillar of leopard moth tunnelling in main stem or in basal part of a branch

Currant clearwing moth

The currant clearwing moth (*Synanthedon tipuliformis* (Clerck)) occurs throughout much of the British Isles but is most numerous in the southern half of England. The caterpillars feed inside the stems of currant bushes (Fig. 25.5) and can cause damage in fruiting plantations; gooseberry is also attacked. Leaves on infested shoots wilt and fruit trusses fail to mature.

Description and life history

The adults are somewhat wasp-like, having blackish bodies with three or four narrow yellow bands and brown-bordered, transparent wings with a span of 15–20 mm. They are active in sunny weather and may be found on or near currant bushes in June and July. They are particularly fond of resting on the foliage of the bushes. Eggs, which are pale yellow, are laid singly on the stem near a bud or wound.

On hatching, the young caterpillar, which is whitish and has a brown head, burrows into the pith to begin feeding. The caterpillar (Fig. 25.5) can burrow either upwards or downwards but usually works towards the younger wood. Caterpillars may be found throughout the autumn and winter months; they become fully grown in April when about 15 mm long. Pupation takes place just below the surface of the stem and, when the adult eventually emerges, the empty pupal case remains projecting from the burrow.

Fig.25.5 Black currant shoot cut open to show mature caterpillar of currant clearwing moth in tunnel

Apple clearwing moth

The apple clearwing moth (*Synanthedon myopaeformis* (Borkhausen)) is usually restricted to old trees and hence is seldom seen in commercial orchards. Apple is the main host plant, but caterpillars have also been found feeding in pear, plum, cherry and *Sorbus*.

Description and life history

The moth (Fig. 25.6) has a slender black body with a red band on the upper side, from which is derived the alternative common name of red-belted clearwing. The wings are dark-bordered and transparent and have a span of about 20 mm. Adults are active by day in June and July and eggs are laid in crevices in the bark.

Fig.25.6 Apple clearwing moth (× 3)

The caterpillars are creamy-white with a pale brown and somewhat flattened head. They live for about two years, feeding in or beneath the bark in the trunk, or near the bases of main branches, where injury by other means has previously occurred. Caterpillars also occur in cankers on smaller branches and in areas of bark occupied by caterpillars of the cherry bark tortrix moth (*Enarmonia formosana*) (Scopoli)). Suitable areas of bark often contain first- and second-year clearwing caterpillars and infestations may persist in a trunk for several years. When fully fed the caterpillars are about 15 mm long. Pupation occurs in the feeding tunnels just below the surface of the bark.

Raspberry moth

The raspberry moth (*Lampronia rubiella* (Bjerkander)) occurs throughout Britain on wild and cultivated raspberry. The caterpillar, which is known as the raspberry borer, burrows into the buds of fruiting canes. It rarely causes much injury to cultivated raspberry in England, but in Scotland it is a more serious pest, especially in the older plantations.

Fig.25.7 Raspberry cane with shoot (on right) damaged by caterpillar of raspberry moth

Loganberry and blackberry are also damaged.

Nature of damage

The injury is characteristic and shows in the late spring when, in addition to healthy side-shoots, others can be seen that are withering and have not developed far beyond the bud stage (Fig. 25.7). Each withered bud contains a tunnel made by a caterpillar of the raspberry moth. The small bright red caterpillar or a brown pupa is usually found inside the tunnel when the withered bud is opened (Fig. 25.8).

Description and life history

The raspberry moth is small, having a wing-span of 9–12 mm. The wings are dark purplish-brown with several conspicuous yellow spots. The moths appear at the end of May and in June and lay their eggs in flowers of the host plant.

After about a week, whitish caterpillars hatch from the eggs and burrow into the plug of the fruit, in which they feed. During this period the damage caused is so slight that it is difficult to tell which fruits have been attacked. When the fruits begin to ripen, the caterpillars leave them and search for places where they can make cocoons in which to spend the winter. These cocoons, which are small and white, usually occur in the soil and rubbish around the stools, or in crevices in canes and supporting stakes.

Early in April the caterpillars emerge from their cocoons, crawl up the canes and bore into the bursting buds. Each caterpillar then feeds within a bud for 4–6 weeks; during this time it becomes bright pink with a blackish head. When mature it pupates within the burrow or a suitable crevice, or amongst the young foliage after tying it together with silk. The moth emerges three or four weeks later.

Currant shoot borer

The currant shoot borer (*Lampronia capitella* (Clerck)) is an uncommon and local species. The caterpillar feeds on *Ribes*, particularly red and white currant. It is mainly a pest in gardens.

Fig.25.8 Injured raspberry shoot cut open to show caterpillar of raspberry moth inside

Description and life history

The moth is active by day during late May or early June. It has a wing-span of about 15 mm and is recognized by its yellow head and purplish-brown fore wings, each with three or four yellowish spots. Eggs are laid in young fruits, frequently two per fruit.

On hatching, each caterpillar bores directly into a seed. In late June or early July, when about 2 mm long and red, the caterpillars leave the fruit and hibernate in white cocoons, usually situated among buds or scales on a fruiting spur. At this stage the only evidence of infestation is prematurely ripened fruit.

On emergence in the spring each caterpillar bores into a bud or shoot and then tunnels along it to feed on axillary buds. As a result, infested shoots wilt and die. As the caterpillar matures it changes from red to green. Pupation takes place either in the tunnel or under loose rind on the branch.

Control measures

Leopard moth

In most years caterpillars are so infrequent that only an occasional older one is found tunnelling in the basal part of a branch or in the trunk. Such individuals can be killed by pushing a piece of wire up the tunnel; a canker paint should then be applied to the damaged area. Alternatively, an insecticide which vaporizes can be introduced at the base of the tunnel and the hole then sealed with grafting wax, clay or putty.

After very hot summers, when young caterpillars are likely to be numerous, younger orchards should be inspected in September–October. Infested shoots should be pruned and the caterpillars killed to prevent entry into older wood.

Goat moth

If the trunk of a tree is badly infested, the whole tree should be cut down and burnt. When an attack is slight the measures suggested for control of leopard moth should suffice.

Currant clearwing moth

Attacked branches should be cut out and burnt. Chemical control is difficult because the caterpillars invade the stems during the fruiting period. However, a high-volume spray of a recommended insecticide applied immediately after harvest may give some control.

Apple clearwing moth

Control measures are rarely necessary and there are no recent recommendations. An old method is to spray or brush a winter wash on to the bark of infested trees in May, but care must be taken to prevent contact with foliage otherwise severe scorching will result.

Raspberry moth

An annual clean out and burning of rubbish from around the stools, together with the old canes and damaged supporting stakes, helps to decrease the number of hibernating caterpillars.

Where the pest is established, serious injury can be prevented by drenching the stools with a winter wash in late February or early March, shortly before the caterpillars are due to emerge from hibernation.

Alternatively, effective control should be obtained by a high-volume application of a recommended insecticide as a spring spray in early April to kill emerging caterpillars.

Currant shoot borer

Infested shoots should be removed and burnt before the beginning of May. There are no current recommendations for the chemical control of this pest.

26
Swift moths

Fig.26.1 Potato tubers damaged by caterpillars of swift moth

Several species of swift moth are found in Britain. The caterpillars of two of them — the ghost swift moth (*Hepialus humuli* (L.)) and the garden swift moth (*Hepialus lupulinus* (L.)) — damage a range of crops by feeding on the roots. Both these species are generally distributed throughout Britain. The caterpillars live in the soil and attack farm and market-garden crops (Fig. 26.1), nursery stock and grass, sometimes causing serious losses. They may also be found beside or tunnelling into the roots of docks, dandelions and other weeds.

Caterpillars of the **ghost swift moth** are perhaps most common in grassland and sometimes damage meadows and lawns. Any crop planted in freshly broken grassland is liable to be attacked by these caterpillars. The amount and duration of the damage depend on the age of the caterpillars when the grassland is ploughed and planted.

Hops are particularly liable to be attacked by this pest. Older plants may not be seriously affected but, if they are grubbed, succeeding crops such as strawberries are likely to be badly damaged.

The **garden swift moth** is a commoner species and attacks many more kinds of plant. Like the ghost swift moth it may cause damage

Fig.26.2 Chrysanthemum stem-base damaged by caterpillar of garden swift moth

Fig.26.3 Male ghost swift moth (× 3)

to hops. Strawberries and lettuces are especially liable to injury; large cavities are eaten in the base of the crowns and hearts respectively, and crops may be completely ruined. The caterpillars also feed on cereals, most root crops and many ornamentals including chrysanthemum (Fig. 26.2), dahlia, gladiolus, iris, lily of the valley, narcissus and peony.

Description and life history

Ghost swift moth

The ghost swift moth is fairly large with a wing-span of 40–65 mm. The wings of the male (Fig. 26.3) are silvery or greyish-white with a reddish-brown edge; those of the female (Fig. 26.4) are greyish-brown or buff and sometimes there are brick-red markings on the fore wings.

The moths fly in June, July, August and occasionally in September. The males can be seen at dusk, hovering over the herbage until found by the females, which are probably attracted to them by a scent.

The female lays 200–700 eggs, which are dropped singly over grass or other vegetation while the moth is in flight; they are spherical and dirty-white when first laid but soon turn black.

The eggs hatch in 12–18 days and the young caterpillars burrow into the soil and begin to feed on the fine rootlets of nearby plants. The caterpillars (Fig. 26.1) are white with a reddish-brown head and several dark dots along the body.

The caterpillars grow slowly and usually take two years to become fully fed, when they are about 65 mm long. In their second summer they may tunnel into larger and thicker roots and, occasionally, into the lower part of stems. When conditions are favourable the life cycle is completed in a year, but in certain unfavourable conditions the caterpillars have been known to spend a third winter in the ground. During the spring fully fed caterpillars change into pupae (Fig. 26.5) in cells in the soil below the plants.

Garden swift moth

The garden (or common) swift moth (Fig. 26.6) is generally smaller than the ghost swift moth, varying from 25 to 40 mm across the wings. The fore wings are brown with white markings and the hind wings are greyish. The colouring is very variable but the female is usually more drab and larger than the male.

Unlike those of the ghost swift, the males of the garden swift do not hover and attract the females but seek them out. The female lays her eggs while in flight but seems to drop them deliberately near suitable food plants. She lays 50–200 eggs, which hatch in 10–14 days. The egg, caterpillar and pupa resemble those of the ghost swift moth, but the caterpillar and pupa are appreciably smaller, and the caterpillar more slender and translucent. The caterpillars burrow into the soil and usually complete their life cycle in one year if plenty of food is available. Feeding continues throughout the winter but is at a peak in February and March. Pupae are formed in the spring and the moths emerge in mid-May and June. In some years a few moths may also be seen in August and September.

Fig.26.4 Female ghost swift moth (× 2)

Natural enemies

Swift moth caterpillars are attacked and occasionally destroyed by parasitic fungi, which give their dead bodies a grey, mummified appearance. They are also eaten avidly by moles and birds.

Control measures

Cultural control

It is important not to plant into land in which the caterpillars are known to be present. Since newly broken grass or waste land is always likely to be infested by these and other soil pests, it is advisable to devote a season to thorough cleaning operations before planting with perennials.

Fig.26.5 Pupa of ghost swift moth (× 2½)

Fig.26.6 Male garden swift moth (× 2½)

Periodic cultivation and digging of the soil help to keep swift moth caterpillars in check by disturbing and injuring them, and exposing them to the attacks of birds. The caterpillars are most harmful to crops such as strawberries and hops growing in soil that is seldom disturbed. Trouble is usually most serious where land is full of couch-grass. Weed control in a perennial crop reduces the risk of egg laying because it is a cover of weeds or grass rather than the actual crop that is attractive to the female moths. Clean cultivation is thus an important means of controlling this pest.

In gardens where swift moths are common, bulbous and herbaceous plants should be lifted and replanted at fairly frequent intervals.

It is important to ensure that loam used for potting is free from caterpillars. Serious losses in chrysanthemums and peonies have been traced to this source.

Chemical control

In districts where these pests are troublesome, growers should seek advice on suitable insecticide treatments.

27
Tortrix moths on apple

Fig.27.1 Apples at harvest showing damage by caterpillars of fruitlet mining tortrix. The fruits are positioned to show the sides which were touching in the cluster

Caterpillars of many species of tortrix moth are found on fruit trees, but the majority feed only on foliage or occur in such small numbers that they are not regarded as important pests. The caterpillars of certain species, however, also feed regularly on the fruit. These include the fruitlet mining tortrix (*Pammene rhediella* (Clerck)), which attacks apples and plums in June and July, the summer fruit tortrix (*Adoxophyes orana* (Fischer von Röslerstamm)), which damages apples and pears in July and August, and the fruit tree tortrix (*Archips podana* (Scopoli)), which attacks maturing apples from late August until harvest.

The fruit tree tortrix is the most common species and occurs in all the main apple-growing areas. The fruitlet mining tortrix is locally common in the Weald of Kent and has also been found in East Anglia and Worcestershire. The summer fruit tortrix is traditionally a pest of north Kent but is extending its range.

Although caterpillars of each species will feed on the foliage of many kinds of fruit plants and other trees and shrubs, apple suffers most injury to the fruit. Damage of economic importance is usually localized. The following descriptions refer mainly to apple. The fruitlet mining tortrix on plum is described on page 124.

Recognition and damage

Moths of most species are active at night and are seldom seen during the day unless disturbed from their resting positions among foliage. An exception is the fruitlet mining tortrix whose adults are active in daytime, especially during sunny periods. As a group, the tortrix caterpillars can be recognized by their characteristic habit of wriggling backwards when disturbed. They are rather slender, from 12 to 25 mm in length when fully grown, and usually green. All live in sheltered positions, usually among leaves which are webbed together or rolled, often at the tips of young shoots.

Caterpillars of individual tortrix species can be recognized only after considerable practice, but those of greatest economic importance injure the fruits and these injuries form the easiest method of identification.

Fruitlet mining tortrix

The whitish caterpillar is found during June and early July at the centre of a cluster of apples which are joined together by silken webbing. In this position the caterpillar feeds on the fruits only; on opening the cluster, a small number of holes, each usually surrounded by a dark ring and often only a few millimetres deep, will be found on the protected side of each fruit. Occasionally a tunnel penetrates deeply into the flesh but rarely reaches the core, so that infested fruits remain on the tree until harvest. Similar damage may also occur to single fruits underneath a leaf which has been webbed to the side. The damage can be distinguished from that caused by apple sawfly (see page 93) or codling moth (see page 109) by the presence of shallow punctures, and also because the deeper tunnels follow a meandering course instead of going straight to the core, as happens when sawfly or codling attacks fruits.

Often the damage is not noticed until harvest. By this time the injured areas have developed a rough, corky surface and are dimpled or misshapen owing to reduced growth around the damaged area (Fig. 27.1).

Summer fruit tortrix

Mature caterpillars, pale green with a pale greenish-brown to dark brown head, occur in late May and June and again from mid-July to September. They are usually found between webbed leaves, especially at the tips of shoots, but may also damage apple and pear fruits in July and August. A large, shallow, irregular area of skin is sometimes removed from the surface (Fig. 27.2). Development of the fruit is too far advanced for malformation to appear

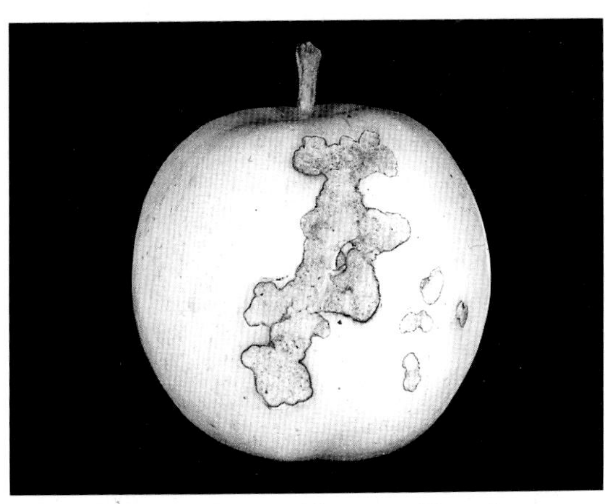

Fig.27.2 Apple damaged by caterpillar of summer fruit tortrix

and the injured area takes on a uniformly russet appearance.

Fruit tree tortrix

Mature caterpillars, which are dark green with a darkish-brown head, occur in late May and June. They usually feed on the foliage but occasionally eat deep irregular cavities in the sides of apple fruitlets. Other caterpillars, such as those of the winter moth (see page 154) and clouded drab moth (*Orthosia incerta* (Hufnagel)), cause similar injury, so these symptoms do not necessarily indicate the presence of tortrix unless the caterpillar is also found.

The important damage occurs from late August until harvest, when young caterpillars attack the maturing fruits and are taken into store. Some caterpillars feed in the eye (calyx) cavity, while others web a leaf to the side of a fruit or live in a silken web where two fruits are touching. Small irregular areas of skin are removed from the fruit (Fig. 27.3) and occasional punctures may extend to a depth of 1–2 mm in the flesh. This injury not only reduces the quality of fruits but allows infection by fungi which may lead to rotting during storage.

Fig.27.3 Mature apples damaged by young caterpillars of fruit tree tortrix. Similar damage can be caused by caterpillars of other tortrix moths

Description and life history

Fruitlet mining tortrix

The moths, which are dark brown with a lighter metallic area at the tip of each fore wing, have a wing-span of 10 mm. They emerge during May and, on calm, sunny days, may be seen flying over the tops of apple trees. The flat, translucent, scale-like eggs are laid singly on the underside of the rosette leaves and begin hatching during the week following petal fall.

The caterpillars are whitish with a black head until the last stage when the head becomes brown. Except when feeding, the caterpillar lives in the silken webbing between fruits or, if fruits are not available, it will tunnel into young shoots. By early July it is fully grown and then makes a strong silken cocoon under loose bark, where it remains until changing into a pupa in the following spring. There is only one generation each year.

Hawthorn is the normal host of this species and is the source of orchard infestations.

Summer fruit tortrix

The young caterpillar hibernates in a silken cocoon underneath a piece of dead leaf or bud-scale which is firmly attached to a spur or small branch. These yellowish to green caterpillars (Fig. 27.4) resume activity between bud break and green cluster in the spring. At first they burrow into the fruit buds, then feed on the blossom truss and later web together leaves on young shoots. The pupa is formed in the webbed leaves.

The moths that emerge are light brown and are present from early June until late July. They are active at night and lay batches of up to 100 or more eggs on the leaves. Each egg mass is yellowish-green and appears as a flat, waxy scale.

Another brood of caterpillars develops in July and August, giving rise to a second generation of moths from mid-August to late

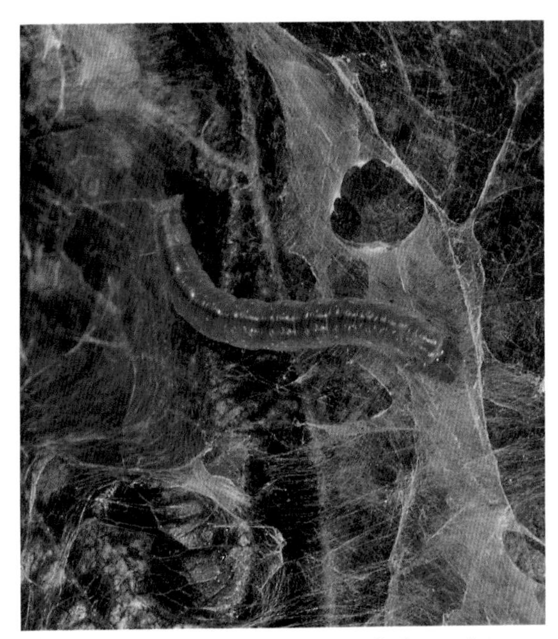

Fig.27.4 Caterpillar of summer fruit tortrix on webbed leaf (× 4)

September. The young caterpillars hatching from eggs laid by these moths moult once or twice before they hibernate.

Fruit injury is caused mainly by the larger caterpillars which occur in July and August.

Fruit tree tortrix

The life cycle is very similar to that of the summer fruit tortrix, except that the caterpillar develops more slowly. Moths (Fig. 27.5), therefore, do not appear until mid-June and are found until mid-August. Eggs are laid in green scale-like masses (Fig. 27.6) which closely resemble the leaf in colour.

The caterpillars disperse after hatching and, while young, each one lives in a silken web beside the midrib on the underside of a leaf. At first they are yellowish with a black head; later the body becomes grey-green and in the final stage (Fig. 27.7) the head becomes brown. Caterpillars of this generation hibernate after moulting once or twice, although in warm summers some may develop to maturity and give rise to a partial second generation of moths in September and early October. The caterpillars hatching from eggs laid by these

Fig.27.5 Fruit tree tortrix moths on apple leaf (×1½). Left: female. Right: male

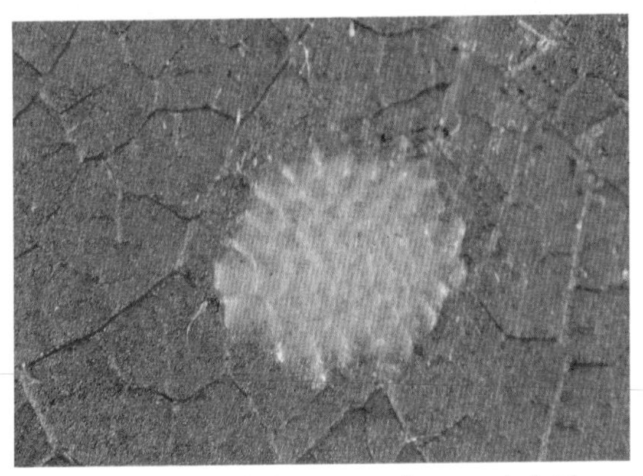

Fig.27.6 Eggs of fruit tree tortrix recently laid on leaf (× 5)

moths feed for a short while, moult once or twice and then hibernate in a silken cocoon attached to a spur.

Caterpillars of both generations may damage fruit.

Pheromone traps

Sticky traps containing caps impregnated with an attractant which is specific to the male moth, i.e. a sex pheromone, are available for summer fruit tortrix and fruit tree tortrix. These can be used to monitor the emergence and flight of these tortrix species and so improve the accuracy of timing sprays for their control. If fewer than 30 fruit tree tortrix moths

Fig.27.7 Final-stage caterpillar of fruit tree tortrix on leaf (× 3)

are caught per trap per week, it is not necessary to spray against this species. A threshold for summer fruit tortrix has not yet been established, but the indications are that it is similar to the threshold for fruit tree tortrix.

Control measures

Special measures to control tortrix caterpillars are necessary only if damage of economic importance occurs to the fruits. Where pheromone traps are not used, the decision on whether or not to spray against tortrix depends on the previous history of damage.

The application of conventional insecticides should be timed to coincide with egg hatch; this generally occurs 10–14 days after egg laying but depends on temperature. If spraying is delayed, caterpillars are less susceptible and probably less accessible.

Insect growth regulators that are ovicidal should be applied to coincide with egg laying.

Fruitlet mining tortrix

As eggs of this species start hatching shortly after petal fall of apple, control measures must be applied at an earlier date than for the other species. A recommended insecticide spray should be applied about 10–14 days after petal fall. Only one treatment is needed.

Reinfestation appears to occur slowly; treatment can therefore be discontinued after one or two years providing an annual survey is made in late June or early July to detect damaged fruits.

Summer fruit tortrix and fruit tree tortrix

The sprays applied at green cluster for control of winter moth are also toxic to tortrix caterpillars, but serious damage to maturing fruit has sometimes occurred after this treatment. Also, reinfestation can occur in the summer as a result of moths flying into orchards from neighbouring hedges or woodland.

Control of codling moth and fruit tree tortrix moth should be considered as a combined operation. Fruit tree tortrix usually begins to appear about 7–10 days later than codling moth, so the codling moth spray should be delayed by about a week if fruit tree tortrix is considered the more serious pest.

The summer fruit tortrix moth flies 2–3 weeks earlier than threshold numbers of codling moth. Therefore, where summer fruit tortrix is a problem, a separate earlier spray may be required.

Owing to the long flight period of tortrix moths more than one spray may be necessary. The use of pheromone traps enables sprays to be timed more precisely and unnecessary treatments to be omitted.

The summer spray programme will result in a great decrease in the number of overwintering caterpillars, hence there should be no need to apply control measures in spring, unless other species of caterpillars, such as winter moth, are present.

28
Tortrix moths on strawberry

Fig.28.1 Strawberry fruits damaged by caterpillar of a tortrix moth

Caterpillars of several species of tortrix moth can cause damage to strawberry. By far the most important in Britain are the strawberry tortrix (*Acleris comariana* (Lienig & Zeller)), the straw-coloured tortrix (*Clepsis spectrana* (Treitschke)) and the dark strawberry tortrix (*Olethreutes lacunana* (Denis & Schiffermüller)), but damage is also caused locally by other species including the carnation tortrix (*Cacoecimorpha pronubana*) (Hübner)) and the flax tortrix (*Cnephasia asseclana* (Denis & Schiffermüller)).

The **strawberry tortrix** is an important pest in many strawberry-growing districts of Britain, especially in the Wisbech area, the Vale of Evesham, south Lancashire and the Dee Valley, which are all low-lying areas where its wild host plant — marsh cinquefoil (*Potentilla palustris*) — occurs. This pest has spread to other districts mainly by the introduction and transplanting of infested strawberry runners.

The **straw-coloured tortrix** (or cyclamen tortrix) also inhabits low-lying areas, but outbreaks of this species are more localized. It feeds on marsh cinquefoil and various other wild plants including willowherb (*Epilobium* spp.) and also black currant, strawberry, hops, grape-vines and several glasshouse plants including cyclamen and rose.

The **dark strawberry tortrix** often occurs in districts where the other species are less common or absent. As with the straw-coloured tortrix, outbreaks occur locally, but the species has caused serious damage to protected strawberry crops in Hampshire.

The **carnation tortrix** is frequently a glasshouse pest in Britain. Since the 1920s it has established itself on outdoor plants in central and southern England, especially privet, spindle and ivy. It is a serious pest of hardy ornamental nursery stock and occasionally causes damage to strawberry.

The **flax tortrix** is a widespread species. The caterpillars occasionally damage strawberry fruitlets.

Recognition, life history and damage

Most tortrix moths are active at twilight so are seldom seen during the day unless disturbed from their resting positions among foliage. As a group, tortrix caterpillars can be recognized by their characteristic habit of wriggling backwards when disturbed. When fully grown, they are 15–20 mm in length and green, grey or brown depending on the species, though colour often varies within a species.

The caterpillars spin the leaves of their host plants together with silk to form shelters where they feed. They often move to new sites (generally to younger leaves) and some species may move into flowers later; fruit may also be damaged (Fig. 28.1).

Fig.28.2 Strawberry leaf damaged by caterpillar of strawberry tortrix

Strawberry tortrix

The moths have wing-span of about 15 mm. They are variable in colour with a pronounced dark mark mid-way between the base and tip of each fore wing (Fig. 28.3). They emerge in mid-June with a second generation appearing from mid-September until late October. Eggs are laid on the upper and lower sides of the leaves. Those laid by second-generation adults overwinter and the young caterpillars hatch in the following spring — from late April onwards.

Caterpillars are present from late April to June and again from July to September. When young, the caterpillars are white with a black head but, when fully grown, they are pale green with a brownish head. They are usually found among webbed leaves, but in spring some may move into the flowers. As the majority of caterpillars feed on the foliage (Fig. 28.2), there is little economic damage to well-established plants unless infestations are very large.

Straw-coloured tortrix

The moths are pale straw-colour with dark markings (Fig. 28.4) and are slightly larger than the strawberry tortrix. They appear in the first week of June, earlier than the strawberry tortrix, with a second generation in August–

Fig.28.3 Strawberry tortrix moth (× 2)

September. All stages of the caterpillar have a black head and an olive-green to brown body with pale warts (Fig. 28.5). In contrast to strawberry tortrix, this species overwinters as young caterpillars in leaf shelters on the host plants. Feeding begins in early spring on the foliage; later the caterpillars move into and feed mainly in the flowers, often capping over the petals in a characteristic manner (Fig. 28.6) and feeding on stamens and stigmas. Considerable economic damage to the crop can occur.

Fig.28.5 Caterpillar of straw-coloured tortrix (× 2)

Dark strawberry tortrix

The life cycle is similar to that of the straw-coloured tortrix, with caterpillars overwintering on the host plant. The moth is beige with darker markings (Fig. 28.7). The caterpillar is similar to the straw-coloured tortrix caterpillar but has shiny black warts and is usually darker (Fig. 28.8). The flowers and developing fruitlets can be seriously damaged by the caterpillars.

Carnation tortrix

The life cycle is similar to that of the preceding two species. The moths, which are basically brown with orange hind wings, appear mainly during two periods of the year — in June and again in September–October. The adult males are unusual in being active during the day. The caterpillar is grey-green and occurs from October to May with a second brood in June–July, although there is often some overlapping of the two generations. This caterpillar feeds on leaves and in the flowers and can cause serious damage.

Fig.28.6 Strawberry flower infested with caterpillar of straw-coloured tortrix. Petals have been spun together to form a roof under which the caterpillar feeds

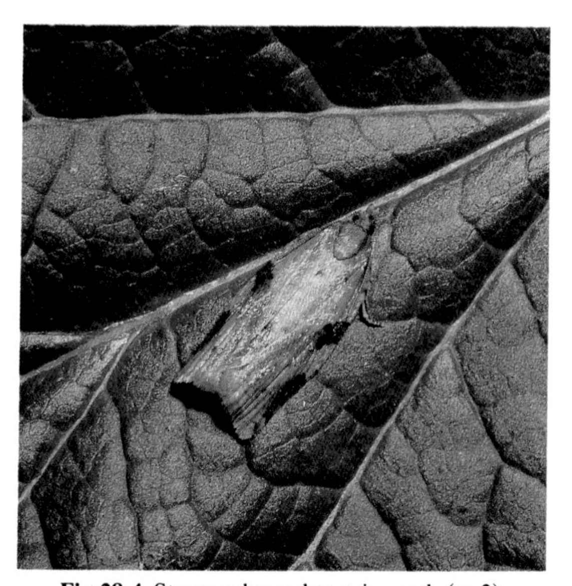

Fig.28.4 Straw-coloured tortrix moth (× 3)

Flax tortrix

The adults are mottled grey. Eggs are laid in various situations, but mainly on deciduous trees, in late summer. On hatching, the greyish caterpillars immediately hibernate without feeding. In the following June there is some wind dispersal of caterpillars to various low-lying plants including strawberry. When strawberry is attacked the fruitlets can be damaged as well as the foliage and flowers.

Natural enemies

Tortrix caterpillars are often parasitized by parasitic wasps and other insects. The chalcid wasp *Litomastix aretas* (Walker) is a particularly important enemy of the strawberry tortrix, parasitism sometimes reaching 90%.This can result in considerable variation in the number of caterpillars from one season to another. Many pesticides kill beneficial insects and excessive treatment may lead to an increase in tortrix numbers.

Fig.28.7 Dark strawberry tortrix moth (× 2½)

Fig.28.8 Caterpillar of dark strawberry tortrix (× 2½)

Control measures

For strawberry tortrix, a spray should be applied as close to the beginning of flowering as possible. Egg hatch is generally complete by then, but in cold springs the hatching period is prolonged, so that some caterpillars may hatch during the flowering period making control sometimes less effective.

Various insecticides are recommended for control of strawberry tortrix and spraying with any of them generally gives good control. Sprays are, however, more effective when applied at high volume (at least 1000 litres/ha) than at low volume.

As caterpillars feeding on the foliage usually have only a slight effect on yield, control of the summer generation occurring in July/August is generally necessary only when infestations are severe.

Other species which overwinter on the plant as caterpillars can be controlled by applying a spray as soon as they have begun to feed on the leaves in the spring. With protected crops this may be as early as March. Damage by these species later in the season is unusual, so that later spray treatments are not normally required.

29
Web-forming caterpillars

Fig.29.1 Webbing of caterpillars of small ermine moth on bird-cherry

Caterpillars of some moths feed gregariously, for at least part of their lives, in communal silken webs or 'tents' which they spin on their food plants. Some species which have this habit and are economically important in Britain are described here. These species are the small ermine moths (*Yponomeuta* spp.), the lackey moth (*Malacosoma neustria* (L.)) and the brown-tail moth (*Euproctis chrysorrhoea* (L.)). The vapourer moth (*Orgyia antiqua* (L.)), whose caterpillars occur together in large numbers though not feeding within webs, is also included.

Small ermine moths

These moths derive their name from the attractively speckled wings of the adults (Fig. 29.2). They are probably best known for the defoliation of trees and shrubs, especially hawthorn, caused by the caterpillars. Outbreaks have been reported from many places throughout the British Isles. Several kinds of small ermine moth occur here: *Yponomeuta padella* (L.) on blackthorn and hawthorn; *Yponomeuta malinellus* Zeller on apple; *Yponomeuta cagnagella* (Hübner) on spindle (both deciduous and evergreen species); *Yponomeuta evonymella* (L.) on bird-cherry; and *Yponomeuta rorrella* (Hübner) on willow.

Nature of damage

Infestations tend to occur regularly in certain localities but are more widespread in some years. During three or four weeks in early summer, a severe attack can turn a pleasant green hedge into a skeleton of bare twigs and a few withered leaves tied up in numerous silken webs. Hedges can be killed by repeated attacks. Damage is rarely seen these days in commercial orchards in which winter or spring sprays are applied.

Description and life history

The moths are about 8 mm long and about 22 mm across the wings. The fore wings are lustrous white or grey dotted with black (Fig. 29.2), while the hind wings are uniformly grey. The moths emerge in July and August and lay their eggs in batches on the branches of their food plants (Fig. 29.3). The eggs are flat and each batch is covered by a hard protective coat which closely resembles the colour of bark. The eggs are thus well camouflaged.

In September tiny caterpillars hatch from the eggs but remain under the protective covering until the spring, usually May, when they emerge to feed on the developing foliage. On some food plants, e.g. on apple but not hawthorn, the caterpillars at first live gregariously in leaf mines, causing brown blisters on the leaves. Later these caterpillars, and the newly emerged caterpillars on plants where no mining occurs, live in colonies between leaves that are webbed together (Fig. 29.1). As these leaves are eaten, the web is extended to include more leaves. When all the foliage on one spur or shoot has been devoured, the colony moves to another site and repeats the procedure. With severe infestations the numerous webs meet, so that whole branches or large areas of hedge may be covered by a mesh of caterpillar silk.

By the end of June or early July, the caterpillars are fully grown, being about 12 mm long and dusky grey with black spots. They then

Fig.29.2 Small ermine moth (× 4)

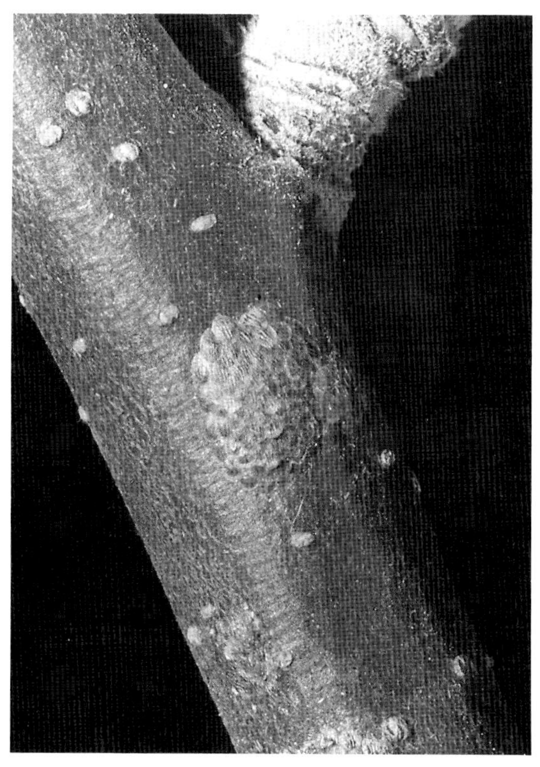

Fig.29.3 Egg batch of small ermine moth on bark of apple tree (× 6)

spin cocoons, often many together, within the web and change into pupae.

Natural enemies

Small ermine moth caterpillars are heavily parasitized by several kinds of chalcid and ichneumonid wasps. They are also eaten by starlings and other birds.

Lackey moth

This moth is common in many parts of southern England but is rare north of the Midlands. It tends to remain unnoticed in a district for years, then during one season caterpillars become numerous and cause defoliation of trees and shrubs. The infestation dies away in the next one or two years, then several more seasons may pass before there is another outbreak.

Nature of damage

Lackey moth caterpillars feed on the foliage of many trees and shrubs including apple, cherry, pear and plum among fruit trees, and alder, blackthorn, elm, hawthorn, hazel, oak, rose and willow among woodland trees and shrubs. Severely infested trees are weakened by the loss of foliage, so growth is poor and fruits fail to mature. The effects may persist for one or even two more seasons. This moth was once fairly common on commercial fruit but is now mostly confined to neglected orchards and to gardens.

Description and life history

The moths are on the wing at night from the end of July to September. They are brown and measure about 35 mm across the wings. After mating, the females lay their eggs on the twigs of the various trees or shrubs on which the caterpillars feed. Batches of 100–200 greyish-brown eggs are laid in the form of a band, 6–14 mm wide, encircling a twig (Fig. 29.4). The moths die in September or earlier and the eggs hatch in the following late April or May.

The caterpillars (Fig. 29.5) are nearly black when they hatch but soon show a colourful pattern: the ground colour is blue-grey and there are white and orange-red stripes outlined in black on the back and sides. The whole body is clothed with reddish-brown hairs.

Lackey moth caterpillars live in colonies for most of their lives. After hatching, they spin a small, silken web between two or three leaves and, as they grow, they build larger nests or tents by spinning together twigs and branches in a thick web of silk (Fig. 29.6). A large colony may form a nest 30 cm long and 15 cm wide — a conspicuous object on the host tree, especially when most of the foliage has been devoured. The caterpillars shelter inside these nests but often

Fig.29.4 Egg band of lackey moth (× 3)

Fig.29.5 Caterpillar of lackey moth (× 1½)

Fig.29.6 Young caterpillars and webbing of lackey moth

rest on the outside, apparently sunning themselves, or lie, many together, along an exposed branch. They feed voraciously so that sometimes a tree is stripped bare of its foliage and is almost smothered in webbing.

When fully fed in the latter half of June or early July, each caterpillar spins a yellow or white silken cocoon, which may be found attached to the bark, between leaves on the tree or amongst low herbage. Inside the cocoon the caterpillar turns into a blackish-brown pupa from which the moth emerges about three weeks later.

Brown-tail moth

In some localities, especially near the coast, caterpillars of this moth often become a public nuisance because contact with their irritant hairs can cause painful rashes on the skin. Although recorded as a major pest and injuring fruit trees at many inland sites in southern and eastern England in the 18th century and the early part of this century, the insect seems for the past 30 or more years to have occurred mainly in the coastal areas of Hampshire, Sussex, Kent, Essex and Suffolk.

Nature of damage

The caterpillars occur on a range of shrubs and trees in hedgerows and on waste land. Plants commonly attacked are blackthorn, bramble, dog rose, hawthorn, sea buckthorn, apple, pear, plum and English elm. Forsythia, oak, sallow and strawberry are also suitable hosts. In some years the caterpillars are abundant and cause trouble by invading gardens, when they may damage fruit and other trees and bushes, as well as inflicting rashes on persons handling them or walking through infested scrub. Their irritant hairs can be blown by the wind and collect on domestic washing hung to dry in gardens, subsequently causing rashes when the clothing is worn. Acute inconvenience has often been experienced at caravan and other holiday sites.

Description and life history

The moths appear in the latter half of July and in early August and may be found resting beneath leaves during the day. They have white wings and, although the head and thorax

Fig.29.7 Caterpillar of brown-tail moth (× 3)

are white, much of the abdomen is dark brown; both sexes, but notably the females, have a conspicuous tuft of brown hairs at the tip of the abdomen. They measure approximately 32–38 mm across the wings. The females lay eggs in several elongate batches on the stems of the food plant or beneath leaves; each batch is then covered with brown hairs detached from the tuft at the tip of the abdomen. The eggs hatch during the last two weeks of August and the first week of September.

The caterpillars are striking in appearance (Fig. 29.7). The body is nearly black with a double row of vermilion marks of varying width along the back of most of the abdominal segments, a row of downy, white scales along the upper part of each side, and long, yellowish-brown hairs which arise in tufts from brownish warts.

On emerging, the young caterpillars at once begin to spin a silken retreat or tent, which may incorporate several leaves and in which they shelter during dull and cold weather. In warm, sunny weather the caterpillars leave the tent by day and feed on surrounding leaves by eating away the upper layers, often lying with their bodies side by side in an arc-shaped 'front' which gradually advances up the leaf. The attacked leaves wither and form a light brown 'feeding area' around the tent. These feeding areas slowly increase in size (Fig. 29.8) and by autumn may reach 45 cm or more in diameter; they become conspicuous in contrast to the rest of the green foliage and are a characteristic sign of the insects' presence. Even in fine weather, there seem always to be some caterpillars inside the tents.

Fig.29.8 Apple leaves damaged by young caterpillars of brown-tail moth before they hibernated

About mid-September the caterpillars moult into their second stage. They continue feeding into October, though to a gradually lessening extent, with fewer and fewer leaving the tent each day. By the end of October feeding has generally ceased. From the outset, silk is continually added to the tents until by the time activity ceases they have become very dense and tough hibernation quarters for the second-stage caterpillars.

In the following April, as soon as warm, sunny weather occurs, the caterpillars begin to reappear outside the tents, though little feeding occurs until early May. During May young shoots are rapidly stripped and, during this period of activity, as the caterpillars forage farther and farther from their winter quarters, they spin a succession of new silken retreats. These, though not so dense as the winter ones, are often extensive and connected by silken 'walk-ways' along the branches. Early in June, when the caterpillars are over 25 mm long, they begin to move away from the tents and become solitary.

In late June the caterpillars begin to pupate in a silken cocoon spun between leaves, either solitarily or gregariously in a large, thick silken mass. The moths emerge in mid-July. When ready to lay eggs, the females seem loath to fly, so that egg laying and subsequent infestations of caterpillars tend to be intensified in sections of hedge; the kind of plant that is attacked probably depends on what is available nearby.

Natural enemies

Three species of ichneumonid wasp and one tachinid fly are known to be among the natural enemies of brown-tail moth, but the extent to which these affect its abundance is unknown.

Vapourer moth

Although caterpillars of this moth do not feed gregariously in webs or tents, this species does produce conspicuous overwintering egg masses and, in summer, large numbers of the caterpillars occur on a single infested tree. It is, therefore, convenient to include this insect here.

Vapourer moth is widely distributed throughout the British Isles. It seems especially to favour urban districts and is often common on trees in the parks and gardens of large towns. Occasionally, localized outbreaks occur on heather moors.

Nature of damage

The caterpillars attack the foliage of many different plants including blackthorn, hawthorn, hazel, heather, rose and most kinds of fruit and woodland trees; they sometimes feed on herbaceous plants in gardens and glasshouses. In most years they are not numerous, but occasionally large numbers appear and cause much damage. Nowadays, infestations are very rare in commercial orchards because of the widespread use of insecticidal sprays for the control of pests such as aphids (see page 19) and codling moth (see page 111).

Description and life history

The moths occur from July to September, or occasionally in October. The sexes are very different in appearance. The *male* measures 25–32 mm across the wings, which are ochreous or chestnut-brown with darker markings on the fore wings. Near the rear angle of each fore wing is a crescent-shaped, white spot. The body is brown. The *female* is fat and sluggish and unable to fly as the wings have been reduced to mere stumps. The body is yellowish-grey and hairy.

Male vapourer moths fly during the day, usually rather high above the ground. Females, however, remain alongside their pupal cocoons. After mating, each female lays up to 300 brownish-grey eggs on and around the webbing of the cocoon (Fig. 29.9), thereby forming a large overwintering egg mass. The young caterpillars do not emerge until the following spring, generally in May or June. The eggs hatch intermittently during a period of two months or more and, as a result, caterpillars of very different sizes, and in the latter part of the year all stages of the insect, may be found at the same time.

Fig.29.9 Egg mass of vapourer moth on cocoon (× 2½)

Soon after emerging, the young caterpillars scatter over the tree and feed on the foliage. Caterpillars may be found feeding from May to August or, rarely, September. They are brightly coloured with a mixture of red, black and yellow markings on a violet or grey background (Fig. 29.10); there are two red warts on each segment of the body. The caterpillars are

Fig.29.10 Caterpillar of vapourer moth (× 1½)

also very hairy and there are four tufts of long yellow hairs on the back and a pencil of longer, dark hairs protruding from each side of the head; other long, dark hairs occur on each side and at the the end of the abdomen. When fully grown the caterpillar is about 25 mm long.

When fully fed, each caterpillar spins a thin but tough silken cocoon (Fig. 29.9) within which it changes into a dark brown pupa. Hairs from the caterpillar are mixed into the webbing of the cocoon, which is attached to a leaf or a twig, or may be fixed in a crevice in the bark of a tree or on a nearby post or fence. The adults emerge in about three weeks. Moths developing from eggs hatching unusually early in the year may give rise to a partial second generation, although this is rare.

Cultural control

Small ermine moths

Where attacks of small ermine moth have occurred on a small scale, it is worth inspecting susceptible trees and shrubs in early June. Small webs with caterpillars can then be removed and destroyed.

Lackey moth

If only a few trees are affected, the nests formed by the caterpillars can be destroyed by hand. This should be done early in May while the caterpillars are still small. If trees are first banded with sticky material, fallen caterpillars will be prevented from crawling back and reinfesting them. Nest destruction may also be used for 'mopping up' any survivors of a chemical treatment (see below).

The egg bands are not difficult to see in winter and they can be collected and destroyed before caterpillars emerge and cause damage; this can be done at the same time as pruning.In small gardens and on small trees no other control measures should be needed.

Brown-tail moth

Cutting out and burning the winter tents between November and the end of March is very effective and, though sometimes impracticable in dense, thorny scrub, is worth doing wherever possible and as an adjunct to other measures. As the winter tents are mostly formed near branch tips they can easily be cut from fruit trees, while hard trimming of hedges will remove a great many.

Vapourer moth

In districts where outbreaks of vapourer moth caterpillars are known to occur, growers or gardeners should look for and destroy the egg masses during the winter while pruning. On a small scale, hand-picking the conspicuous caterpillars during the spring or summer may be enough to prevent serious damage.

Chemical control

Winter washes

Attacks of small ermine moths can be prevented by the application of a winter wash to kill the young overwintering caterpillars. However, this is worth doing only in winters following a severe attack.

All deciduous trees and shrubs must be dormant when sprayed with a winter wash, December and January being the best months for treatment. Winter washes may be used on all plums except Myrobalan but should not be applied after mid-January; they must not be used on evergreens. Whenever used on deciduous trees or shrubs, winter washes must be applied thoroughly so that all the twigs are wetted; ground crops need to be protected from the spray. Application should not be made in frosty or windy weather or when plants are wet.

Overwintering eggs of the lackey moth are resistant to winter washes, and winter spraying is not suitable for control of brown-tail moth infestations. Winter washes will, however, give good control of vapourer moth eggs and some control of winter moth eggs (see page 156).

Spring and summer sprays

Small ermine moths To control small ermine moth infestations the best time to spray is usually during May, when the caterpillars have just left their first feeding webs and are beginning to feed more extensively on the hedge or tree. Good control of the small caterpillars can be obtained by spraying at high volume with a recommended insecticide at the rate of 25 litres per 100 m of hedge. Fine nozzles should be used to obtain efficient penetration of a hedge. To avoid killing bees, insecticide should not be sprayed on to trees or hedges in flower and should not be used where drift on to flowering plants can occur. Also, to avoid harming other non-target insects and other wildlife, the least harmful insecticide that is available should be chosen and the spray limited to as small an area as possible.

Lackey moth Although lackey moth caterpillars can be controlled with sprays applied in spring or summer, the treatment is ineffective if the host tree is enveloped in webbing. Best results are obtained by application of an insecticide while the caterpillars are young. High-volume sprays should be applied as suggested above for control of small ermine moth caterpillars.

Brown-tail moth Brown-tail moth caterpillars can be controlled by spraying in sunny weather about mid-September, when all the eggs have hatched but before the winter tents have thickened appreciably. A possible difficulty with this timing is that in areas where brambles are present the public would have to be warned against picking blackberries for a few days after spraying. An alternative is to spray in mid-May, though there is probably then a greater risk of some caterpillars inside the winter tents escaping treatment. Spraying in early June when the caterpillars become solitary seems to be reasonably effective, but as they increase in size the insects are probably less susceptible to insecticides. At all times, high-volume sprays should be used as these are more likely to penetrate the silk webbing. Insecticides must be carefully selected where sites are frequented by the public.

Vapourer moth Extensive or heavy infestations of vapourer moth caterpillars can be controlled by spraying as recommended for small ermine moth caterpillars.

30
Winter moths

Fig.30.1 Apple buds and foliage damaged by caterpillars of winter moth

The caterpillars of several species of moth damage fruit trees in the spring. These include tortrix caterpillars (see page 136) and caterpillars of the winter moth group. Tortrix caterpillars spin the leaves together tightly and, if disturbed or held in the hand, wriggle violently. By contrast, caterpillars of the winter moth group web the leaves together more lightly, or hardly at all, and crawl with a characteristic looping action (Fig. 30.5). The most important of the winter moth group are the winter moth (*Operophtera brumata* (L.)), the mottled umber moth (*Erannis defoliaria* (Clerck)) and the March moth (*Alsophila aescularia* (Denis & Schiffermüller)). They attack practically all standard and bush fruits as well as most woodland trees and bushes. Formerly, they were regarded as among the most serious pests of fruit, but a suitable spray programme prevents them from causing appreciable loss.

Nature of damage

The caterpillars are first seen during April and continue feeding until the end of May. The characteristic method of feeding of the caterpillars of the winter moth and the March moth is to spin two leaves together very loosely — or a leaf to the side of a blossom truss — and then to eat holes in the leaves, attacking the edges and centres more or less indiscriminately (Fig. 30.1). They may also eat the flower buds (Fig. 30.1) and later bite holes in the young fruitlets. The damaged area of a fruitlet heals and at harvest time appears as a flat, concave or convex area with a corky surface (Fig. 30.2). Deep bites which reach the core result in a malformed fruit with a deep cleft indicating the site of injury. If the leaves are not open when the eggs hatch, the young caterpillars burrow into the unopened leaf and fruit buds and cause much harm.

The caterpillars of the mottled umber moth feed on the foliage in a similar manner but do not spin leaves together so often, nor are they so inclined to feed on the trusses of fruitlets or blossom.

In bad attacks, everything green on the tree may be devoured by the end of May and the orchards appear as leafless as in winter. Fresh leaves are produced later, but the trees are crippled for the year and production of fruit in the following year is seriously affected.

Fig.30.2 Apples damaged by caterpillars of winter moth

Description and life history

Winter moth

The male is greyish-brown and measures about 25 mm across the fore wings. The female (Fig. 30.3) cannot fly, having greatly reduced wings, and looks very different from the male. The female's vestigial wings and large body make the legs appear long so that superficially the female resembles a spider.

The first moths emerge from pupae in the soil at the end of October and peak emergence occurs about a month later. Very few moths are seen after the end of December. The females crawl up the trunks of trees to the smaller branches and shoots to lay their eggs. Males may commonly be seen flying after dark and are often attracted to artificial lights.

The eggs, light green when laid but soon changing to orange, are placed singly or in small groups among bud clusters, in crevices in the bark, or at the cut ends of shoots. Each female lays 100–200 eggs.

On apple, hatching begins about the time of bud break of Cox's Orange Pippin and is generally complete by the late green-cluster stage, but in late springs hatching may not be complete until pink bud or even later. On other fruit trees, hatching occurs during the same period but the stages of bud development are different.

The youngest caterpillars are olive-green with black heads and are extremely small. They can be blown by the wind from one tree to another; by this means spread may occur from neighbouring woods and hedgerow trees into orchards. As the caterpillars grow they become a brighter green, and creamy-white stripes develop along the back and sides (Fig. 30.4). By the end of May most caterpillars are

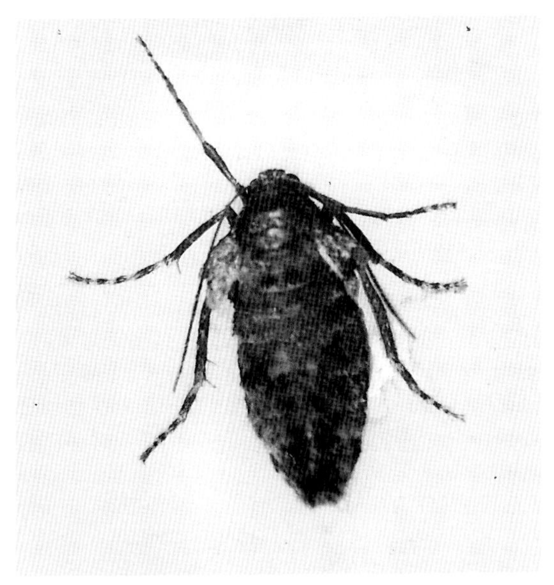

Fig.30.3 Female winter moth (× 7)

fully fed; they then descend to the soil on a silken thread and change into pupae in the soil beneath the tree. Moths emerge from these pupae in the following autumn and winter.

Northern winter moth This species (*Operophtera fagata* (Scharfenberg)) is closely related to the winter moth and difficult to distinguish from it. However, the male northern winter moth is paler than the male winter moth and has a slightly larger wing-span, while the caterpillars have dark rather than pale stripes along the back and sides. The habits of the two species are similar, but the northern winter moth is most likely to occur in orchards close to birch trees.

March moth

The March moth appears in March or sometimes earlier. The male measures about 32 mm across the rather narrow fore wings, which are grey-brown with transverse bands. The female is wingless and is brown with a tuft of hairs at the hind end of the body.

The eggs, which are brown, are laid in rows or bands around twigs and are covered with hairs from the female.

The caterpillar (Fig. 30.5) is yellowish-green

Fig.30.4 Caterpillar of winter moth (× 2)

with a darker green line edged with yellow down the back and pale yellowish lines along the sides. It can be distinguished from the caterpillars of the winter moth and the northern winter moth by its more slender appearance and the presence of a pair of vestigial prolegs on the fifth abdominal segment. The pupal stage is spent in the ground.

Mottled umber moth

The mottled umber moth is rather larger than the winter moth and the northern winter moth. The male measures nearly 40 mm across the fore wings, which are pale brown or brownish-yellow, banded and mottled with dark brown.

Fig.30.5 Caterpillar of March moth (× 2½)

The female is wingless and has a yellow-brown body with two or more dark spots on every segment.

The moths normally appear from the first week in October to the middle of December, but occasionally they are seen in January or even February. The eggs are rather larger than those of the foregoing species and are straw-coloured. As many as 400 are laid by one female.

When fully grown the caterpillar is much larger than those of the winter and March moths and distinct in coloration, being chestnut-brown (sometimes yellow-brown) above with a wavy dark stripe on each side and yellow on the sides and underneath. When fully grown it descends to the ground to form a pupa.

Natural enemies

Small mammals, e.g. shrews, and carabid beetles prey on winter moth pupae. Hymenopterous parasites also attack the pupae. The combined effect of predation and parasitism could significantly reduce populations of the moth.

Control measures

Caterpillars of the winter moth occur much more commonly on fruit trees than do those of the other three species, and it is against this pest that control measures are directed.

In orchards

Partial control of eggs can be obtained with winter washes, but the most effective treatments are sprays applied against the young caterpillars as soon as hatching is complete in the spring.

Spring sprays On **apple**, the optimum time for spraying is at late green cluster to early pink bud. At this stage nearly all the eggs have hatched and the fruit buds have opened, so that the early hatching caterpillars that burrowed into the developing buds are now in a more exposed position and are more likely to be killed by the spray. By green cluster, caterpillars of the fruit tree tortrix (see page 138) have resumed activity after hibernation and can, therefore, be controlled at the same time. In addition, this is the most suitable time to control aphids (see page 19) and apple sucker (see page 14), and a fungicide is usually applied then to control apple scab. Thus, a single spray containing suitable insecticides and a fungicide can be used to control winter moth and tortrix caterpillars, aphids, apple sucker and apple scab.

On **other fruit trees**, the recommended times for spraying are petal fall for pear and white bud for plum and cherry.

In late springs when hatching may not be complete until pink bud or later, or when winter moth attacks are particularly severe, a second spray may be necessary at petal fall on apple, cherry or plum.

In gardens

Grease banding By placing grease bands around the trunks of trees, the wingless female moths can be trapped as they crawl up the trunks to lay their eggs.

The grease-banding material may be applied directly to the bark or on paper bands. Direct application is preferable on old trees with rough bark, since paper bands cannot be made to fit sufficiently tightly to prevent the females from crawling underneath. The bands should be placed 1–1.5 m from the ground, and the grease should form an unbroken ring 12–15 cm wide. The bands should be in position early in October and should remain tacky until April. They should be scraped frequently to maintain a fresh sticky surface.

Spraying If a spray is preferred, a recommended insecticide should be applied at the same stages of bud development as mentioned above.

BEETLES

Coleoptera

31
Apple blossom weevil

Fig.31.1 Apple truss with 'capped' blossoms

Before the introduction of DDT in 1946, the apple blossom weevil (*Anthonomus pomorum* (L.)) was a common and troublesome pest of apple and, occasionally, pear. Until recently, damage had not occurred in orchards in which sprays of either DDT or HCH had been used. However, following the withdrawal of DDT and the less frequent use of HCH, the pest has reappeared in a few places. Fruit trees near woodland are more likely to be attacked.

Nature of damage

At petal fall it may be noticed that some of the blossoms remaining on the tree have failed to expand, the petals having died and become brown (Fig. 31.1). These are known as 'capped' blossoms; underneath the brown petals there will be a whitish grub which has eaten away the base of the flower and so prevented further growth. Later, a pupa or a newly emerged weevil will be present. In severe attacks a large proportion of the blossom may be damaged and much of the crop lost.

Description and life history

The adult weevil (Fig. 31.2(a)) is a small beetle, about 5 mm long, with a long rostrum (snout). It is black or brownish and is dotted with fine grey hairs; on the back is a greyish or yellowish mark in the form of a V.

Weevils have started moving on to the trees from hibernation sites by the time the buds are breaking (usually about mid-March) and the number of active weevils increases until the green cluster stage (mid-April). They feed mainly on the blossom buds, where the developing anthers provide a rich supply of protein. Before bud burst the weevils can obtain food only by boring through the inner bud scales which still ensheath the fruit bud; later the sides of the exposed blossom buds are pierced directly.

Eggs are laid from bud burst until green cluster. The female bores a hole through the side of a blossom bud with her rostrum and then, by means of her ovipositor, inserts an egg among the anthers. Each female lays 40–50 eggs, but only one is laid in each blossom bud.

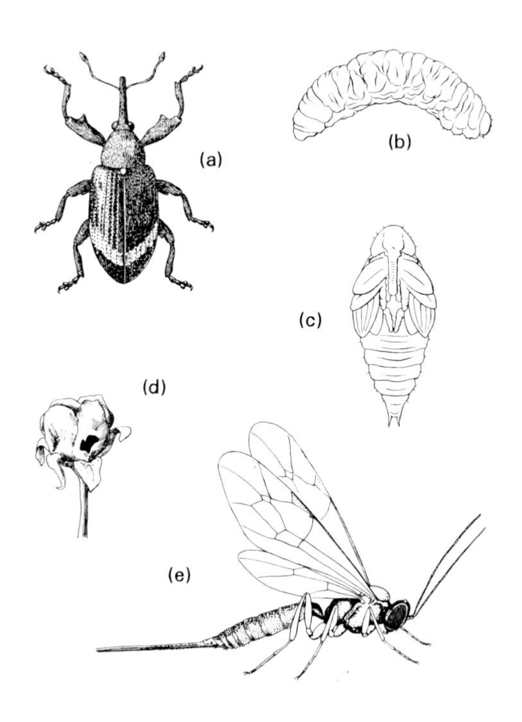

Fig.31.2 Apple blossom weevil and parasite. (a) Weevil (× 6). (b) Weevil grub (enlarged). (c) Weevil pupa (enlarged). (d) Attacked blossom bud showing emergence hole of weevil. (e) Adult parasite with wings extended (× 6)

Hatching begins at the green cluster stage and is complete by pink bud (late April). The grub (Fig. 31.2(b)), which is creamy-white with a brown head, feeds on the stamens and the bases of the unopened petals and is mature after 3–4 weeks. Within the protection of the capped blossom it then changes into a cream-coloured pupa (Fig. 31.2(c)). Transformation into the adult occurs after about two weeks, and a few days later the newly formed adult escapes by gnawing a hole through the dead petals (Fig. 31.2(d)). It then feeds on the underside of leaves, removing small areas.

Adult activity ceases during the latter half of June when the weevils seek shelter under loose bark or move to neighbouring hedges or woodland. A resting period follows until the next spring.

Natural enemies

The apple blossom weevil is eaten by small birds such as tits. It is also attacked by insect parasites; for example, the grubs of the ichneumon *Scambus pomorum* (Ratzeburg) (Fig. 31.2(e)) feed on the weevil grubs and pupae in the capped blossoms, sometimes destroying large numbers.

Control measures

In orchards

Routine sprays applied at green cluster against caterpillars help to control any apple blossom weevil grubs that are present; numbers do not, therefore, build up in sprayed orchards. Where apple blossom weevil is a serious problem in unsprayed orchards, a single spray of a recommended insecticide should control the adult weevils. The timing of the spray is important as the object is to kill the adults before eggs are laid in the blossom buds. Therefore, for the best results a spray should be applied just before bud burst.

In gardens

Many weevils can be trapped if folded sacking bands are placed around the trunks. These provide suitable hibernation sites and will also trap caterpillars of the codling moth (see page 111). Sacking bands are most effective if alternative shelter is absent, so rough bark should be removed from the trunks and main branches of older trees before the bands are applied. This can be done easily with a narrow strip of wire netting, which is pulled to and fro against the far side of the trunk. The bands should be in position before mid-June and can be removed after the crop has been picked; they should then be burnt or placed in a pail of boiling water.

Insecticides are also available for home garden use.

32
Chafer grubs

Fig.32.1 Grubs of garden chafer in damaged pasture

The grubs of five species of chafer beetle are pests of local importance in Britain. The cockchafer or maybug (*Melolontha melolontha* (L.)) occurs mainly in southern England; the garden chafer (*Phyllopertha horticola* (L.)) in western Scotland, parts of Wales, the Lake District and the southern counties of England; the summer chafer (*Amphimallon solstitialis* (L.)) mainly in the southern half of England; while the Welsh chafer (*Hoplia philanthus* (Füessly)) and brown chafer (*Serica brunnea* (L.)) are very local pests in western and northern districts. Adults of a foreign pest species — the Japanese beetle (*Popillia japonica* Newman) — are sometimes found in aircraft arriving from N. America.

Nature of damage

Adult chafer beetles feed mainly on leaves, flowers and fruit of deciduous trees and shrubs, rarely causing any serious damage in Britain; they are probably most widely known as intruders into lighted buildings on warm evenings in May, June and July.

Fig.32.2 Grub of cockchafer (× 3)

Chafer grubs feed on plant roots and are normally associated with upland grass or deciduous woodland. Damage to upland pasture is more prevalent on the southern slopes of warmer, sheltered areas adjoining deciduous woodland. Large populations of grubs can build up in neglected permanent grassland. Injury in pasture, lawns and golf courses results in poorly growing patches that brown readily in dry weather; the grubs can be found immediately below the surface (Fig. 32.1). Severely damaged areas are sharply defined and, because the roots are severed, the turf can be rolled up like a carpet. Frequently, much damage occurs when birds find an infestation of chafer grubs in permanent grass: the birds rip up the grass while searching for the grubs.

Potatoes planted after old pasture are liable to attack by grubs of some chafers, usually the cockchafer. Small numbers of grubs may also be found in arable ground in the districts where the beetles are common; they sometimes cause serious injury to nursery and garden plants, e.g. lettuce, raspberry, strawberry and young trees. Injury to the roots or rootstock causes small saplings, or tender tap-rooted plants like lettuce, to wilt suddenly; stronger plants show stunted growth and a tendency to lose their leaves prematurely.

Description and life history

Chafer beetles are medium to large insects belonging to a group of beetles distinguished by the antennae, which have tips consisting of a series of thin plates. In most chafer species the wing-cases do not entirely cover the abdomen; this feature is particularly noticeable in the cockchafer (Fig. 32.3). Adult chafers are active in late spring or summer and the females lay their eggs in the soil under close plant cover.

The grubs (larvae) hatching from the eggs have brown heads, strong biting mandibles, three pairs of legs, and whitish bodies that normally lie in a characteristically comma-like position (Fig. 32.2). They are sometimes known as white grubs. The mid and hind pairs of legs are similar in length; this feature distinguishes chafer grubs from some closely related beetle grubs in which the hind pair of legs is shorter than the mid pair. Although their legs are well developed, chafer grubs do not move far from where the eggs were laid, movement being mainly up and down in the soil. The grub stage lasts 9–36 months depending on the species.

In summer or autumn the grubs form cells deep in the soil, where they develop into adults after passing through a soft whitish pupal stage. The adults burrow up from the cells in the following spring or summer.

Cockchafer

The cockchafer is the largest British species. The adult (Fig. 32.3) is about 25 mm long with a black head and thorax and reddish-brown wing-cases. The grub (Fig. 32.2) may be as long as 44 mm when fully grown. This species is associated with deciduous woodland and the grubs are pests in nearby gardens and nurseries.

Adult cockchafers emerge from the soil in May and early June; they are nocturnal in habit, feeding at night on foliage and resting on trees during daylight. Grubs hatch 5–6 weeks after the eggs have been laid, but they are not fully grown until the end of the third summer after hatching. Although young grubs cause little damage, small numbers of the larger grubs can cause considerable losses in row-planted crops.

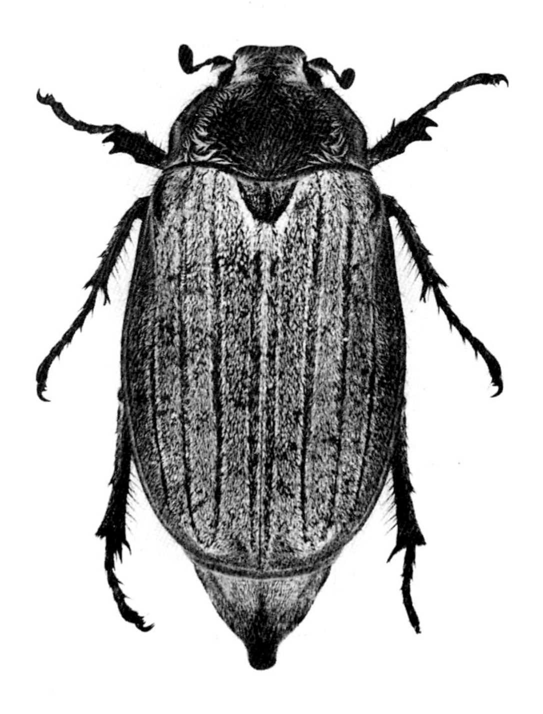

Fig.32.3 Cockchafer (× 3)

Garden chafer

The garden chafer is probably the most abundant species. The adult (Fig. 32.4) is about 9 mm long with a metallic-green head and thorax and reddish-brown wing-cases; it is markedly hairy. The grubs (Fig. 32.1) are white and about 18 mm long when fully grown.

This species is the most widespread of the group and the most troublesome on grassland. It is also found in nurseries growing hardy stock, particularly in rarely disturbed soil around trees and shrubs. In a few districts the grubs are a persistent pest of upland pasture, favouring the steeper slopes with friable, free-draining soil and patches of bracken. Occasionally, lowland pasture and lawns are severely damaged by grubs. The adults chew holes in the young fronds of bracken and the flowers of herbaceous plants, trees and shrubs, e.g. the opening buds of rose and the flowers of rhododendron. The fruits of apple and pear can also be damaged.

The life cycle of the garden chafer occupies

Fig.32.4 Garden chafer (× 5)

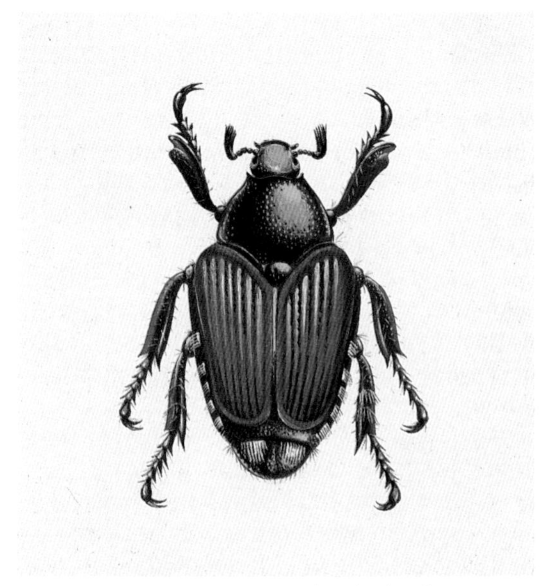

Fig.32.5 Japanese beetle (× 4)

one year. Adults emerge in late May and early June, the majority within a period of 10 days. At emergence time swarms of males may be seen on warm days over infested pasture, looking for emerging females. After mating, each female burrows back into the ground immediately and deposits her eggs. Thus, infested pasture is regularly reinfested each summer, although the areas with obvious damage are often different each year. About five weeks after the eggs have been laid, the grubs hatch and feed on the grass roots until late autumn, when they burrow further down to make earthen cells in which they hibernate. No more feeding occurs before pupation. The grubs pupate in their cells in the following spring and, after about four weeks, the adults slowly dig their way out of the soil.

It should be noted that garden chafer grubs do not cause any damage in the spring or early summer, unlike grubs of those species with a longer life cycle.

Summer chafer

The adult summer chafer (Fig. 32.6) is about 16 mm long, bright brown to dull yellow and distinctly hairy. It flies in June and July at dusk. The grub reaches a length of 30 mm and is a local pest in pasture. The life cycle is completed in two years.

Welsh chafer

The Welsh chafer infests grassland in some districts of Wales and the north-west of England. The adult is about 7 mm long with a black head and thorax and reddish-brown wing-cases. The grub is similar to that of the garden chafer but may be distinguished from it by the larger number of hairs on the upper part of the abdomen which give it a rusty appearance.

The life cycle is completed in a year. The adults emerge in June and swarm and mate on warm days. Eggs are laid in midsummer. After hatching, the grubs feed in the autumn and spring; pupation occurs before the beginning of June.

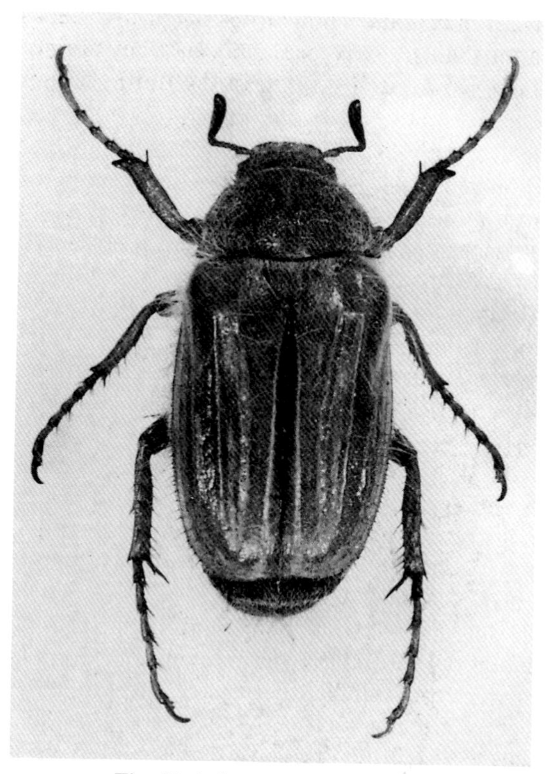

Fig. 32.6 Summer chafer (× 4)

Brown chafer

The brown chafer is common is a few well-wooded localities, especially on lighter soils. The adult is about 9 mm long and yellowish or reddish-brown with a dark head. It emerges from June to August. The grub stage lasts about two years.

Other chafers

Adults of the **rose chafer** (*Cetonia aurata* (L.)) are often noticed in the south of England because they fly in full sunlight and are brilliant metallic golden-green. They are about 18 mm long. The grub has transverse rows of reddish hairs on the body and, when fully grown, is 25–32 mm in length. The adults appear in late May and early June. Eggs are laid in early summer. The grubs hatch and feed for 2–3 years before pupating in late spring. The grubs of this species are not pests, but the adults attack rose flowers.

The **Japanese beetle** was introduced into

North America from eastern Asia. It now occurs widely as a pest in the USA and southern Canada, where the adults feed on deciduous trees and herbaceous plants and the grubs feed mainly in grassland. This pest is not yet established in Europe, but a large area of southern Europe lies within its known climatic range and heavy infestations could be expected to develop once the beetle was introduced. Adults have been discovered in aircraft arriving in the UK after periods of intense beetle activity near US airfields in June and July. The live beetles were killed by insecticide treatment.

It is important that any Japanese beetle found in Britain is recognized and reported quickly. The adult (Fig. 32.5) closely resembles the garden chafer (Fig. 32.4). The main difference is that the Japanese beetle has small distinct tufts of white hair round the margin of the body; these tufts are absent in the garden chafer. In addition, the legs of the Japanese beetle are thicker than those of the garden chafer and green instead of brown. Any suspected Japanese beetle found in England or Wales should be placed in a strong box or tin and sent to the nearest office of the Ministry of Agriculture, Fisheries and Food; any found in Scotland should be sent to the Scottish Office Agriculture and Fisheries Department.

Natural enemies

Chafer grubs are parasitized by the larvae of two tachinid flies. Some birds eat chafer grubs and some eat the adults. Bacterial ('milky') and fungus diseases and nematode parasites of chafer grubs have been known for some time on the continent of Europe.

Control measures

In pasture

Intensively managed grassland, in which there is no formation of surface mat and where regular treading maintains a compact surface, is rarely subject to chafer infestation. Chafer grubs are also little or no trouble where the ground is disturbed by arable cultivations. The destruction of the protective mat renders the ground unfavourable for reinfestation, and the eggs, grubs and especially the pupae are easily killed by such operations as discing. Therefore, any farm policy that brings infested pasture into cultivation, even for a season or two, will greatly reduce chafer grub injury. Damaged areas that cannot be ploughed can sometimes be surface-cultivated and reseeded. Because most female chafer beetles lay their eggs very close to the place of their emergence, dispersal is limited and reinfestation of treated areas will be slow.

Persistent soil insecticides have been extensively used abroad to control chafer and similar grubs in pasture. Little work has been done on chemical control in Britain, but treatment has been successful when applied in late July or August before the appearance of damage symptoms. Applications made in September or after damage has been seen (usually September or October) have been less effective.

Where damage occurs frequently, farmers should assess the need to apply an insecticide by digging out small areas of turf to a depth of 3–5 cm and counting the number of grubs in the soil. If more than about 70/m^2 are found, damage is likely and treatment is probably justified.

Damage caused by birds searching for grubs may be reduced by using bird scarers. To prevent permanent damage to pasture, the worst affected areas should be rolled, where possible, to encourage re-rooting. The application of grass seed, fertilizer or farmyard manure should also be considered.

In ornamental and sports turf

The use of insecticides is occasionally necessary to control chafer grubs in lawns, golf courses and recreation grounds. Certain persistent soil insecticides are recommended for chafer grub control in ornamental and sports turf and should be effective.

On arable land

On farm crops Chafer grubs do not occur in significant numbers in arable farmland unless it has become neglected. Trouble usually follows the ploughing of infested grass; almost any crop is then likely to suffer root injury. Continual disturbance of the soil is unfavourable to the eggs and grubs and does not attract the egg-laying females. Early ploughing of grassland, discing and rotary tilling are especially effective.

In nurseries and market gardens Injury to perennial crops such as stawberries, young trees and ornamentals sometimes occurs. Chafer grubs may be controlled by persistent soil insecticides applied against other soil pests of these crops.

In gardens Chafer beetles often occur in private gardens in 'chafer districts'. A few well-grown grubs in an arable plot can cause annoying losses of plants such as lettuce, strawberry and some ornamentals. When the plants are in rows, wilting or stunted attacked plants occur in succession and the offending grubs will be found in the soil near the last one to be attacked. Grubs should be removed by hand and killed when they are found. Thorough cultivation will generally ensure that plant losses are minimal.

33
Colorado beetle
(A crop pest not yet established in Britain)

Fig.33.1 Colorado beetle on potato leaf (× 4)

The Colorado beetle (*Leptinotarsa decemlineata* (Say)) is a serious pest of potato. It was first recorded in 1824 in the western USA feeding on a wild plant. However, when potato was introduced by the early settlers in 1855 the beetle attacked this new food voraciously and spread across the USA and into Canada. It was first reported in Europe — in Germany — in 1877. In the same year one was found at Liverpool docks on a ship carrying Texan wheat and this prompted Britain to introduce its first plant health legislation: The Destructive Insects Act 1877. Britain's first outbreak of Colorado beetle occurred in 1901, and between 1941 and 1952 137 breeding colonies were detected. By using drastic measures all these colonies were exterminated, so that the pest has been prevented from establishing itself in the British Isles. Single breeding colonies were found and eradicated in 1976 and in 1977.

Fig.33.2 Eggs of Colorado beetle on underside of potato leaf (× 2)

Fig.33.3 Larva of Colorado beetle (× 6)

Description

The adult (Fig. 33.1) is about 10 mm long with alternate black and yellow stripes running from front to back along each wing-case; the head and thorax are brown with variable black markings like Arabic script. The eggs are yellow, cylindrical and about 2 mm long (Fig. 33.2). The larvae, which move about freely, are at first orange-brown but later become carrot-red with two rows of black dots on each side (Fig. 33.3). The pupae are bright orange and about 10 mm long.

There is no similar insect in Britain that feeds on potato leaves. Ladybird pupae may be found on potato foliage: these are similar in shape and colour to Colorado beetle larvae but are immobile (see page 172).

Life history (Fig. 33.4)

The beetle spends the winter buried in the soil, usually 25–50 cm below the surface. In the spring or early summer it works its way to the surface and, on warm sunny days, it flies or crawls in search of potato crops. Flights are normally short, not exceeding two or three kilometres (one or two miles), but under certain conditions associated with high temperatures and often stormy weather, large numbers may take to flight at the same time and be carried by wind over longer distances.

After reaching a crop, the beetles start to eat the foliage and soon the females begin to lay eggs in batches of 40–70 on the underside of the leaves. Daily egg laying continues throughout the summer and one female may produce as many as 4000 eggs. After 6–8 days the eggs hatch into larvae, which usually remain on the same plant and feed until fully grown. The larval stages last 32 days at 16.5 °C.

The fully grown larvae burrow into the soil to pupate. The pupal stage lasts about 18 days at 16.5 °C. The final change to an adult beetle then takes place.

During July and August these summer beetles come to the surface, feed and, if the weather is warm, lay eggs which can produce a further generation of beetles before the potato haulm (branched stems) dies off in the autumn. At the end of the summer the beetles burrow

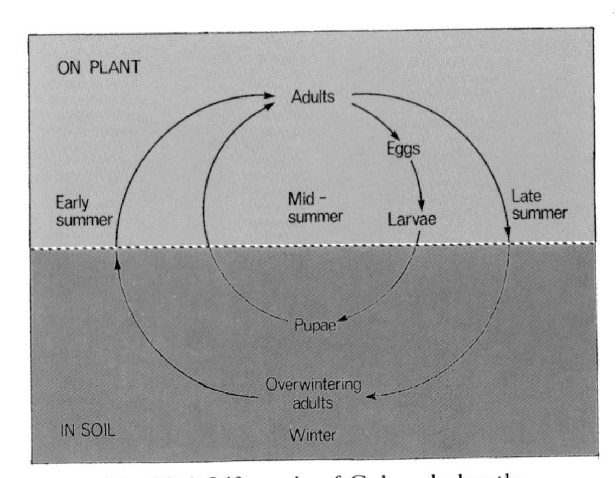

Fig.33.4 Life cycle of Colorado beetle

down into the soil again and stay there for the winter.

Nature of damage

Figure 33.5 shows a field of potatoes badly attacked by Colorado beetle. The injury started by the adults is carried on by the larvae eating in the same way until the haulm is stripped of leaves (Fig. 33.6). Another characteristic sign of feeding by both adults and larvae is the black and rather messy excrement that is left on the leaves and stem. Although normally only the foliage of the plant is

Fig.33.5 Potato crop in France defoliated by Colorado beetle

Fig.33.6 Potato plant with upper part defoliated by larvae of Colorado beetle

attacked, the yield of tubers is reduced. Exceptionally the tubers are also eaten, but this is more likely to happen in store.

Adults and larvae of the Colorado beetle also feed on other plants of the potato family (Solanaceae) such as tomato, egg-plant, capsicum, black nightshade and woody nightshade.

Precautionary measures

The Plant Health (Great Britain) Order 1987 requires occupiers of land to report to the Ministry of Agriculture, Fisheries and Food any suspected outbreak of Colorado beetle. This Order also *prohibits* the keeping of any live Colorado beetles and the spraying or other treatment of any affected crop without the authority of the Ministry. Inexpert treatment might aggravate the situation. When a breeding colony is discovered, the Ministry takes immediate action to eradicate it.

In certain years, as a precautionary measure, there has been large-scale spraying of potato crops in areas where outbreaks have occurred in previous years. Since 1955 precautionary spraying has been needed only in 1977.

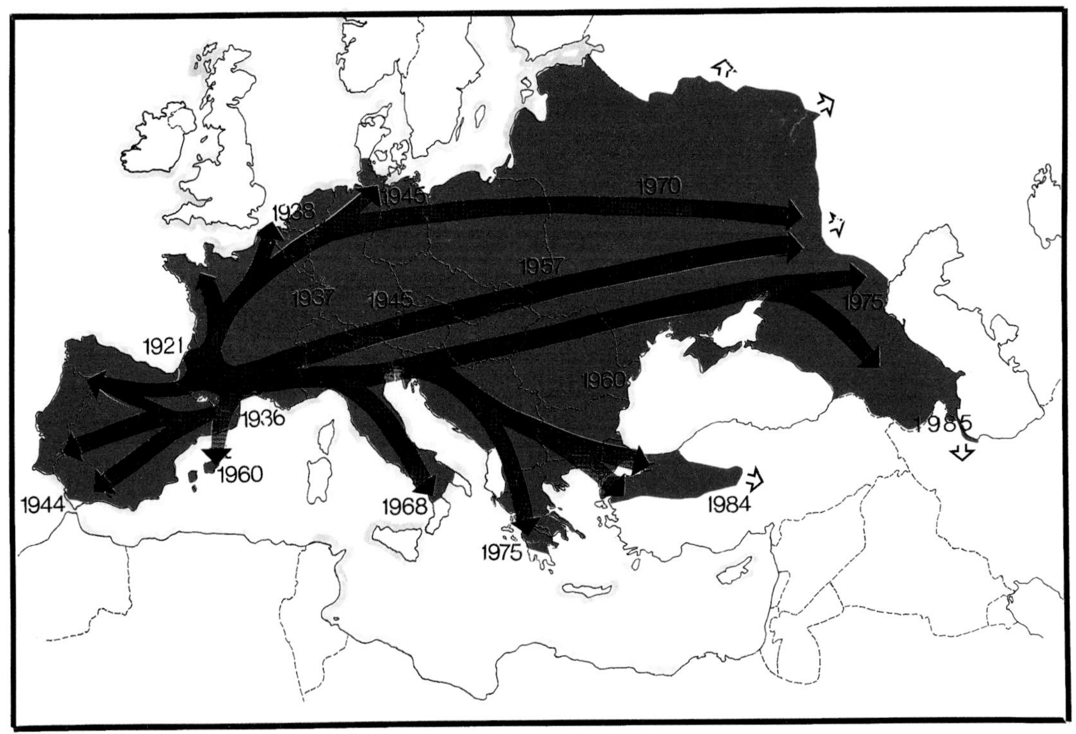

Fig.33.7 Approximate lines of advance of Colorado beetle in Europe from 1921 to 1985. (Compiled from reports issued by the European Plant Protection Organisation and from other sources)

Spread

The beetle was originally restricted to semi-desert regions of the western USA, where it fed on *Solanum rostratum* – a wild plant of the potato family. Following the introduction of the cultivated potato the beetle spread rapidly, reaching the Atlantic coast of the United States and south-eastern Canada by 1874. Although successful in colonizing new areas, official eradication prevented its establishment in Europe until 1922 when it gained a foothold near Bordeaux, France. Since then, particularly during the Second World War when counter measures could not be applied effectively, the insect has spread and become a pest throughout most European countries. The map (Fig. 33.7) shows the distinct routes of this invasion. Colonization of new areas continues. The physical extremes of the beetle's original habitat have predisposed it to adapt to different environments and to different methods of potato cultivation. It was also one of the first plant-feeding insects to become resistant to insecticides and is able to develop resistance to newly introduced insecticides.

The Channel has, so far, proved an effective barrier to the direct flight of beetles from the Continent to Britain, but they are easily brought over with plant produce or other merchandise, or in cars, ships or aircraft. Most of the beetles found in Britain can be traced to imported horticultural or agricultural produce, particularly leafy vegetables and grain, and to imported timber. The finds on leafy vegetables usually occur soon after the spring emergence of adult beetles where a non-host crop has followed potatoes in the previous year. These incidents are often associated with unseasonably mild weather in the exporting country and have occurred from mid-February onwards in some years.

The Plant Health (Great Britain) Order 1987 aims to prevent destructive pests and diseases, including the Colorado beetle, from becoming

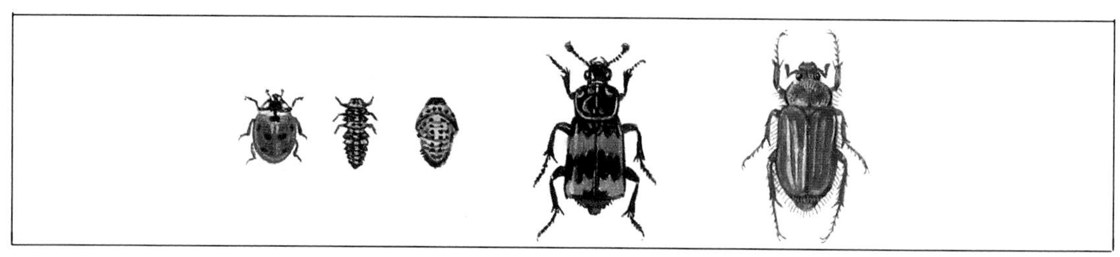

Fig.33.8 (a) Ladybird: adult, larva, pupa. (b) Burying beetle. (c) Summer chafer. (All life size)

established in Britain. When potatoes, and other vegetables with which the pest is likely to be carried, are imported from countries where Colorado beetle is established, the Order requires field control measures to have been taken in those countries; it also places packing restrictions on the vegetables being imported.

Danger to Britain

If the Colorado beetle were allowed to establish itself, the yield of potatoes would be reduced; loss of yield would be greatest in hot, dry years. To avoid losses caused by the beetle, growers would need to spray their potato crops. This spraying would add to the cost of potato growing and increase the risk of the environment being contaminated with insecticides.

Therefore, it is in the interests of all, not only farmers and gardeners, to do everything possible to assist the Ministry in keeping Britain free from this pest.

Suspected beetles

Beetles or larvae suspected of being Colorado beetle, found in England or Wales, should be placed in a strong box or tin and sent to the nearest office of the Ministry of Agriculture, Fisheries and Food. Those found in Scotland should be sent to the Scottish Office Agriculture and Fisheries Department.

Beetles mistaken for Colorado beetle

The insects most frequently confused with the Colorado beetle are ladybird (coccinellid) larvae and pupae, burying (or sexton) beetles (*Nicrophorus* spp.), and the summer chafer (*Amphimallon solstitialis* (L.)) — see Fig. 33.8.

Ladybirds

The adults are marked with black spots on a red, orange or yellow background: they are never striped. The larvae are very active, rather flattened and violet-grey or blackish with orange-yellow markings — not humped and bright red like the Colorado beetle larva. The pupae are yellow or orange marked with black; although similar in colour and size to the larva of a Colorado beetle, they are attached at the hind end to a leaf and cannot move around freely like a larva. (The pupa of a Colorado beetle is in the soil.)

Ladybirds and their larvae feed on aphids and *should never be killed*, for they are most useful insects.

Burying beetles

These beetles are much larger than the Colorado beetle and are marked on the back with orange-red and black bands. The markings run from side to side across the wing-cases, not from front to back as in the Colorado beetle. Burying beetles feed on carrion, so performing a valuable function.

Summer chafer

This beetle is yellowish-buff and is considerably larger than the Colorado beetle. There are ridges down its wing-cases which to some people suggest stripes, but they are very different from the black and yellow stripes of the Colorado beetle. The larva of the summer chafer lives in the soil and damages grass and herbaceous plants by attacking their roots (see page 163).

34
Flea beetles

Fig.34.1 Small striped flea beetles on turnip seedling showing leaf damage (× 2)

Growers of cabbage, kale, turnips and similar crops will be familiar with flea beetles (also known as turnip 'fly'), which can cause serious damage to the young plants, sometimes destroying them in the seedbed. These beetles were a major pest of brassica crops until the mid-1950s when attacks became much less common and less severe. There are many species of flea beetle and a range of crops may be attacked. The species described here are those attacking Brussels sprout, cabbage, cauliflower, kale, radish, swede, turnip, beet, mangel, flax and occasionally watercress. Beet and mangels are rarely damaged as seriously as brassicas.

The cabbage stem flea beetle, whose larvae tunnel in the stems of overwintering brassicas, is not included here. For information on this pest, see pages 180 and 185.

Nature of damage

The main damage is done by the adults eating holes in the leaves and stems of seedling plants (Fig. 34.1), starting just before the seed leaves appear above ground. Plants are still attacked after the first true (rough) leaf has been produced but, unless growth is checked by drought or frost, or the beetles are very numerous, attacks become progressively less damaging. Loss is usually greatest in a dry spring when the seedlings make little headway after germination; two or three sowings may be destroyed unless preventive measures are taken. In some seasons severe attacks requiring treatment occur when swede, turnip and late-planted brassica crops have reached the 6–8 rough-leaf stage.

Although most damage occurs in April and May, considerable damage sometimes occurs in summer when large numbers of beetles migrate downwind after brassica seed crops have been cut.

Description

Because they are small and often start an attack on or below the surface of the soil, flea beetles are not easy to see, although their presence soon becomes obvious owing to the characteristic damage to the leaves (Fig. 34.1). As their name implies, they have the distinctive habit of jumping when disturbed.

Small striped flea beetle
This species (*Phyllotreta undulata* Kutschera) is one of the most common flea beetles. The adult (Fig. 34.1) is very small, being 2.5 mm long, and black with a yellow stripe down each wing-case.

Large striped flea beetle
This flea beetle (*Phyllotreta nemorum* (L.)) is similar to the above species but slightly larger as it is 3 mm long.

Other cruciferous flea beetles
Other common species are *Phyllotreta atra* (Fabricius) and *Phyllotreta cruciferae* (Goeze). The adults of both are very small and black though that of *P. cruciferae* has a green or blue lustre.

Mangold flea beetle
This species (*Chaetocnema concinna* (Marsham)) is responsible for the damage to beet and mangels. The adult is 2 mm long and black with a bronze-coloured lustre; there are rows of deep punctures on the wing-cases.

Flax flea beetles
There are two species which damage flax and linseed. Both are very small. The adult of the flax flea beetle (*Longitarsus parvulus* (Paykull)) is about 1 mm long and black, while the adult of the large flax flea beetle (*Aphthona euphorbiae* (Schrank)) is 1.5–2 mm long and metallic green. The wing-cases of both species are finely punctured.

Life history

Most kinds of flea beetle spend the winter hidden under the bark on trees, at the bottom of hedges, in refuse heaps, stacks, etc. Large numbers, for instance, may be found at the bottom of a straw bale or haystack near a turnip or oilseed rape field.

The beetles emerge in the spring and at first feed on any suitable plants that are available. The cruciferous flea beetles then feed on crops of the cabbage family and weeds such as charlock, whereas the mangold flea beetle feeds on beet, mangels, rhubarb, docks, knot-grass and *Polygonum*. Little is known of the feeding habits of the flax flea beetles, but in addition to flax and linseed they may feed on related weeds, grass, clover, wood spurge and fruit trees. On warm days flea beetles fly and probably travel considerable distances, collecting on newly sown host crops as soon as these appear above ground.

Eggs are laid on or in the soil near the plants during May and June. After hatching, the behaviour of the larvae varies with the species: the larvae of the large striped flea beetle crawl up the stem and tunnel into the leaves, forming bladder-like mines, while larvae of the other species feed on the plant roots. The fully fed larvae pupate in the soil and after two or three weeks change into adults which start feeding on the foliage of the plant.

Development from egg to adult takes 6–8 weeks. The adults survive until July or August of the following year when beetles of the next generation take their place.

Control measures

Cultural control

Cultural methods can do much to decrease plant losses caused by flea beetles and other pests. A seedbed of good tilth and moist, adequately manured soil will help plants to grow quickly through the susceptible stage. However, in a bad season, or on land which is near coppices or other suitable overwintering sites of the beetles, it is necessary to rely on insecticides.

Chemical control

The three methods of using insecticides for flea beetle control are:

(1) treating the seed and then, if a severe attack develops, treating the growing crop later in the season;
(2) spraying the crop when the rows become visible, if close inspection shows flea beetles to be present;
(3) applying certain granular insecticides primarily for control of other pests.

Seed treatments These are now usually applied as a routine and are most important for precision-drilled crops, which are particularly at risk because of the low seed rates. A seed treatment only protects the crop from moderate attacks between germination and the first true-leaf stage, after which a spray treatment may be necessary.

Seed supplied by a seed merchant is often treated with an insecticide and a fungicide, so that the crop is protected from seedling diseases as well as flea beetles.

Alternatively, growers can treat their own brassica seed with a recommended insecticide powder. To improve the adhesiveness of the powder, 15 ml of paraffin or vaporizing oil can be mixed with each 2 kg of seed before mixing in the powder. Seed to which the sticker has been added should be sown within a week of treatment.

Sprays Excellent control of severe or late attacks of flea beetles on brassica crops can be obtained by spraying with a recommended contact insecticide. Usually only a single spray is necessary.

Granules Some of the granular insecticides recommended primarily for the control of other pests, such as cabbage root fly on brassicas and various seedling pests of sugar beet, give incidental control of flea beetles.

Various methods of application are recommended. Granules may be broadcast before sowing, or applied at drilling (i) by the bow-wave method or (ii) to the moving soil in front of the seed coverers or (iii), for insecticides unlikely to damage the seedlings, in the furrow with the seed.

In gardens Various dust and spray formulations of insecticides are available for use in the garden. Treatment should be applied as soon as damage is seen.

35
Insect pests of brassica seed crops

Fig.35.1 Mustard flower buds damaged by pollen beetle: blind stalks at raceme tips have been caused by larvae, those lower down by adults

Brassica seed crops are attacked by a succession of insect pests from the early seedling stage until the pods begin to ripen in summer. Yields of seed can be very seriously reduced by any one or a combination of these pests. The most common are pollen (or blossom) beetles

(*Meligethes* spp.) the cabbage seed weevil (*Ceutorhynchus assimilis* (Paykull)) and the cabbage stem weevil (*Ceutorhynchus quadridens* (Panzer)); each causes specific and easily recognizable damage. These and other pests frequently found on brassica seed crops are listed in the table on pages 180–181. Those also occurring on oilseed rape are included on pages 182–189.

Description

Pollen beetles (Fig. 35.2) are metallic greenish-black, about 2.5 mm long, and are most numerous in the crop just before and during flowering.

The weevils are seldom as numerous as pollen beetles and may therefore be less well known. Adults of both weevil species are about 2.5 mm long with a pronounced slender snout (rostrum) (Fig. 35.3). The cabbage seed weevil is lead-grey with a faint lighter grey stripe down the middle of the back; it is usually found with pollen beetles on the upper parts of the plant. The cabbage stem weevil is more mealy-grey with a white spot in the middle of the back; it is seldom numerous and rarely found on the upper parts of the plant.

Life history and nature of damage

These insects have similar life cycles but differ in their behaviour and the damage that they do. The adults overwinter in sheltered places (copses, dykesides and farmyard litter), near buildings, beneath tree bark, and in the soil beneath or near to the previous year's seed crops. They fly actively and invade brassica seed crops in spring, chiefly on warm, sunny and calm days.

The first **pollen beetles** to become active in the spring are usually found on early flowering dandelions (*Taraxacum* spp.). The adult females bite tiny slits in the base of small flower buds of brassica plants and lay their eggs inside. There is one generation each year. Both the adults and their larvae can damage other buds and flower parts. Damaged buds wither and die, and the number of pods that set is decreased (Fig. 35.1). Occasionally, a distorted or stunted pod forms where a flower bud was only slightly damaged.

Cabbage stem weevil adults lay their eggs in the leaf stalks of brassica plants from about the sixth or seventh broad-leaf stage onwards. The larvae (grubs) tunnel into the stem (Fig. 35.4) during May and June. The lower leaves fall prematurely and the plants lose vigour but, unless some other problem is affecting the crop or stem weevil larvae are very numerous, there is usually little effect on yield. The larvae make exit holes in the stems, often near a leaf scar, fall to the ground and pupate in an earthen cell a few centimetres below the soil surface.

Cabbage seed weevil adults lay their eggs in young pods. Each larva (grub) eats several seeds and damages others during its development. The number of seeds eaten by each larva depends mainly on seed size, which varies with the type of brassica and with the season. When fully fed the larvae bore exit holes in the pod

Fig.35.2 Pollen beetle (*Meligethes aeneus* (Fabricius)) on turnip flower (× 8)

walls; these holes are easy to see during June, July and August. The larvae fall to the ground and pupate in earthen cells similar to those of the stem weevil.

Crops attacked

Pollen beetles attack all brassica seed crops. On overwintered brassicas they are usually most serious on crops with a long flowering period and less serious on crops that produce buds, flowers and pods quickly. Generally there are a few pollen beetles during early April; numbers then increase until a peak is reached in late May or early June. Crops that produce buds and flower early in this period suffer less damage than later-flowering crops.

Although the **cabbage stem weevil** readily attacks overwintered crops, it is usually more serious on spring-sown brassicas and worst on crops sown between late April and late May. Of the spring-sown crops, white mustard (*Sinapis alba*) is usually less affected than brown mustard (*Brassica juncea*), although recently introduced varieties of white mustard seem more susceptible to attack than the older, hollow-stemmed varieties.

The **cabbage seed weevil** attacks all spring-sown and overwintering brassicas with the exception of white mustard. For all practical purposes this crop is immune to attack by seed weevil larvae, although adults are often found on the crop.

Cultural control

Early drilling is of great value in minimizing pest damage both for spring-sown and over-wintered crops, but there is a risk of frost damage if overwintered crops are allowed to get too forward in autumn. Vigorous plants can often withstand considerable pest damage and make a partial recovery. For example, plants can compensate to some extent for buds destroyed by pollen beetles by producing pods from buds that would not otherwise have developed. However, it is most unlikely that the plants could compensate for damage by either of the two weevil species.

Chemical control

Threshold infestations for spraying

There is not a consistent relationship between the number of adult pests present in the crop in spring and subsequent seed losses. It is therefore very difficult to give precise recommendations for chemical treatment that will suit all crops and all seasons. However, crops with small numbers of plants per hectare usually withstand larger populations of insects on individual plants than do denser crops in which the individual plants are smaller.

On overwinterd crops (except Brussels sprouts and winter cauliflower) chemical treatment is usually worth while if there are more than 15 pollen beetles or one cabbage seed weevil per plant at any time during April or May. *Frequent and early inspection of crops is necessary to determine when these critical levels have been reached* and it is most important to time treatments correctly. Very tall and dense crops can only be treated from the air, but aerial treatment is costly and perhaps difficult to justify if the crop area is small.

There are usually fewer plants per hectare in Brussels sprouts and winter cauliflower seed crops than in other overwintered brassicas, and

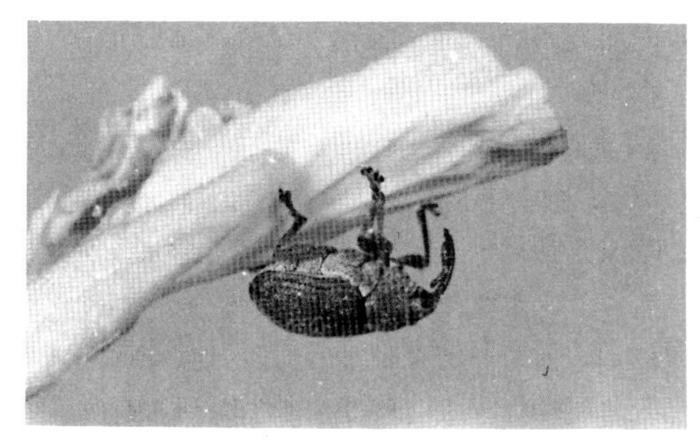

Fig.35.3 Cabbage seed weevil on turnip flower (× 10)

Fig.35.4 Mustard stems cut open to show (centre and right) damage by larvae of cabbage stem weevil compared with undamaged stem (left)

30–50 pollen beetles or 2–4 cabbage seed weevils per plant during April usually justify treatment.

Mustard is always grown at higher plant densities and should be treated if three pollen beetles are counted on each plant during the second or third weeks in May. Brown mustard and other susceptible spring-sown brassicas should be treated for cabbage seed weevil if there is a count of one or more per plant during the last fortnight in May or the first week in June.

Timing of sprays

The precise timing of sprays should be geared to the stages of flower-bud development; the numbers of insects and times of year given above should be used as a guide only.

Fig.35.5 Mustard pod damaged by larvae of brassica pod midge

During April and May numbers of adult pollen beetles increase in individual brassica seed crops because of constant immigration. In early spring the beetles migrate to crops and flowering weeds from their overwintering quarters, but later there is much movement within and between crops. This migration may give the false impression that early treatments have been ineffective, particularly on spring-sown crops like mustard. The treatment at green-bud stage (see Table 35.1) should delay the build-up for 7–10 days, by which time the crop is usually ready for a second spray. Pollen beetle adults cause damage from the early green-bud stage onwards and larval attack is most serious during the late yellow-bud stage and during the week or so after flowers have first opened. Because of this prolonged attack, and because cabbage seed weevils do not begin to lay eggs until pods have set, it is not possible to control damage by pollen beetle adults and larvae *and* damage by seed weevil larvae with a single pre-blossom spray. For pollen beetles two sprays are needed: one at the early green-bud stage and one at the early yellow-bud stage. Where cabbage seed weevils are numerous a third spray is often necessary. On spring-sown crops this should be applied as late as possible in the yellow-bud stage, but *before any flowers are open* in order to avoid danger to

Table 35.1 Main pests of brassica seed crops (in order of occurrence during the life of the crop)

Pest	Nature of damage	Time of damage	Crops attacked	Control
Cabbage stem flea beetle	Larvae tunnel in leaf stalks and stems.	From autumn through to spring.	Overwintering crops.	Spray with a recommended insecticide in October–November against adults or in November–March when larvae are found in plants. Isolate from previous year's crop.
Flea beetles (page 173)	Adults eat holes in and destroy seed leaves.	April and May.	All brassicas. Damage worst in dry seasons on spring-sown crops.	Combined insecticide/fungicide seed-treatment. If attack is prolonged or season very dry, supplement with a spray of a recommended insecticide.
Cabbage root fly (page 237)	Larvae mine and destroy lower part of stem, and roots.	May and June.	All brassicas.	Seldom necessary on seed crops.
Cabbage stem weevil	Larvae tunnel in leaf stalks and stems.	Late April, May and June.	All brassicas, but older white mustard varieties least susceptible.	Flea beetle seed-treatment gives some control. Spray with a recommended insecticide as for cabbage seed weevil at seven broad-leaf stage (spring-sown crops) or in early May (later-flowering overwintered crops).
Turnip gall weevil (page 207)	Larvae live in galls on roots and lower parts of stem.	Most months of the year.	All brassicas.	Flea beetle seed-treatment gives some control, but special control measures are unnecessary on seed crops.
Pollen beetles	Adults and larvae destroy flower buds.	April, May and June.	All brassicas.	Spray with a recommended insecticide between green-bud and first yellow-bud stages. Do not spray crops in flower.
Cabbage seed weevil	Larvae destroy seeds in developing pods (usually one larva per pod). Adults make feeding holes through which pod midge can lay eggs.	June and July.	All brassicas except white mustard, which is immune.	Spray with a recommended insecticide at late yellow-bud stage. The best control, with least harm to beneficial insects, is achieved by a post-flowering treatment with an appropriate insecticide capable of penetrating the pod wall.

Table 35.1 Main pests of brassica seed crops (in order of occurrence during the life of the crop)

Pest	Nature of damage	Time of damage	Crops attacked	Control
Brassica pod midge	Several small white/cream larvae inside each pod. Pods ripen prematurely and shed seed (Fig. 35.5).	June and July	All brassicas, but over-wintered crops worst affected. Always associated with seed weevil.	Post-flowering treatment against cabbage seed weevil also gives some control of midge. Isolate from previous year's crop.
Cabbage aphid (page 27)	Colonies of mealy-grey aphids cover leaves and branches.	July and August.	All brassicas.	Effect on yield doubtful. Control usually impracticable except by use of aircraft.
Cabbage white butterflies (page 97)	Caterpillars eat leaves.	June and July.	All brassicas.	Rarely serious enough on seed crops to justify treatment.

pollinating insects. On winter crops a post-flowering spray will give the most effective control of seed weevil.

Application of sprays

In most years and on most spring-sown crops two treatments repay their cost. On overwintered crops the decision is often more difficult to make, chiefly because treatment of tall crops with tractor-drawn spraying machines inevitably results in some damage to the crop from wheels and draw-bars. This damage can be minimized by fitting large-diameter land wheels with narrow treads, by using special high-clearance machines, or by using aircraft; the cost of the latter is usually the deciding factor. With aerial application it is very important to apply only insecticides that are approved for this use.

Danger to bees and other beneficial insects

All insecticides are poisonous to flying bees, but their deposits vary in the length of time they remain poisonous. Care should be taken in the choice of insecticide to be used on crops close to flowering. It is essential not to spray while there are open blossoms in the crop. It is good policy to spray in the early morning or late evening — times less hazardous to honey bees than the middle of the day — but wild bees and other pollinating insects will be killed whenever crops with open flowers are treated with insecticide.

A complete spray programme against pollen beetles and cabbage seed weevil almost certainly affects their natural enemies and may have undesirable long-term side effects. If the recommendations given here and on the manufacturers' product labels are followed carefully, better yields of higher quality seed should be obtained with minimal effects on beneficial insects.

36
Insect pests of oilseed rape

Fig.36.1 Flower buds of oilseed rape damaged by pollen beetle

Many insect species, both harmful and beneficial, colonize oilseed rape. As with many other crops, the incidence and importance of insect pests tend to increase with the area and frequency of cropping. Where rape has been grown regularly on a farm or in the vicinity, damage by pests is likely to occur.

The major pests can be grouped according to their feeding habits: (i) pollen (or blossom) beetles (*Meligethes* spp., mainly *M. aeneus* (Fabricius)), whose adults feed on unopened buds (Fig. 36.1) and whose larvae feed on young pods, (ii) cabbage seed weevil (*Ceutorhynchus assimilis* (Paykull)) and brassica pod midge (*Dasineura brassicae* (Winnertz)), whose larvae feed in pods (Figs 36.2 and 36.3), and (iii) cabbage stem weevil (*Ceutorhynchus quadridens* (Panzer)), cabbage stem flea beetle (*Psylliodes chrysocephala* (L.)) and rape winter stem weevil (*Ceutorhynchus picitarsis* Gyllenhal), whose larvae bore first into leaf stalks and thence into the stem (Figs 36.4 and 36.5). The cabbage aphid (*Brevicoryne brassicae* (L.)) and cabbage leaf miner (*Phytomyza rufipes* Meigen) are minor pests: the aphid rarely causes yield loss and the leaf miner has not been known to cause economic damage. The time of attack, the parts of the plant attacked and the relative importance of the pests on winter and spring rape are shown in the table below. (See pages 176–181 for these pests on brassica seed crops.)

Description, life history and damage

Pollen beetles (blossom beetles)

There are several species of pollen beetle, all very similar in appearance. The adults are about 2.5 mm long, oval, greenish-black and very shiny. In April they emerge from hibernation in sheltered places such as hedgerows and chew into the buds, laying two or three elongate creamy-white eggs in each. The larvae are also creamy-white, with a distinct head and three pairs of legs. The eggs and newly emerged larvae are very small and not easily seen without a hand lens. Fully grown larvae are about 4–5 mm long. Pupation occurs in the soil. Adults of the new generation are active later in the season and eventually hibernate. There is only one generation per year.

Feeding by adults in unopened buds (Fig. 36.1) and by larvae in flowers and on young pods causes loss of pods and 'blind stalks'. Slightly damaged buds may produce distorted pods or succumb to fungus disease. The development to maturity of pods that would otherwise have shrivelled and fallen can compensate for those lost by pest damage. This applies particularly to winter crops, which are usually in flower when most of the adults arrive. However, very backward crops or areas grazed by wood pigeons may benefit from spraying at the green bud stage, if threshold numbers are reached (see page 188). The attack on spring rape starts before the buds open and the crop is then much less able to compensate for pod

Table 36.1

Insect	Time of attack	Part of plant damaged	Winter rape	Spring rape
Pollen beetles	April May June	Unopened buds Flowers Young pods	(+)	+
Cabbage seed weevil	June July	Pods	+	+
Brassica pod midge			+	+
Cabbage stem weevil	May, June	Leaf stalks and stems	(+)	+
Cabbage stem flea beetle	Autumn & winter		+	0
Rape winter stem weevil	Autumn & winter		(+)	0
Cabbage leaf miner	Autumn		–	–
Cabbage aphid			(+)	+

+ often damaging
(+) occasionally or locally damaging
– of little or no importance
0 not present

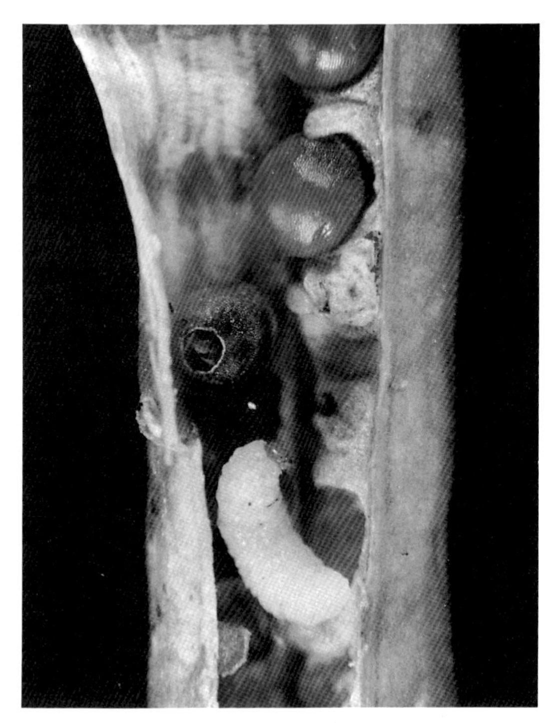

Fig.36.2 Oilseed rape pod opened to show larva of cabbage seed weevil and typical chewing damage to seed (× 7)

loss; the threshold for spraying spring rape at the green bud stage is, therefore, much lower.

Cabbage seed weevil

The adults are 2–3 mm long, dull ash-grey, with a distinct long narrow snout. They emerge from hibernation in sheltered places from late April onwards and tend to congregate initially on headlands, where several may be seen on the upper part of each winter rape plant. They disperse over the field within a few days. The adults do little direct damage to the buds, but their feeding and egg-laying punctures in the pods of both winter and spring rape provide ideal sites for brassica pod midges to lay their eggs (see below). Thus adult seed weevils provide points of entry for infestations of pod midge. There is only one generation of the weevil per year but egg laying continues for several weeks. Normally, there is a single egg in each infested pod, two being unusual and three exceptional.

The creamy-white legless larva (grub) has a distinct yellow-brown head and is, when fully grown, about 4 mm long and easily seen on opening the pod (Fig. 36.2). The larva damages three or four seeds in an infested pod, the remainder developing and ripening normally. The mature larva cuts a pinhead-sized hole in the pod and wriggles out to pupate in the soil; the exit hole is easily visible without magnification and can be seen near harvest. Seed weevil eggs and larvae develop more rapidly on spring rape than on the winter crop because flowering is later and temperatures tend to be higher.

There is generally a good relationship between the number of seed weevil adults in a crop and subsequent damage: the adult threshold level of infestation causing damage is one weevil per plant (see page 188). Dissection of pods has shown that parasites sometimes kill a large proportion of seed weevil larvae so that few survive to become overwintering adults.

Brassica pod midge

The adult is minute, delicate and very difficult to distinguish from other midge species found in the crop. Eggs are laid inside the pods. The midge seems to be dependent for successful egg laying on previous damage to the pod — usually the feeding/egg-laying punctures made by cabbage seed weevil (see above), although it also uses other damaged areas, e.g. fungal lesions. Numerous minute, elongate, transparent eggs are placed in clusters, often beside a seed weevil egg, and give rise to large numbers of larvae in each infested pod (Fig. 36.3).

Newly hatched larvae are tiny, transparent and very difficult to see, but the later stages are white and easily visible. Fully grown larvae are about 2 mm long. Infested pods become yellow, sometimes swollen and distorted ('bladder pods'), and the seeds collapse and shrivel. The pods split open prematurely, the mature larvae escape and the seed is shed. Pupation occurs in a cocoon in the soil. On rape, at least two generations of midge occur per year.

This pest is usually more common on headlands, so that a cursory examination can give a misleading impression of the degree of

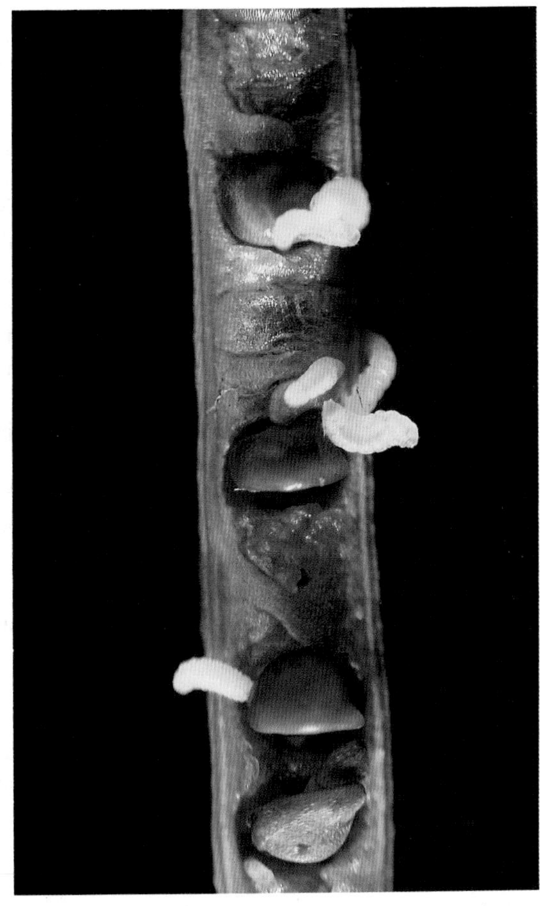

Fig.36.3 Damaged oilseed rape pod opened to show larvae of brassica pod midge (× 8)

infestation. National surveys have given an average figure of about 3% infested pods. However, much heavier infestations causing serious crop loss occur, especially in the southern counties of England, and pod midge is a serious threat to intensive rape growing. Fortunately, the insect flies weakly and isolation of crops from the sites of previous years' oilseed rape crops should reduce the level of infestation. A distance of even 0.5 km between crops in successive years should provide an effective barrie

Cabbage stem weevil

This pest is important only on spring rape. The adult is similar in size and shape to the seed weevil, but the more prominent body scales make the surface rougher in texture. The stem weevil is grey-brown and there is a small patch of white scales in the centre of the back. Adults emerge from hibernation about the same time as pollen beetles and seed weevil. Numerous eggs are laid in small groups just below the lower surface of the leaf stalks and veins and in the stems of spring rape.

The larvae (grubs) tunnel through the plant tissue and eventually all are located in the stem, where they destroy the pith. The mature larvae, which are 4–6 mm long and resemble those of seed weevil but are more elongate and slender, leave the plant through small but obvious exit holes in the lower half of the stem. Entry holes in the lower leaf scars are often enlarged and re-used. Pupation occurs in the soil.

Stem weevil is widespread on spring rape in southern counties of England and virtually every plant in a field may be infested. Plants are not killed and because infestations are uniform the farmer fails to notice them. However, the damage can reduce plant vigour and yield. The larvae may also infest winter rape but their effect on the crop is unknown.

Cabbage stem flea beetle

This pest breeds only on autumn and winter brassicas. The adult is about 5 mm long and usually shiny greenish- or bluish-black, but a bronze form is common. Those emerging from mature rape fly readily in July and August in the vicinity of the crop and are often found in

Fig.36.4 Oilseed rape plants damaged by larvae of cabbage stem flea beetle

large numbers among the harvested seed (where they do no harm). On the ground they exhibit the typical leaping movements of flea beetles. After their dispersal flight in late summer they become quiescent for a few weeks and then in autumn and early winter they feed on seedling winter rape and other new brassica crops. They may seriously damage or destroy young crops, especially in dry weather when plant growth is slow. Eggs are laid in the soil over a period of several weeks.

Newly hatched larvae may enter the plants from October to early April. Egg development and larval activity are inhibited by temperatures below about 3 °C. The larvae are predominantly white with numerous very small darker dots on the back, and three pairs of darker legs. The head and the large plate on the upper surface of the hind end are pale yellow in newly moulted larvae but turn black as the skin hardens. When fully grown the larvae are about 6 mm long.

An infested plant may contain 30 or more larvae, which tunnel first in the leaf stalks of the lower leaves and then cause extensive damage to the stems (Fig. 36.4), tunnelling in the pith and destroying the growing points of primary and secondary racemes before pupating in the soil. Damaging infestations appear to be restricted to winter rape in Berkshire, Buckinghamshire, Cambridgeshire, Essex, Northamptonshire, Oxfordshire and parts of Warwickshire, although the pest now occurs on the crop in many other rape-growing areas.

Rape winter stem weevil

The life history is very similar to that of cabbage stem flea beetle. Adults emerge in the summer from pupae in the soil of infested rape fields and disperse to woods and hedges, where they enter a resting phase before moving into young rape crops in the autumn. They are 3–4 mm long, shiny black with reddish feet and a typical weevil snout. Unlike cabbage stem flea beetle they do not leap when disturbed, are difficult to find and cause little direct damage. Eggs are laid in any suitable crevice in the rape plants or in punctures made in the leaf stalks.

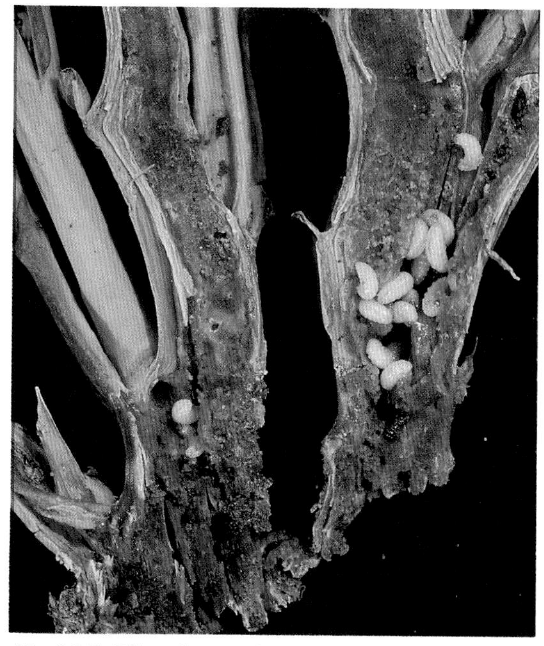

Fig.36.5 Oilseed rape plant opened to show larvae of rape winter stem weevil

The eggs hatch throughout the autumn and winter. The larvae (grubs) are legless, have a white fleshy body and a distinct light brown head. They tunnel into the leaf stalks (Fig. 36.5) and the crown of the plant, which may be completely hollowed out. When fully fed (about 6 mm long) in March or April the larvae leave the plants and pupate in the soil.

Plants attacked in early autumn may be killed. Those which survive may have the terminal shoot destroyed, resulting in the development of many lateral shoots, delayed flowering and loss of yield.

Until 1982 this weevil was an uncommon pest, but in the 1980s it became fairly widely distributed in eastern counties of England from Essex to South Humberside. Economic damage has occurred, particularly in Essex, Lincolnshire and Suffolk.

Cabbage aphid (see also page 27)

Both spring and winter rape are colonized in summer by winged aphids, which in suitable weather quickly produce colonies of grey 'mealy' (waxy) aphids, first in the growing point and then extending down the stem and on to the pods. Because of the similarity in colour, cabbage aphid infestations have occasionally been mistaken by farmers for seed weevils, but the latter never occur in static groups or colonies. Although heavy infestations of aphids on individual plants may be obvious, the proportion of such plants in the crop is often small. In very hot, dry summers damaging populations may occur, especially on spring rape.

Cabbage leaf miner

This insect occurs in autumn on young winter rape plants. The adult is a small and inconspicuous fly about 4 mm long with a grey body and brownish legs. As the larvae tunnel in the veins and leaf stalks of the lower leaves, they may be mistaken for larvae of cabbage stem flea beetle (see above). However, leaf miner larvae are typical fly maggots, being slender, creamy-white and without an obvious head or any appendages; the internal black mouth-hooks at the head end and the two breathing apertures at the hind end are just visible. Although the lower leaves of infested plants turn yellow and drop, no case of significant damage has been recorded and no control measures are advocated.

Inspection of crops

Crops should not be sprayed as a routine but only when careful examination of at least 20 plants at random on a traverse across the field (not just on the headland) has shown that pest infestations have reached threshold levels (see page 188).

Crops should be inspected on warm, dry days from late April onwards, when pollen beetles and cabbage seed weevil become active on the upper parts of the plants and are easily seen. They are more easily and accurately counted, especially on winter rape, if each plant is shaken over a tin lid, dish or tray of convenient size (about 30 × 25 cm), the numbers recorded and the insects discarded before the next plant is examined.

Only pollen beetles and cabbage seed weevils can be assessed in this way. The latter tend to lie immobile in the counting tray for a few seconds with their legs folded beneath them. Pollen beetles are more active but can still be counted easily. Cabbage stem weevil adults seem to spend much time on the ground and are less readily found when plants are examined. It is extremely difficult to distinguish brassica pod midge adults from numerous other species of small insects, some of which are beneficial parasites.

If large numbers of cabbage stem flea beetle are seen during harvesting of oilseed rape, damage to neighbouring crops is likely in the autumn. Crops should be inspected for adult feeding damage in September and October and for larvae from mid-October onwards.

Growth stages of winter and spring rape

Oilseed rape may be of the winter type, sown in autumn (about 90% of the present crop area), or of the spring-sown type. When invasion by pests occurs in May the two types of crop are at very different growth stages. In mid-May winter rape is usually about 130 cm tall and in flower. By contrast, spring rape at this time is usually only about 10 cm tall with

Fig.36.6 Field of winter oilseed rape at right stage for spraying post-flowering

four to six true leaves and the primary bud cluster beginning to form; under drought conditions plants of this size may already have the primary bud cluster and three to four axillary clusters. These differences affect the importance and severity of damage by pollen beetles, cabbage seed weevil and brassica pod midge and the timing of control measures against them.

Control measures for winter rape

Threshold infestations for spraying

Pollen beetles Infestations of pollen beetles may occasionally be present at the green bud stage. If so, a population of 15 or more beetles per plant is considered worth spraying. On very backward crops a threshold of five beetles per plant may be used. It is stressed, however, that damage by this pest to winter rape is very rare.

On no account should sprays be applied against pollen beetles during flowering: spraying is then extremely hazardous to bees and is also uneconomic as large numbers of beetles usually do no harm to the crop at this stage.

Cabbage seed weevil The main invasion of the crop usually occurs from mid- to late May. The infestation should be assessed when the crop is well into flower. A population of one weevil or more per plant is worth spraying but, if brassica pod midge infestations were obvious on adjacent fields in the previous year, it is probably advisable to spray smaller populations of weevils.

Brassica pod midge There are no established threshold infestations for spraying against this pest. If a post-flowering spray is applied for seed weevil, then a large proportion of immature midge larvae in the pods at the time of treatment will also be killed.

Cabbage stem flea beetle No threshold has been established for adults, but infestations damaging a large proportion of young crops should be controlled. Infestations of five or more larvae per plant in autumn or winter are thought to justify treatment.

Timing of sprays

Pollen beetles If control of pollen beetles is necessary, a spray of a recommended insecticide should be applied only at the green or yellow bud stages. However, if backward areas of a crop are sprayed, there is likely danger to bees which may be working the rest of the field or flowering weeds in the green area. An insecticide with minimal toxicity to bees should, therefore, be given preference in these circumstances.

Cabbage seed weevil and brassica pod midge The beginning of the emergence of seed weevil larvae from eggs inside pods of winter rape is well synchronized with the end of the flowering period. If applied at this time, a spray of a recommended insecticide with pod-penetrant properties will kill larvae of both seed weevil and pod midge. Such a spray should be applied *at the end of flowering* (Fig. 36.6), i.e. as soon as the field is *predominantly green overall* (with virtually all pods set), although *a few late flowers* may still be present on secondary racemes. As an alternative, certain contact insecticides are recommended for application against adult weevils before eggs are laid; best results are usually achieved from sprays applied between 20 pods set and 80% petal fall.

Cabbage stem flea beetle In areas where damage occurs regularly, a recommended systemic granular insecticide applied at, or shortly after, drilling will give good control of adult flea beetles and their larvae. Alternatively, a recommended contact insecticide spray can be applied against adults and again as necessary against larvae.

Rape winter stem weevil When applied against cabbage stem flea beetle recommended insecticide granules or sprays should give some control of winter stem weevil. Early sprays against flea beetle adults should also kill adult weevils;

later sprays give some control of winter stem weevil larvae before they enter the plants.

Control measures for spring rape

Threshold infestations for spraying

As damage by pollen beetles, cabbage seed weevil and cabbage stem weevil begins in the very early stages of growth, i.e. as soon as the first green buds have formed, crops should be examined regularly and carefully. The adults of pollen beetles and seed weevil will be easily visible in the growing point or in the bud clusters. Adults of stem weevil are less readily seen. If three or more pollen beetles or one seed weevil per plant are seen, a spray should be applied before flowering. If the pests again reach these levels before flowering, the spray should be repeated but not after the yellow bud stage.

Timing of sprays

For optimum control of pests and to safeguard bees, all sprays should be applied to spring rape from early green bud to late yellow bud, i.e. *all spraying* of this crop should be done *before flowering*. 'Early green bud' means the appearance of the first cluster of buds in the growing point when plants usually have four to six true leaves and are about 10 cm high. Green buds are of course present until all the flowers have opened but, if spraying is delayed beyond the stage indicated, considerable damage and loss of flowers and pods will already have occurred.

Several insecticides are recommended for control of pollen beetle and seed weevil adults in spring rape. Some of them should also give partial control of cabbage aphid.

Insect activity continues during flowering, but insecticide applications then are extremely hazardous to bees and must be avoided. There are no recommendations for post-flowering spraying of spring rape.

Aerial spraying

Some insecticides may be applied by aircraft to oilseed rape: it is most important to use only those that are approved for this use.

Danger to bees

Although honey bees and wild bees have little effect in pollinating oilseed rape or increasing yield, they are extremely important pollinators of other crops. They are often attracted in large numbers to flowering oilseed rape and, unfortunately, several insecticides used are toxic to them. Indeed, some are highly toxic at the time of application and for at least 24 hours afterwards.

To avoid harming bees, great care should be taken when applying all pesticides. Adherence to the following rules should minimize 'bee incidents'.

(1) Spray only if it is known that pest infestation levels have reached the suggested thresholds.
(2) Use an insecticide of low toxicity to bees.
(3) Never spray crops in flower unless there is a specific recommendation on the product label to do so.
(4) Do not spray crops of uneven maturity which are partly in flower, or which contain flowering weeds, when bees are active unless the product used is known to be safe.
(5) Apply post-flowering sprays to winter rape in the evening (for preference) or, failing this, in the early morning before bees are active.
(6) Give beekeepers adequate warning of the intention to spray (at least 24 hours) to enable them to close or move hives as they think fit. Most Beekeepers' Associations have a communication system for this purpose, but it is usually impracticable to close hives (except for a very short time) or to move them; therefore, the other rules are extremely important.

37
Pea, bean and clover weevils

Fig.37.1 Field bean leaf damaged by adult *Sitona* sp.

The pea and bean weevil (*Sitona lineatus* (L.)) and the clover weevils (*Sitona hispidulus* (Fabricius) and *Sitona sulcifrons* (Thunberg)) are common and widespread insects found on leguminous crops. The adult weevils feed on the leaves, whereas the grubs (larvae) feed on the roots or root nodules. As their names indicate, the different species have distinct host preferences, but all of them live on several different crop plants.

Nature of damage

The characteristic leaf damage shown in Fig. 37.1 may be seen every year. Older plants and those that are growing well can tolerate quite severe leaf-notching without appreciable loss of yield. Seedling crops are more susceptible, most damage occurring when the growth of the crop is checked by poor soil or cold weather. In experiments in which pea plants were artificially defoliated to simulate injury by adult weevils, removal of 12.5% of the total leaf tissue at the four-expanded leaflet stage led to an eight per cent loss of yield. Only occasionally is pea and bean weevil responsible for this amount of damage, so control measures are not usually worth while.

The grubs can eat a considerable proportion of the root nodules of leguminous crops and this sometimes reduces the yield and protein content of peas, broad beans and field beans. It is most likely to happen when the nitrogen content of the soil is small, or unfavourable conditions such as a drought prevent nitrogen uptake. If growing conditions are good, nodule damage is less important.

Adult weevils transmit two viruses, namely broad bean stain and broad bean true mosaic, in broad bean and field bean crops. Both viruses cause leaf malformation with light and dark green mottling or yellowish blotching. Early infection seriously reduces yield; stain virus also discolours broad bean seeds ('Evesham stain'), which may be serious in crops for canning. The pea and bean weevil can transmit both viruses but is a much less important vector than the bean flower weevil (*Apion vorax* Herbst) (see page 192). The viruses are also transmitted through some of the seed set by infected plants, so it is important to ensure that crops grown for seed are not seriously infected.

It should be noted that the conspicuous holes sometimes seen in the seeds of peas and beans are not made by the pea and bean weevil nor by clover weevils but by pea and bean beetles (*Bruchus* spp.).

Fig.37.2 Female pea and bean weevil (× 8)

Description and life history

Pea and bean weevil

This weevil (Fig. 37.2) is a common species in England and is fairly typical of the group in appearance and habits. The adult is about 5 mm long and light brown with faint creamy-yellow stripes.

The adult weevils spend the winter among dead leaves, tufts of grass, etc. They can also overwinter in fields of clover, lucerne and other leguminous crops. The weevils become active when the weather turns warm in spring and at once they begin to feed, attacking mainly peas and beans. The first sign of feeding is generally seen near the margin of the field. It is at this time, when the plants are just emerging, that the worst damage is done.

The female weevil lays eggs in the soil near the plants. The grubs, which are white with a brown head and are legless, soon hatch and begin to feed on the root nodules that are characteristic of leguminous plants. The grubs are fully fed at about the end of June, when they pupate about 5 cm down in the soil.

The adults begin to emerge two to three weeks later, often to find that the peas have already been cut or that the foliage of peas and beans is drying out and unpalatable. The weevils, therefore, tend to move out of the pea and bean fields in search of fresh growth. This emigration of weevils — usually in late July and early August — often coincides with the cereal harvest. Consequently, some weevils become mixed with the grain and may subsequently be mistaken for storage pests. More important, the young clover in undersown leys may suddenly be exposed to weevil attack at a critical stage of growth. Undersown clovers bordering on pea or bean fields should be watched for the first two weeks after cutting the cover crop. If there are signs that damage is increasing, the new ley should be treated at once. Weevils can seriously affect the establishment of clovers in the seedling stage. The weevils feed on the clover until the cool autumn weather, when they seek shelter for the winter.

Clover weevils

These species look very like the pea and bean weevil and their life histories are generally similar. However, there are some variations: for example, the species most injurious to clover, *Sitona hispidulus*, lays eggs from autumn to spring. These weevils may be abundant in clover leys in the winter and may subsequently infest nearby crops of peas or beans.

Bean flower weevil

This is a narrow, pear-shaped weevil, about 2.5 mm in length, with a long thin snout and long legs; it is black except for the wing-cases which have a dark blue metallic sheen. It is found on beans, tucked down in the growing point, but not on peas or clover. It is very active and makes small holes in the young leaves (Fig. 37.3) and flower parts, but this damage is probably of little direct significance. However, the adult is an important vector of certain viruses infecting broad and field bean crops (see page 191). It appears to overwinter in woodland and hedgerows on farms where field beans have been grown intensively. This weevil is sporadic in occurrence and easily overlooked, readily falling from the plants when disturbed and apparently spending much of its time on the ground.

Fig.37.3 Field bean leaves damaged by bean flower weevil

Control measures

On field crops

Cultural control Good tilth is important: peas sown on a fine tilth, on land that is in good heart, generally escape serious injury when crops on poor, cloddy soil are damaged.

Chemical control Only exceptionally is damage to the foliage by these pests severe enough to reduce the yield of peas or beans. However, spraying round the edges of the field can sometimes be worth while. Clovers and lucerne have smaller leaves than peas or beans and failure to establish is more likely. With these crops, therefore, it is more important under poor growing conditions to spray at least the edges of the field when damage is noted. There have been some worthwhile increases in yield of peas and beans following the reduction of nodule damage by the application of granular insecticides to the soil, but such increases do not always occur.

Several insecticides are recommended for use as a spray on peas and beans, but treatment is likely to be justified only when weevils are very active and plant growth is slow.

In those areas where pea and bean weevil damage to root nodules occurs regularly and is thought to limit yields, the use of a recommended granular insecticide applied at sowing will reduce damage.

In gardens and allotments

If young crops of peas or beans are badly attacked, the rows should be kept well hoed and the plants dusted with a recommended soil insecticide.

38
Pests of grass and clover seed crops

Fig.38.1 Heads of S48 timothy damaged by maggots of timothy fly

Grass and clover seed crops are attacked by a wide range of pests, although relatively few cause serious loss of yield. Described here are (i) those pests which affect seedling establishment, (ii) pests of the leaf and stem, and (iii) pests of the flower head and seed. Pest situations in reseeded grassland are not dealt with.

Pests affecting seedling establishment

The main seedling pests are slugs, frit fly, clover weevils and stem nematode. Clover cyst nematode may also attack the roots of young plants.

Slugs (see also page 467)

Damage On heavy soils in wet seasons, slugs, particularly the field slug (*Deroceras reticulatum* (Müller)), can severely thin newly germinated crops, especially those which follow trashy crops or stubbles. Seedlings are attacked below ground and at soil level, usually in the spring and autumn.

Cultural control Fine, firm seedbed conditions are essential for good establishment of grass and clover seedlings. Poor tilth or inadequate consolidation will increase the likelihood of slug damage.

Chemical control Test-baiting with small quantities of molluscicide pellets can be used to assess the need for treatment before sowing. When slugs are readily found, molluscicide pellets broadcast on the soil surface a week before sowing will decrease numbers and will be more effective than when applied with the seed or after sowing. Treatment should be timed to coincide with moist conditions suitable for slug activity.

Frit fly (see also page 258)

Damage The maggots (larvae) of the frit fly (*Oscinella frit* (L.)) burrow into the shoots of most of the grasses grown for seed. Very young plants are killed outright, while the growth of older plants is weakened. Grasses most likely to be attacked are Italian and perennial ryegrasses and the fescues, but cocksfoot is rarely infested. Damage by the maggots of the first generation occurs in late spring and the third-generation maggots attack early autumn sowings. In some years, however, egg laying on grasses is continuous because of an overlap of generations.

Fig.38.2 Red clover plants (right) infested with stem nematode and showing stunting and basal swelling of shoots compared with uninfested plants (left)

Cultural control Early spring sowings of grasses undersown in barley are unlikely to be damaged by frit fly.

Chemical control In those areas prone to frit fly attacks local warnings will indicate the risks to late summer sowings. Where damage is expected, prompt treatment with a recommended insecticide is advised.

Clover weevils (see also page 190)

Damage Clover weevils (*Sitona hispidulus* (Fabricius) and *Sitona sulcifrons* (Thunberg)) and the related pea and bean weevil (*Sitona lineatus* (L.))

are common and widespread insects feeding on leguminous crops; the adult weevils feed on the leaves, and the grubs (larvae) on the roots or root nodules. The species most injurious to clover is *S. hispidulus*, the adults of which can occasionally cause severe damage to seedlings soon after emergence.

Cultural control A good firm seedbed will encourage rapid growth of the seedlings and reduce the risk of severe weevil damage.

Chemical control At present there is no recommendation for chemical control of clover weevils on clover.

Stem nematode (see also page 440)
Red and white clover are each attacked by their own specific race of stem nematode (*Ditylenchus dipsaci* (Kühn) Filipjev). The race which attacks white clover is probably common, but often overlooked, in wild clover. Stem nematode appears to be widely distributed in sown swards of white clover.

Damage There is some evidence that stem nematode is a causal factor in reducing vigour, growth and competitive ability of white clover plants, especially in Wales and western areas of England.

Red clover seedlings may be killed at an early stage and severe infestations often lead to thin or bare patches in the crop. Infested plants are stunted and the stems and petioles are often swollen (Fig. 38.2). The flower heads become infested and the seed contaminated, which is an important means of spreading the pest. Stem nematode attacks on red clover are much less common than formerly.

Cultural control Crop rotation will minimize the risk of a soil-borne nematode infestation damaging the newly sown crop. Fields where attacks have occurred should be given several years' break from red clover. In addition, the use of a variety resistant to stem nematode is recommended.

Chemical control Nematode infestations in red clover seed can be controlled by fumigation.

Fig.38.3 Ryegrass damaged by grass aphids

Under the statutory certification scheme, if stem nematode is detected during field inspection of a crop grown for Basic or Higher Voluntary Standard grade, the seed must be fumigated or downgraded to Certified (minimum standard).

Clover cyst nematode
Clover cyst nematode (*Heterodera trifolii* Goffart) is widely distributed in Britain, especially in Wales, and often occurs with stem nematode on white clover crops (see above). Like the stem nematode, it may be a factor in white clover decline, but more work is needed to determine its status as a pest.

Leaf and stem pests

Pests included in this category are aphids, the cereal rust mite, which transmits ryegrass mosaic virus, and clover leaf weevils.

Aphids (see also page 36)
Several species of aphid colonize grass seed crops. The main ones are the fescue (or grass) aphid (*Metopolophium festucae* (Theobald)), the grain aphid (*Sitobion avenae* (Fabricius)) and the bird-cherry aphid (*Rhopalosiphum padi* (L.)). All three species are likely to be more abundant following mild winters; the grass aphid in particular multiplies rapidly in warm, dry weather, spreading quickly throughout the crop. Timothy, fescues, cocksfoot and

ryegrasses are all attacked.

Direct damage When present in large numbers, aphids cause wilting (Fig. 38.3), stunting and death of shoots throughout the crop and in some years they cause serious loss of yield.

Virus spread Of particular importance is the ability of the bird-cherry aphid to transmit barley yellow dwarf virus (BYDV), which can seriously decrease yields of ryegrass seed crops.

Chemical control Aphids can be controlled effectively by sprays of a recommended aphicide. Spraying is, however, unlikely to be economic unless aphids are present in very large numbers in the spring and early summer.

Cereal rust mite

Ryegrass mosaic virus is transmitted by this eriophyid mite (*Abacarus hystrix* (Nalepa)), which lives in the grooves on the upper surface of leaves. The mite is very small, being almost invisible without magnification. Populations build up to a peak in late July before harvest and the mites disperse in the wind to infest newly sown spring crops. Dispersal is complete by late October.

Ryegrass mosaic virus This virus is widespread and severe in midland and southern counties of England, affecting both seed and herbage yields. Italian ryegrass is more susceptible than perennial species, while crops sown in the spring are more seriously affected than those sown in the autumn. Pale green streaks develop on the upper leaf surfaces of infected plants (Fig. 38.4) and eventually become dark brown as the leaves mature. The virus causes a reduction in tillering, and infected plants are usually stunted with yield losses of up to 30% in the first harvest year. The virus also occurs in cocksfoot and fescues, but the effect on yield is unknown.

Fig.38.4 Italian ryegrass affected by ryegrass mosaic virus

Cultural control Autumn-sown crops in their first harvest year are often free from both the mite and the virus because the mites have frequently dispersed before the emergence of the crop. Grasses sown in the spring are more heavily infested during their first year than those sown in the autumn. Mite numbers seem to be closely related to the intensity of grazing: early grazing, especially by sheep in the autumn, reduces the mite population and virus infection in the following year. Some varieties are more tolerant of virus infection than others.

Chemical control There is no recommendation for chemical control.

Clover leaf weevils

The two main species of leaf weevil (*Hypera postica* (Gyllenhal) and *Hypera nigrirostris* (Fabricius)) are common but affect the yield of seed crops in very dry seasons only.

Description, life history and damage Adults of both species emerge from hibernation, mainly from hedgerows, in March and April. *Hypera postica* is a greyish-brown weevil, while *H. nigrirostris* is a shiny blue-green. The adults feed on the leaves, making small holes, but this damage is of little significance. Eggs are laid in the leaf and stem tissue. The larvae, which hatch within about two to three weeks, are curved legless grubs with dark heads; they are unusual for weevil grubs in that they are exposed feeders with protective coloration. The grubs of both species are green, but those of *H. postica* have a white dorsal stripe. After hatching, the grubs migrate to the buds where they feed for 20–60 days, destroying both leaf and flower buds. They pupate in the foliage or in the soil. The adults emerge two to three weeks later, feed on the foliage for several weeks and then hibernate.

Cultural control Clover seed crops should be grown away from hay and silage crops and well separated from other seed crops, especially those attacked the previous year, as these weevils readily move from field to field.

Chemical control There is no recommendation for chemical control.

Flower head and seed pests

Pests included in this category are clover seed weevils, clover seed midges, cocksfoot moth and timothy flies.

Red clover seed weevils

Two species of weevil (*Apion trifolii* (L.) and *Apion apricans* Herbst) can cause damage to red clover seed crops, more commonly in southern England and Wales, resulting in considerable loss of seed in some seasons.

Description, life history and damage The adults are black and pear-shaped, about 2 mm in length, with long snouts. They emerge from hibernation in late spring and feed on the clover leaves, making small jagged holes, but the damage is of no significance. Eggs are laid during May in the developing flower heads. The white legless grubs hatch four to eight days later and burrow into the florets where they eat the developing ovules. Feeding continues for about 18 days until the fully grown grubs pupate in the flower heads. Adult weevils emerge six days later and lay more eggs. The second-generation grubs continue to feed on the developing seeds. The adults from the second generation feed on the foliage until the end of the summer, when they move off the crop to hibernate in the hedgerows.

Cultural control Some control occurs when the early June-flowering broad red clovers are cut for hay, destroying many of the first-generation grubs; this reduces damage to the seed crop later in the year.

Chemical control There is no recommendation for chemical control.

White clover seed weevil

The white clover seed weevil (*Apion dichroum* Bedel) is one of the most important pests of white clover seed crops in south-east England, where it can decrease seed production by 25–30%.

Description, life history and damage Unlike the red clover seed weevils, this species has only one generation a year. The adults are shiny black and 3 mm long; they emerge from hibernation in early May, colonizing clover fields until late June. At first the adults feed on the leaves, making small jagged holes, but this damage is of no significance. Egg laying begins at the end of June; the female weevil bores into the flower to obtain pollen and to form an egg-laying hole into which she lays a single egg. After 10 days the eggs hatch and the white legless grubs feed on the developing seeds for about three weeks before pupating. Ten days later, from about early August onwards, the adult weevils emerge and migrate to surrounding hedgerows and woodland to hibernate.

Chemical control There is no recommendation for chemical control.

Clover seed midges

The clover seed midge (*Dasineura leguminicola* (Lintner)) is an important pest of red clover seed crops, whereas the white clover seed midge (*Dasineura gentneri* Pritchard) causes occasional damage to both white and red clover seed crops.

Damage Infested heads bloom irregularly and turn brown prematurely. Seed from attacked heads is light and will not germinate.

Description and life history The adults are pale red and about 2 mm long; they appear when red clover is in bud. Eggs are laid in clusters in the green heads by the clover seed midge and in open flowers by the white clover seed midge. The larvae, which are pink, hatch 3–6 days later and migrate to the ovaries, where they feed on the developing seeds for up to six weeks. When fully fed, the larvae drop to the ground and pupate; adults emerge two to three weeks later. There are usually two generations a year. The larvae from the second generation overwinter in cocoons in the soil and pupate the following spring.

Cultural control These pests are effectively controlled by correct management of seed and surrounding hay crops. Seed stands can be grazed or cut early in the season, so that flower heads are removed just before peak adult emergence in late May. If the first generation is removed by cutting the crop, the second generation will cause little economic damage. Crops cut after the main flowering period should be made into silage rather than hay because larvae can still survive on cut heads left to dry.

Chemical control Control by insecticides cannot be recommended because of the difficulty of timing applications.

Cocksfoot moth

The cocksfoot moth *Glyphipterix simpliciella* (Stephens)) is widely distributed. It rarely causes economic loss of yield, although occasionally it decreases the yield of older crops.

Description, life history and damage The moth is small and pale grey. It emerges in late May and lays eggs at the base of the seeds. The caterpillars, which are pale green, hatch in mid-June and feed on the seeds for about four weeks, leaving empty glumes with a characteristic small hole near the base (Fig. 38.5). After leaving the seed head, the caterpillar bores into the stem (Fig. 38.6) just above the soil. Pupation occurs in the stem and the adults emerge during the following May.

Cultural control If old seed stands become infested, removing and burning the straw and stubble will give complete control by destroying the pupae. New sowings should be sited well away from old, infested crops.

Fig.38.5 Glumes damaged by caterpillars of cocksfoot moth

Chemical control Use of insecticides is unnecessary.

Timothy flies

Timothy flies (*Nanna* spp.) are sporadic pests and only occasionally cause loss of yield to seed crops in England and Wales, although severe attacks have occurred in Scotland.

Damage Attacked ears have a characteristic appearance with the spike lacking spikelets, usually on one side, and showing a dark streak where the maggots (larvae) have fed (Fig. 38.1). In a severe attack about 35% loss of seed can occur.

Description and life history The flies are dark grey and 5 mm in length. They lay their eggs in

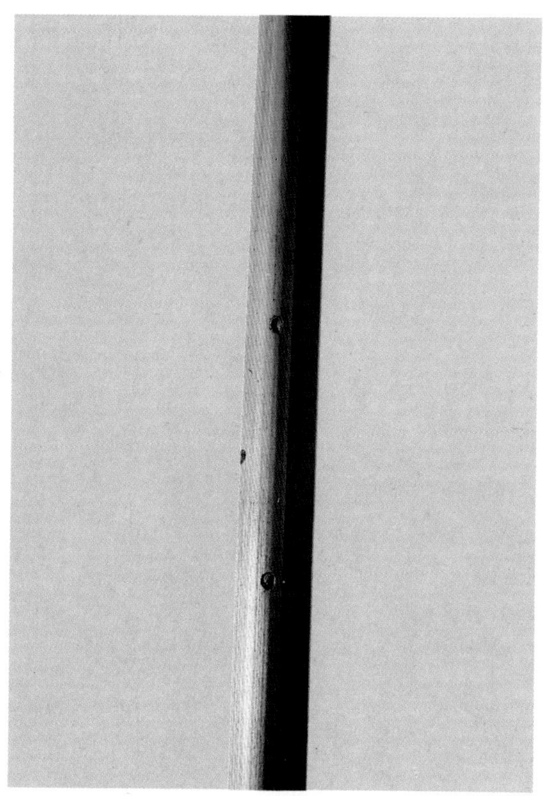

Fig.38.6 Grass stem showing holes made by caterpillars of cocksfoot moth

April and May at the base of the leaf sheaths. After hatching, the small yellow maggots bore into the shoots, where they feed on the developing spikes for about two weeks. By the time the heads are ready to emerge, the maggots are fully grown and fall to the ground where they overwinter as pupae. The adults emerge in the following April.

Cultural control Early-flowering varieties are less susceptible to attack. Similarly, well-grown and forward plants in mature seed crops are rarely heavily infested because their flowering period does not coincide with the main emergence period of the flies.

Chemical control There is no recommendation for chemical control.

39
Pygmy beetle

Fig.39.1 Beet plant showing characteristic foliage damage caused by earlier feeding of pygmy beetles on folded heart leaves

The pygmy beetle or pygmy mangold beetle (*Atomaria linearis* Stephens) can usually be found in any beet or mangel field if sought diligently. However, because of its small size it may be overlooked and the injury it causes attributed to other pests. Feeding on host plants starts in March and continues well into the summer, chiefly below ground, but also above ground in showery weather. Pygmy beetle attacks sugar beet, mangel, red beet, fodder beet and spinach; it can be induced to feed on many weeds but seldom does so in the field.

Before 1935, when beet was often grown after beet, pygmy beetle was the most important pest of sugar beet because it overwinters in beet fields and attacks the germinating seedlings. After 1935 its importance diminished following the British Sugar Corporation's enforcement of a rotation, primarily to control beet cyst nematode (see page 397). In recent years, however, the importance of pygmy beetle has increased, probably because of the use of monogerm seed sown at wide spacing. The wider the seed spacing the greater the risk of seedling damage, especially in a period of slow growth.

Nature of damage

The feeding of the beetle on the stem, at and below ground level, and on the roots produces characteristic pits that blacken within a few hours (Fig. 39.2). When many pygmy beetles are present, every stem and root has numerous bites and many seedlings wither and die. Once past the cotyledon stage, attacked plants usually survive, but they are stunted and yield less at harvest. Fresh bites can be found on the tap root at a depth of 3–15 cm as late as July.

Damage to the foliage results from the beetle feeding on the still-folded leaves in the heart of the plant; it is usually much less serious than bites in the young stem just below ground level. As the heart leaves expand, the holes and notches become apparent, often symmetrically distributed about the midrib and with the edges healed and usually reddened. As these leaves expand further, the holes enlarge and tear (Fig. 39.1), producing the tattered leaves often characteristic of fenland crops. Even at the late stage illustrated the beetles can usually be found feeding in the heart of the plant.

Pygmy beetle damage should not be confused with damage caused by flea beetle (see page 173), which is confined to the cotyledons and outer true leaves of the seedling. Flea beetles feed on either surface of the leaf, producing circular pits that leave the other surface intact.

Fig.39.2 Beet seedling showing characteristic oval, blackened pits caused by pygmy beetle on stem below soil level

Description and life history

The adult pygmy beetle is brown and about 1.5 mm long (Fig. 39.3). Eggs are laid in the soil around beet plants from April onwards. The larvae feed on the roots but are not known to affect their growth. Young adults, pale at first but soon darkening, emerge from the pupal stage in the soil from early summer onwards. The earliest adults to emerge mature in about three weeks, mate and lay eggs from which a second generation may arise.

Because egg laying is protracted and the generations overlap, new beetles continue to emerge throughout the summer and autumn and then spend the winter in the soil of old beet fields. More beetles emerge in the following spring, by which time great numbers — perhaps up to five million per hectare — have accumulated in some old beet fields.

Before dispersing from the old beet fields, many beetles feed on old beet crowns which have survived the winter. These are often infected with virus diseases but, as far as is known, pygmy beetles do not transmit viruses.

Fig.39.3 Pygmy beetle (× 40)

Fig.39.4 Outline map showing relative incidence of pygmy beetle damage in each sugar factory area in England (1968–80)

Dispersal

Mass flights from the old beet fields begin with the first spells of fine, spring weather when the air temperature approaches 20 °C. The first long-range flights can occur in early April but, if the weather remains cold, may be delayed until late May. Normally most flights occur in May and June but, if the weather in these months is colder than average, many beetles fly in July and early August also.

After the start of dispersal flights, the beetles can readily be found in the soil close to the seedling stems in the new beet fields and the characteristic injury soon begins to appear. In beet crops all the pygmy beetles tend to congregate round the beet seedlings and there are virtually none between the rows.

Incidence of damage

Complete loss of crop is more likely where beet follows beet or other host crops. In such circumstances the beetles are present and begin to feed as soon as the seed germinates; few seedlings emerge and many of these are subsequently killed. If, by error, there is any overlapping of sugar-beet or mangel cropping in partly cropped fields, the area where seedlings are destroyed often coincides to within a metre of the site of the previous beet or mangel crop.

In some years in some crops grown in rotation, severe thinning or even crop failure occurs, usually in the more intensive beet-growing areas and especially in fields where the seedlings grow slowly.

The incidence of damage varies considerably

from year to year and follows no obvious pattern: 1960, 1968, 1971 and 1976 were years of severe damage,whereas damage was particularly slight in 1966, 1969 and 1975. The reasons for these fluctuations appear to be associated with the effects of the climate on the life cycle of the beetle, but the extent of damage from a given number of beetles per seedling depends on the stage and growth rate of the seedlings.

Damage is most severe in the fenland regions (Fig. 39.4), probably because of the intensity of beet-growing in that area.

Control measures

Cultural control

The best way of avoiding the most severe damage by pygmy beetle is not to grow beet after beet or mangels.

Chemical control

In those areas subject to damage, wide seed-spacing, and especially planting to a stand, is hazardous. Where the pest is a regular problem, a seed treatment effective against pygmy beetle can be used, or a recommended granular insecticide applied in the seed furrow during drilling.

On crops being damaged, some control may be achieved by application of a recommended contact insecticide: a high-volume spray should be applied in a band along the rows.

40
Raspberry beetle

Fig.40.1 Larva of raspberry beetle on damaged raspberry fruit

The yellowish-white larvae so often found damaging raspberries, hybrid berries (e.g. loganberries, tayberries) and blackberries are the young of the raspberry beetle (*Byturus tomentosus* (Degeer)) and are a serious pest of these crops. In plantations where no control measures are applied, this beetle can multiply rapidly and crops can be so badly attacked as to become valueless. Infested fruit (Fig. 40.1) is unsuitable for processing or for the fresh fruit market, where the presence of the larvae makes consignments unattractive to buyers.

Description and life history

The main emergence period of adults from the soil occurs in late April or May in England and mid-May to early June in Scotland. Small numbers also emerge in Scotland between mid-July and mid-August. The adults (Fig. 40.2) are 4 mm long and when newly emerged are golden brown, later becoming greyish, sometimes tinged with green. At first they tend to frequent any open flowers and are often common in hawthorn and apple blossom, but they are soon attracted to the buds and opening flowers of raspberry and hybrid berries. In Scotland, where often there are no alternative flowers present at emergence, adult beetles will feed on the terminal leaflets of young primocanes (young or vegetative canes) before migrating to feed on the developing white buds. In sunny weather the beetles are very active and fly readily, but they become sluggish when it is cool or wet.

Mating begins in June, and during June and July the female beetles lay small, creamy-white, shiny eggs in the blossoms of raspberry and hybrid berries; eggs are laid on blackberry much later. The beetles prefer to lay their eggs in blossoms that have just set fruit. The eggs may be attached to stamens or pistils or to the centre of the flower; they hatch in about 10–12 days.

Fig.40.2 Raspberry beetle on damaged flower bud (× 12)

The larvae have a brown head and a yellowish body with darker markings (Fig. 40.1). At first the tiny larvae feed on the surface of the young fruit but soon move to the calyx region and feed on the basal drupelets. When the fruit begins to ripen, they burrow into the plug and continue to feed on the inner surface of the drupelets. Sometimes the first attacked fruit is abandoned and another fruit is entered. Occasionally, small larvae feed in the unopened leaf buds on the tips of young canes.

The mature larvae, which are about 8 mm long, leave the berries and burrow into the soil to a depth varying from 2 to 30 cm. There they form small cells and, after about a month, change into pupae. After a further four or five weeks, i.e. in early autumn, the pupae change into adults which remain in the cells until the following April or May. In Scotland some larvae remain through the winter and emerge as adults in July or August, or even in the following May.

Nature of damage

In the early part of the season the adults can cause considerable damage to the buds. In flowers they feed on the nectar and stamens and often cause black scars to form on the ring of green tissue round the base of the stamens, but this damage does not affect the quality of the fruit. The most serious damage is done by the larvae feeding in the developing fruit.

Primocane varieties, even though flowering later than traditional varieties, may still be attacked by raspberry beetle because of the long period of activity of the adults.

Control measures

Damage by raspberry beetle larvae can be effectively prevented if chemical sprays are properly timed. There is a choice of insecticide, but it is important to observe the minimum permitted intervals that should elapse between spraying and picking the fruit.

The leafhoppers that transmit the virus causing rubus stunt disease are also controlled by some of the insecticides recommended for control of raspberry beetle.

Timing and number of sprays

On raspberry One spray, applied when the first pink fruit is seen, usually gives adequate control. Growers who specialize in the production of very high quality dessert fruit sometimes apply two sprays, as recommended below for loganberry, to avoid the slight damage which may otherwise occur to the basal drupelets of the earliest fruits. (Such damage can be more severe in Scotland.) The risk of damage by raspberry beetle is greater in plantations containing mixtures of early and later flowering raspberries; this can be reduced if all varieties are treated with an insecticide within a week of each other.

On hybrid berries with a long flowering period (e.g. loganberry, boysenberry) two sprays are required to protect the whole crop. The first should be applied when about 80% of the blossom is over (usually mid-June) and a second when the first fruit is colouring, which is usually about two weeks later.

On hybrid berries with a short flowering period (e.g. tayberry, sunberry) one spray post-flowering at least two weeks before harvest should give adequate control unless flowering is unusually protracted or the fruit is produced for the high quality dessert market or infestations are severe (see below).

On blackberry A single spray, as soon as the first flowers are open, will kill the beetles before any eggs have been laid and will prevent fruits from becoming infested.

Severe infestations Occasionally, omission of control measures in previous years results in such large populations of beetles that they damage the unopened flower buds and the growing tips of the young canes, especially on raspberry. When this occurs, an application of a recommended insecticide should be made at the white bud stage and followed by a further treatment, as recommended above, at first pink fruit.

Beneficial insects All the insecticides currently recommended for control of this pest are harmful to bees and other beneficial insects and should, therefore, never be applied during the main part of the blossom period. If any blossoms are open, the risk to bees can be reduced by spraying in late evening, particularly on a dull day.

Application of sprays

To kill the young larvae before they penetrate the fruit, a good spray cover must be obtained because the larvae seldom move from individual fruits. Therefore, a spray must be applied in a large volume of water.

41
Turnip gall weevil

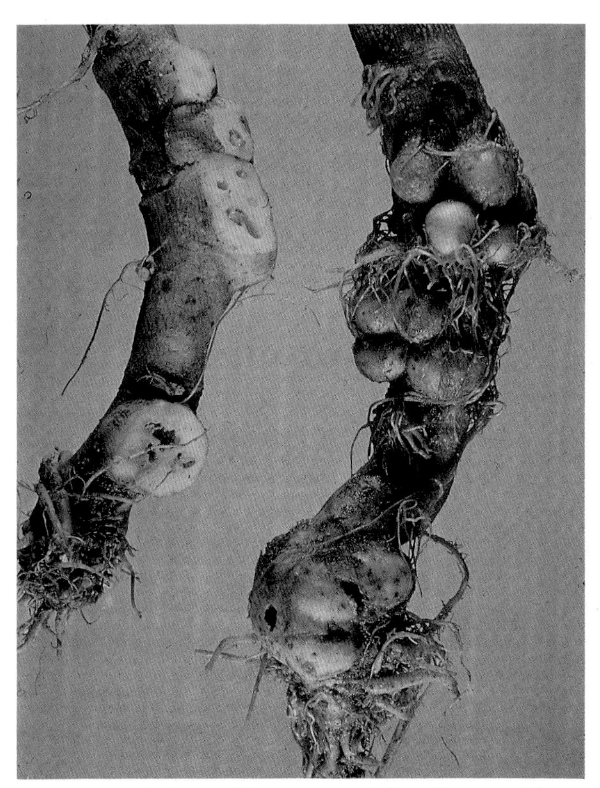

Fig.41.1 Spring cabbage roots with galls caused by feeding of grubs of turnip gall weevil. Galls on left root cut open to show larval chambers

The rounded galls of the turnip gall weevil (*Ceutorhynchus pleurostigma* (Marsham)) occur on the roots or stem-bases of cruciferous crops throughout Britain, but they are most common in west Cornwall. Turnip, swede, cabbage, savoy, winter cauliflower, Brussels sprout, kohlrabi, charlock, mustard and rape are all attacked, but damage is most pronounced in late-sown brassica crops which are small at the time of attack.

Although the presence of weevil galls (Figs 41.1 and 41.2), especially on seedlings or transplants, causes understandable alarm, normally crop growth is not seriously affected and the effect on yield is slight or negligible. Seedlings and newly transplanted plants may suffer a setback in growth, but the effects are usually transient unless other conditions are unfavourable, e.g. a drought soon after transplanting or a soil of low fertility. However, galled turnips and swedes may be ummarketable for table use.

Several other species of weevil are harmful to turnip, cabbage, etc., but none of these causes galls. (For stem and seed weevils, see page 177.)

Recognition of galls

The weevil galls superficially resemble hybridization nodules and the swellings caused by club root — a common and serious disease of cruciferous plants which is caused by the fungus *Plasmodiophora brassicae*. The weevil galls are rounded and contain chambers (Fig. 41.1) which the grubs occupy until they are fully grown. Club root swellings, especially those on the smaller roots, are solid and often elongated. Another difference is that weevil galls rarely rot until the insects leave them in late winter or early spring, whereas fungus swellings rot much more quickly. Hybridization nodules are smaller than galls and they are solid and do not rot.

Description

The adult turnip gall weevil (Fig. 41.3) is a small beetle about 3 mm long with a long snout or rostrum. Without magnification it appears to be black above and greyish on the underside but, viewed through a lens, the upper surface is seen to be sparsely dotted with grey and white.

The larva is a small, shiny white or yellowish grub with a brown head. It is legless and is found curled inside the gall.

Fig.41.2 Swede with typical galls caused by grubs of turnip gall weevil

The pupa is white and lies in the earth in a cell formed of particles of soil glued together with a sticky material secreted by the grub.

Life history

The adult weevils emerge in spring and summer. The females lay their eggs singly, just beneath the skin of a root of a food plant. The eggs are inserted in holes which the female makes with her snout. Frequently, several eggs are laid in one root. Most eggs are laid in August–September.

On hatching, the young grub feeds in the root, which reacts by forming a gall (Figs 41.1 and 41.2). At first the gall is small, but it grows gradually and may reach the size of a small

Fig.41.3 Turnip gall weevil (× 20)

marble; frequently several galls form a single, large outgrowth in which each chamber is occupied by a grub. When fully fed, the grubs bite their way out of the galls and enter the soil, where they pupate. Although most grubs leave the galls in early spring, the length of time spent in the different stages varies so much that both adults and grubs may be found throughout most of the year.

Control measures

Cultural control

The trash from most commercial brassica crops is ploughed or rotavated into the soil — a technique that favours survival of the gall weevil. On farms or market gardens where turnip gall weevil has been troublesome, the stumps should be lifted as early as possible in spring and stacked in heaps to dry because the grubs are unable to emerge from the dried roots.

Crop rotation should be practised where feasible. Any galled seedlings seen at transplanting should be discarded.

Chemical control

Insecticide seed treatments will give partial control of turnip gall weevil attack and will also protect the seedlings from attack by flea beetles (see page 175) and cabbage stem weevil (see page 180).

Some of the insecticides used to control cabbage root fly may also give some control of turnip gall weevil.

42
Wingless weevils

Fig.42.1 Cyclamen plant damaged by grubs of vine weevil

'Wingless weevils' is a convenient name for certain medium-sized to large weevils which are incapable of flight. They are found throughout Britain, occurring naturally in heaths and woodland. Several species are pests of cultivated plants. Strawberry roots and crowns are damaged by grubs of the vine weevil (*Otiorhynchus sulcatus* (Fabricius)), red-legged weevil (*Otiorhynchus clavipes* (Bonsdorff)), strawberry root weevils (*Otiorhynchus ovatus* (L.), *Otiorhynchus rugifrons* (Gyllenhal), *Otiorhynchus rugosostriatus* (Goeze) and *Sciaphilus asperatus* (Bonsdorff)) and strawberry fruit weevils (*Barypeithes* spp.). Stawberry leaves and occasionally the fruits are damaged by adults of these species. Other fruit (tree, bush and cane) and hops, especially in young plantations, are damaged by adult weevils which gnaw grafts, stems, buds and leaves; the clay-coloured weevil (*Otiorhynchus singularis* (L.)) is usually responsible for this. Black currants are also damaged underground by grubs of the vine weevil. The main damage to hops is caused by grubs of the hop root weevil (*Plinthus caliginosus* (Fabricius)). Pot plants under glass and ornamental plants in containers and outdoors are attacked by the vine weevil: the roots are injured by grubs, the leaves and stems by adults.

Description and life history

The general pattern of the life cycle outdoors is as follows.

Adult weevils become active as temperatures rise in the spring. The first weevils to emerge are those which completed their development during the previous year and hibernated in litter in woodlands, hedgerows, fields or gardens. These weevils are succeeded by those which overwintered as grubs. After emerging, the weevils, which are nocturnal, feed on their host plants for several weeks before laying eggs in the soil near the plants. The eggs are round and white at first but turn brown two to three days later.

After an incubation period of 8–24 days, the grubs hatch and begin feeding on the roots of their host plants. The grubs are legless with a brown head and whitish, plump body which is curved like a typical weevil grub.

The fully grown grubs burrow down in the soil to form earthen cells in which they pupate (Fig. 42.2). A few of the grubs become adult by the autumn, but the majority do not develop into adults until the following spring or summer. The time needed for development from egg to adult is between 9 and 18 months, but less in glasshouses and under other protection.

Vine weevil

The vine weevil occurs widely outdoors and is the species usually found under glass. The adult female is about 9 mm long and black with roughened wing-cases on which there is fine yellow speckling (Fig. 42.3). The male weevil is unknown so reproduction is always by unfertilized eggs.

Outdoors the adults emerge from April onwards. Their feeding produces leaf notches which are often obvious on 'naturalized' shrubs. Eggs are laid from May until October. In gardens and nurseries the grubs injure various herbaceous perennials such as primulas and alpines; they also attack strawberries.

In warmed glasshouses, where cyclamen and camellia are favoured host plants, adult weevils may emerge in the autumn. Egg laying in pots

Fig.42.2 Earthen cell broken open to show pupa of vine weevil inside (× 6)

or beds containing host plants starts in the following spring and continues at intervals during the summer. Grubs from the first-laid eggs can severely damage plants by late autumn and they become adult in December. All stages of the weevil may occur at the same time.

Clay-coloured weevil
The clay-coloured weevil breeds throughout Britain in deciduous hedges and woodland and in fruit plantations, e.g. currant and raspberry. The adult is 6–8 mm long and coloured brown with lighter markings on the wing-cases, which are often obscured by adhering soil. The adults emerge from hibernation as early as February, feeding first on buds or grafts, later on new growth. They injure young forest trees as well as fruit trees. The grubs feed on the roots of these plants without causing any noticeable injury.

Red-legged weevil
The red-legged weevil occurs locally in chalk or limestone districts in the Midlands and south of England. The adults are about 12 mm long with shiny black bodies and reddish legs. In south-west England the first adults appear in late April and early May during warm weather. They feed on the buds, stalks and leaves of their food plants, injuring fruit plants in the same way as the clay-coloured weevil. The first eggs are laid in May or June, up to 500 being produced by one female. The grubs, which begin to hatch in June, can seriously injure strawberries and were formerly a major pest of this crop in the Cheddar district; they also attack pot plants under glass.

Fig.42.3 Vine weevil on cyclamen leaf (× 4)

Strawberry root weevils
There are four species that sometimes injure strawberry crops on the lighter soils in the eastern and south-eastern counties of England, but all occur locally elsewhere from the Midlands southwards in woodlands, heaths and hedgerows. *Otiorhynchus ovatus* and *O. rugosostriatus* usually overwinter in the grub stage, the adults emerging in June and July. Eggs are laid mostly in August and September.

Strawberry fruit weevils
There are two species that attack strawberry and raspberry. The adults are 3–4 mm long, shiny and brownish-yellow to black. They emerge from hibernation in the spring and feed on the foliage of strawberry and raspberry. The adults of *Barypeithes araneiformis* (Schrank) also attack the developing fruits. The grubs of both species feed on the roots of various plants.

Hop root weevil
This weevil occurs in hop gardens in the West Midlands and south-eastern England. The adult is 5–9 mm long, black and usually encrusted with soil. It can be distinguished from the clay-coloured weevil by the narrower body and parallel-sided wing-cases. The adults may occur at any time of the year but are most numerous in the late summer and autumn. They are sluggish and hide among the bases of the shoots or in the soil. Eggs are laid in the epidermis of the underground part of the bine or in the rootstock. The grubs tunnel in the underground stems, feed for 9–18 months and pupate in the burrows.

Crop damage and its control

On plants in containers, pots or beds

Damage Almost all plants are liable to attack but infestations are commonly found on azalea, begonia, camellia, *Cissus*, conifers, cyclamen, hydrangea, *Kalanchoe*, pelargonium, primula (including polyanthus), rhododendron and rose.

The adults can be found resting during the day under the 'lip' of pots, beneath containers and in debris, or feeding on the plants at night. Holes eaten out of the leaf margins and petals are a useful indication that weevils are present; the pinnae of ferns may be notched or bitten allowing pieces to fall. Plants damaged by adult weevils are not necessarily hosts of the grubs.

Unless a special search is made an infestation may pass unnoticed for months after the grubs have hatched. By this time the grubs are well grown and are attacking the main feeding roots, causing obvious symptoms such as flagging growth, leaf yellowing or wilting (Fig. 42.1); the plant will then have been damaged beyond recovery. The actual injuries are severed roots, barked rootstocks and excavations in corms and tubers.

Hygiene Before a glasshouse or growing area is re-stocked with plants that are susceptible to attack by vine weevil, it should be thoroughly cleared.

Clean, uninfested areas can be protected by installing sticky glue barriers, which will prevent immigration by adult weevils.

Stocks of plants brought in should be examined carefully for both adults and grubs. Adults are easily shaken off. Small grubs may not be seen at first but should be visible by the time the plants are ready for the first re-potting.

Biological control Plants under protection can be treated with nematodes (*Heterorhabditis* spp.) that are parasitic on insects; several applications may be necessary. This control method would complement the integrated pest management programmes now being used in many nurseries.

Chemical control Incorporation of a recommended persistent insecticide in the growing medium used for filling containers or beds, whether in the open or under glass, is the simplest way of protecting plants from attack.

When plants are infested with grubs they are often beyond recovery, but the attack may be halted by treatment with a recommended insecticide. If watered on, the insecticide does not always penetrate under the crown of the plant. A better method is to pour the liquid into a hole dug sideways through the soil under the crown; this hole can be made with a pencil. With cyclamen, it is particularly important to apply sufficient liquid to reach under the corm.

On field-grown ornamentals

Damage In the light-soil districts where wingless weevils are common, injury by grubs to primulas and various alpine herbs and shrubs may occur and leaf notching by adults is frequent on azaleas and rhododendrons.

Control The soil should be thoroughly cultivated before planting. Young azalea, rhododendron and camellia plants can be protected when planting out by dusting the dibble holes with a recommended insecticide dust.

On strawberry

Damage Strawberries suffer most from attacks by weevil grubs on the roots. The worst injury is associated with the lighter soils and with intensive fruit production, as in parts of Cambridgeshire, Kent, Sussex, Hampshire and the Cheddar district. Runners planted on infested land may be quickly destroyed. The chief species attacking commercial strawberries are the vine weevil and strawberry root weevils.

The injury caused by the adults may be seen from late April onwards. It consists of holes eaten out of the leaf margins (Fig. 42.4); this is rarely serious by itself but may be the only obvious indication of the presence of weevils. Adults of *Barypeithes araneiformis* also bore into the young fruitlets, making small cavities. Adult weevils can be found under the straw or polythene mulch and can be seen by torchlight on fine, warm nights.

Signs of grub injury appear either in the autumn when affected patches of plants die prematurely or in the spring when they wilt, just before flowering, owing to the main

Fig.42.4 Strawberry leaves damaged by red-legged weevil

feeding roots having been cut by nearly fully grown grubs. Affected plants are easily lifted; grubs will be found at the roots and often excavating in the base of the crown.

Cultural control Close rotation and poor isolation suit the weevils. Where weevils are common, growers should consider increasing the isolation between old and new plantings and lengthening the rotation for strawberries.

Ploughing and cultivation of old beds increase the mortality of grubs, while clean ditches discourage the movement of adults into and out of the fields. Burning or trimming off after picking decreases the amount of cover available for weevils and improves the efficiency of insecticide applications.

Chemical control Chemical treatment of the plants is not recommended when an attack of grubs is discovered during flowering or fruiting. Chemical control should be used to eradicate an infestation and prevent it from spreading to adjoining crops. A recommended insecticide should be applied to the crowns and surrounding collar of soil after cropping and preferably after burning or trimming off. Such treatments should be applied on damp soil and lightly irrigated in.

When strawberry fruit weevils are present, damage to the fruits can be prevented by applying a recommended insecticide at the end of flowering.

Fig.42.5 Raspberry lateral damaged by clay-coloured weevil

On other fruit

Damage The important injury to tree, bush and cane fruit is that caused by the adult, usually the clay-coloured weevil (see page 212), in young plantations, especially those immediately following grubbed fruit. The worst injury occurs when numerous clay-coloured weevils emerge early in the year and feeding is concentrated on buds, grafts and young growth. In well-established plantations the injury is rarely noticed.

On fruit trees growth of newly planted and grafted trees can be seriously impaired following barking of the graft and young stems; buds are also destroyed and leaves bitten.

On black currant young leaf stalks are bitten and notched, causing leaves to wilt. In addition to this injury by adult weevils, grubs damage the main roots and can cause severe stunting of the bushes.

On raspberry injury begins on the young canes and fruiting laterals in late March or April, i.e. before flowering. The rind of the canes is gnawed and later the fruiting laterals are bitten so that they fracture and wilt (Fig. 42.5). Grubs occasionally infest the roots.

Control On nurseries and in orchards, young trees and grafted stock can be protected from the adults by grease-banding as for winter moth (see page 156). It is also possible to trap the weevils in rolled pieces of sacking or corrugated paper placed on the ground beneath and near infested plants. The traps should be examined every few days and the weevils shaken out and destroyed.

On hops

Damage The main damage to hops is by grubs of the hop root weevil, which weaken the bine and reduce the vigour of the whole plant. Adults of the clay-coloured weevil sometimes gnaw extensive areas of the bark of the bine and also eat holes in the foliage. In addition, they may nip out terminal buds of the young bines, resulting in the growth of laterals.

Control Control measures under conditions of no cultivation have not been studied.

43
Wireworms

Fig.43.1 Base of wheat plant showing frayed area caused by wireworm feeding

Wireworms are some of the best known soil-inhabiting crop pests; they occur in all kinds of soil and attack many farm and garden crops.

Wireworms are the larvae of a group of beetles called click beetles, so named from the audible click made as they flick themselves into the air to right themselves after falling on their backs. About 60 species of click beetle are known in Britain, but only a few of these can be classed as pests. The majority of attacks are caused by one or more of the three common species of *Agriotes* (*Agriotes lineatus* (L.),

Agriotes obscurus (L.) and *Agriotes sputator* (L.)). Their larvae are abundant in pasture and frequently attack arable crops. Damage may also be caused by a species of *Athous* (*Athous haemorrhoidalis* (Fabricius)), which is widely distributed, and by species of *Ctenicera* which occur in some upland areas. For practical purposes the pest species can be considered together.

Description and life history

Adult click beetles (Fig. 43.2) may be found in grass or running about on the soil surface from April to July. Adults of the pest species are mostly very dark brown or black, but certain species of *Ctenicera* have brilliant metallic hues. The female beetles lay their eggs in May and June, singly or in small clusters just below the soil surface. They prefer grassy or weedy ground for egg laying: in bare soil the eggs are more likely to dry up and die.

After about a month the young wireworms hatch; at first they are transparent white and only about 1.3 mm long. As they grow, their colour darkens to the typical shiny golden brown (see Fig. 43.3) of fully grown wireworms, which are up to 25 mm in length. Wireworms have a distinct, dark brown head, powerful biting mouthparts, a cylindrical body that is firm with a tough skin, and three pairs of short walking legs just behind the head. They grow very slowly, spending four or five years in the soil before they are mature.

Young wireworms feed on both living and dead plant material in the soil, but older ones feed mostly on the underground parts of living plants. Individuals feed for two or three short periods each year, feeding being most intense from March to May and during September and

Fig.43.2 Click beetle (× 6)

Fig.43.3 Wireworms (× 3)

October. Wireworms usually reach maturity in July and August, when they burrow deeper into the soil and hollow out small cells 10–25 cm down. Three to four weeks later they become adults. These generally hibernate in the cells but, if disturbed in the autumn, they move upwards and hibernate elsewhere.

Damage to crops

Wireworms will attack most crops; the seriousness of the damage depends on the species of plant, its stage of growth and vigour when attacked, and the plant density, i.e. the number of plants per hectare.

Permanent grass is the natural home of wireworms and supports the heaviest infestations, though there is considerable variation from field to field. Grassland rarely appears to be damaged, except occasionally when newly sown, but when infested pasture is broken up the subsequent arable crops are especially prone to injury, the risk of failure being greatest in the first two years after grass. For some time after grass has been ploughed, the wireworms feed on the dying turf. In these circumstances crop damage may be delayed until the second season after ploughing. Populations decrease quickly under arable cultivation and it is unusual for large numbers of wireworms to persist for more than three or four years after grassland is ploughed. However, in certain soils, particularly clays over chalk, wireworms may persist in arable rotations for many years.

Fig.43.4 Potato tubers damaged by wireworms

The return of arable land to either ley or permanent grass does not inevitably mean a rapid increase in wireworm numbers. This appears unlikely to happen unless a moderate number of wireworms are already present; even then it is likely to be at least five years before the increase becomes obvious.

On all crops, except potatoes, the most serious injury is caused to young plants, usually in spring and early summer, as this is probably the main feeding period.

With the widespread use of protective seed treatments and broadcast applications of persistent insecticides from the mid-1950s, the incidence of wireworm damage decreased greatly. In recent years, however, wireworm damage has become more noticeable, although still at a relatively low level. It is not known if this reflects a slow recovery of populations, following the reduced usage of persistent insecticides, or the lesser use of protective seed treatments.

Cereals

Both autumn- and spring-sown crops of oats and wheat are more susceptible to attack than barley or rye. Maize is likely to be damaged and this can be important because of the low plant density of this crop.

A common sign of damage is the presence of small numbers of adjacent plants wilting and dying. When these are lifted, the stems are found to be chewed and frayed just above the old seed (Fig. 43.1). A wireworm will usually be found at the next normal or slightly wilted plant. Symptoms of attack are similar to those of other pests damaging cereal shoots at or below soil level, so that for correct diagnosis of wireworm attack it is necessary to find numerous wireworms in the soil. The effect of an attack on the crop is most serious on fluffy, unconsolidated, light soils subject to drought or when there is other crop stress such as disease.

Potatoes

Wireworms tunnel deeply into the tubers leaving small round holes on the surface (Fig. 43.4). Although this damage does not affect yield, it causes a serious loss in quality and provides access for slugs, millepedes and other soil organisms. Slugs and cutworms also attack healthy potatoes but, unlike wireworms, they often produce enlarged cavities within the tuber (see pages 471 and 113, respectively).

Wireworms also attack seed potatoes, but this seldom affects the growth of the plant and does not necessarily lead to serious damage to the tuber crop. Indeed, examination of seed potatoes after planting gives little indication of wireworm density or subsequent damage to ware potatoes.

Early crops are little affected but, since wireworm damage increases with time, later lifting results in more injury. If serious damage is expected, crops must be checked for signs of injury in September and consideration given to lifting earlier than usual.

Root crops

Young sugar beet, fodder beet and mangel plants are particularly susceptible to wireworm attack. Injury to beet seedlings shows as small wounds that soon blacken on the stem below ground level. The wound is small but is usually enough to make the seedling wilt and ultimately die. Fortunately, seeds of the beet group can carry sufficient insecticidal dressing to protect the young plants effectively against moderate attack, but this is probably not so with small seeds such as those of turnip. Where seedlings show signs of appreciable damage, thinning should be delayed. With the advent of precision drilling and the consequent small seed rates, good protection is more important than formerly.

Injury to fully grown roots takes the form of small, blackened pits where feeding has occurred below ground level. This injury occurs during the period of autumn feeding and does little harm because the wireworms do not burrow into the flesh of the root as they do into potato tubers.

Horticultural crops

Wireworms can be serious pests on horticultural land and will attack brassicas, dwarf French beans, lettuce, onions, strawberries, tomatoes, anemones and other crops. On tomato the mode of attack is unusual because the wireworms bore into the base of the stem and then upwards in the pith.

Resistance to attack

Certain plants seem to be unpalatable to wireworms. The most notable example is linseed (or flax), which can be successfully grown in very heavily infested fields; other crops that appear to resist attack well are peas, field beans, clovers and lucerne.

Natural control

Wireworms are attacked by various birds, particularly rooks, and by soil insects and fungi.

Drought has an adverse effect on the eggs and small wireworms; this may partially explain the fact that numbers are usually smaller on light soils.

Cultural control

A good firm seedbed with adequate manuring is essential to encourage rapid growth of the young crop. Where an attack is expected, seed rates should be increased slightly to compensate for any likely loss of plant. Good consolidation hinders wireworm attack and cereals can, therefore, benefit from post-emergence rolling when soil conditions permit. Early nitrogen top-dressing can also aid crops to overcome the effects of wireworm attack where rates of earlier nitrogen application to the seedbed were restricted.

Continual disturbance of the soil decreases wireworm populations, as is shown by the decrease in numbers following the ploughing of old grass and by the smaller numbers found in arable land.

By ploughing in February or March and fallowing, a much greater decrease in wireworm numbers is obtained than by ploughing later, say in July. (The wireworms move down into the moister soil during dry periods and may be missed by cultivations.) Good inversion of crop and weed residues is essential and, where possible, pre-ploughing incorporation of these in shallow soil, using disc or rotary cultivators, is advisable. On light land, especially under dry conditions, cultivations to give a firm tilth without loss of moisture are a necessary pre-planting requirement where wireworm attack is likely.

Where wheat bulb fly is a menace and the next crop is winter wheat, a summer fallow may not be advisable because bare fallows are favourite egg-laying sites for the flies (see page 316). By contrast, in areas of the country where leatherjackets are a problem (see page 274), grass cover in August and September is an attractive egg-laying site for crane flies. Ploughing in mid-August may be a suitable compromise in districts where both pests are prevalent.

If grassland is to be ploughed in autumn for winter wheat or oats, there is a risk of both wireworm and frit fly damage to the young crop. A frit fly attack (see page 258) will be most likely if the grass includes much ryegrass and if ploughing is near to sowing time; on the other hand, wireworm damage will be less likely if the land is ploughed then. Where there is a risk of attack by both pests, the best solution is to plough early (at least five weeks before sowing), to allow time for the frit fly larvae to die before the new crop is established, and to use an insecticide against the wireworms. A long interval between ploughing and drilling will also reduce the risk of infection by barley yellow dwarf virus transmitted by aphids migrating from the buried grass (see page 39).

Chemical control

Recommended insecticides should be applied either to the seed before sowing, to protect the seedling plants from attack, or directly to the soil in larger quantities to destroy large populations of wireworms.

Seed treatment

The cost of seed treatment is small, but it only protects the plants during the earlier stages of growth and from moderate attack. A population of 1¼ million per hectare is about the upper limit for seed treatment. Except when applied to a spring-sown crop, seed treatment has little effect on the wireworm population; therefore a subsequent potato crop may require broadcast treatment.

All sugar beet seed supplied under contract is treated with a fungicide and an insecticide. Cereal seed is normally treated with a fungicide; some products may include an insecticide to control wireworms.

Treating seed requires special care and equipment and is a service readily provided by seed merchants. If growers wish to treat their own seed, they should seek advice on methods of seed treatment.

Soil treatment

Soil treatment is generally used to clear land of heavy infestations before susceptible crops are sown or planted, or where seed treatment cannot be used. Once a general soil treatment has been applied for the control of wireworms, it need not normally be repeated until the field has been under grass for several years.

There is some evidence that potatoes are best protected by applying an insecticide to the soil and working it in thoroughly the season before the potatoes are planted, e.g. before sowing the preceding cereal crop. This gives the insecticide time to become thoroughly mixed with the soil by the cultivations on the previous crop and to exert its greatest effect on the wireworm population. Alternatively, the insecticide can be applied immediately before ploughing in the autumn preceding the planting of potatoes.

Where wireworm populations are large, sugar beet or maize may be protected by incorporating insecticide granules in the seed furrow. Such treatment will also control a range of other soil pests in sugar beet and frit fly in maize.

Persistent soil insecticides should not be applied without careful consideration. Wherever possible, seed treatments should be used instead of broadcast applications.

Attacks in progress

Where attacks are noticed at an early stage, the recovery of the crop may be aided by an early top-dressing of nitrogen. Application of an insecticide to the growing crop cannot be expected to stop further damage because the chemical will not reach the pest in the soil.

Spring oats or wheat should not be sown after the failure of a winter crop due to wireworms, unless the soil has first been treated. Alternatively, one of the more resistant crops can be sown.

Control under glass

The risk of wireworms in glasshouses is greatest with new or mobile houses. For these it is advisable to treat the soil during the year before the first crop is planted. Wireworms are killed by steam sterilization or fumigation of soil.

FLIES

Diptera

44
Apple and pear midges

Fig.44.1 Pear shoots with leaves damaged by larvae of pear leaf midge

Four species of gall midge are known as pests of apple or pear in Britain, but nowadays they occur infrequently in commercial orchards. Three of them are probably more important as pests of nursery trees than of fruiting plantations. Damage is caused by the feeding of the larvae; the part of the tree attacked varies with the species.

The pear leaf midge (*Dasineura pyri* (Bouché)) attacks the young foliage of pear, causing the edges of the leaves to remain rolled upwards and inwards towards the midrib (Fig. 44.2). All varieties can be infested but some appear to be more susceptible than others. Heavy infestations may cause some shortening of extension shoots; this is more important in young trees and nursery stock than in mature trees.

The apple leaf midge (*Dasineura mali* (Kieffer)) causes similar damage to apple. All varieties can be attacked though they differ in susceptiblity.

The pear midge (*Contarinia pyrivora* (Riley)) attacks the fruit of pear. However, as this pest appears mainly during the middle part of the blossom period, early and late flowering varieties usually escape damage.

The red bud borer (*Resseliella oculiperda* (Rübsaamen)) is an important pest of nursery trees as the larvae can cause the death of buds on budded stocks. Apple, pear, plum, peach, apricot and rose can be affected by this species.

Pear leaf midge and apple leaf midge

Nature of damage

Pear leaf midge The margins of very young pear leaves are normally rolled inwards, the two rolls meeting along the midrib. If the leaves become infested with a few larvae of pear leaf midge, they open out as they grow but their margins remain partially rolled inwards (Fig. 44.1). If larvae are numerous the leaves may fail to expand at all, remaining tightly rolled (Fig. 44.2). In the early stages of infestation an affected leaf may be distinguished from an unaffected one by a slight, irregular puffiness or 'lumpiness' of the rolled portion; later the rolled portion may become partly or entirely reddened. The affected leaf is more brittle than a healthy leaf and may be down-curved so that in side view it is sickle-shaped. The reddened areas eventually turn black; severely affected leaves which remain closely rolled may become entirely blackened and drop from the tree.

Early in the season infested leaves occur only at the tips of shoots. As the shoot extends, the young leaves at its tip may in turn be attacked by later generations of midges, so that affected leaves may be found at several levels along the shoot.

Apple leaf midge This midge produces similar symptoms on young apple leaves. Water shoots in the middle of the tree are especially liable to attack.

Description and life history

Pear leaf midge The adults are dark brown flies, 1.5–2 mm in length. Because they are so small and also short-lived (1–3 days) they are difficult to observe. There are three or four generations, with three or four flights of midges, per year. The times of appearance of each generation overlap and vary considerably from one season to another. The adults of the first flight usually begin to appear in late April, though the time varies from mid-April to early May, and the flight lasts until late May or even late June.

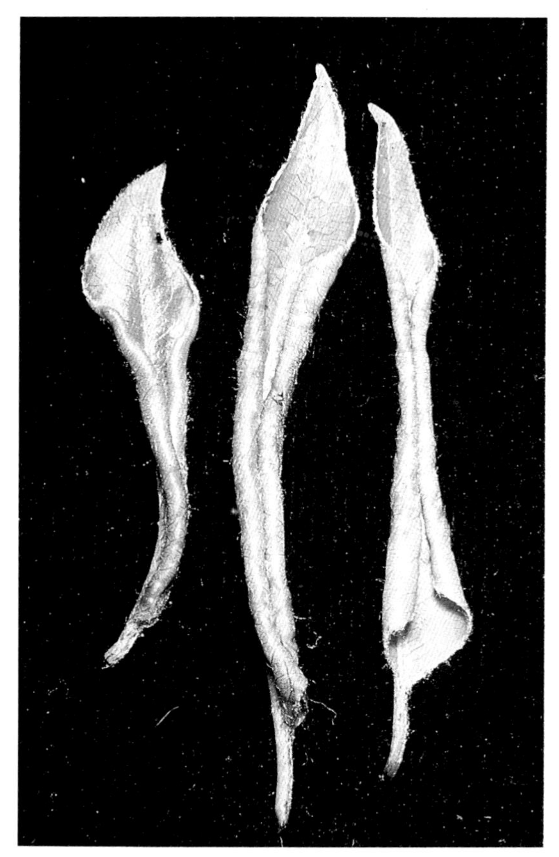

Fig.44.2 Upper surface view of young pear leaves infested with larvae of pear leaf midge

The midges lay their eggs, which are red, within the rolled margins of the young leaves as soon as these emerge from the bud. Usually several eggs are laid per leaf; up to 35 have been found. The larvae, which are whitish, hatch after about four days and their feeding on the leaf surface prevents the blade from unrolling as it grows. Some of the larvae eventually drop to the soil to make cocoons in which to pupate, while others pupate in the rolled leaves.

The whole life cycle normally takes 25–30 days, except that the larvae of the last generation remain in the soil during the winter and pupate in the following spring. Some of the larvae of previous generations may also remain in the soil and not pupate until the spring; the extent to which this happens varies from year to year and influences the size of the first flight of midges. The number of generations during the year is probably partly determined by the length of the season in which new shoot growth occurs.

Apple leaf midge The adult is dark brownish with, in the female, a reddish abdomen. Like the pear leaf midge, it is 1.5–2 mm long and lives for a half to three days. There appear to be three overlapping generations in a year.

The eggs, which are reddish, are laid within the rolled margins of young apple leaves and hatch in 3–5 days. The larvae are creamy-white, turning orange or orange-red as they grow older. When fully fed, after about 20 days, they drop to the ground to pupate. Occasionally some larvae pupate in the rolled leaf or under loose bark.

The complete life cycle takes about 35–40 days, except that the larvae of the last generation remain in cocoons in the soil until the following spring.

Fig.44.3 Pear shoot showing (left) normal fruitlets and (right) fruitlets infested with larvae of pear midge

Pear midge

Nature of damage

Infested fruitlets can be recognized a few days after fruit set because they are larger and rounder than sound fruitlets of the same age (Fig. 44.3). Sometimes they are misshapen though this is not necessarily due to pear midge. Where the pest is present, small yellowish-white larvae will be found inside the fruit (Fig. 44.4). Infested fruitlets start dropping when they are about the size of marbles. Newly fallen ones are often disfigured by black patches which crack open; inside them the larvae lie in a blackened masss of pulp and excreta. Later the dead fruitlets become entirely black and are hollow inside: these are

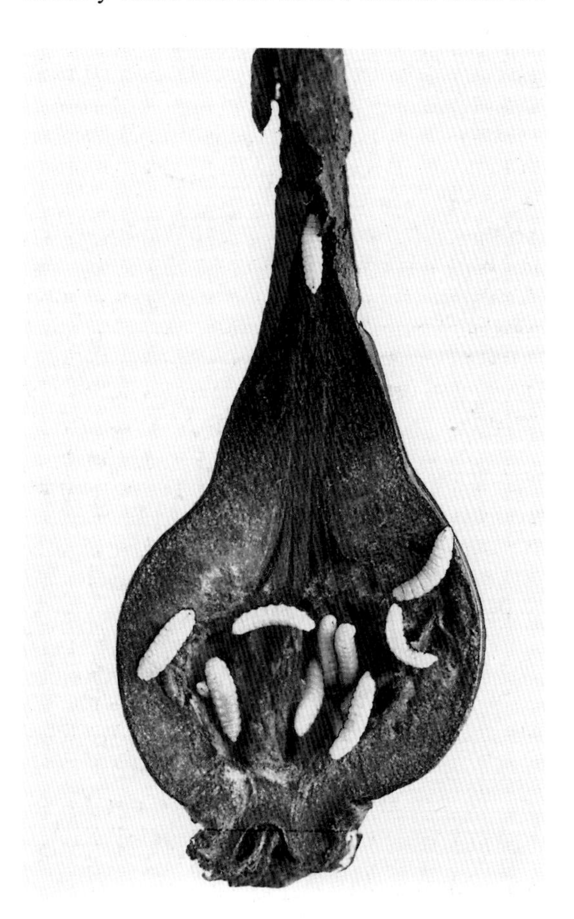

Fig.44.4 Pear fruitlet cut open to show larvae of pear midge inside (× 4)

recognizable signs of an attack even after the larvae have gone.

Description and life history

Pear midge adults are also very small (2–3 mm long) and blackish, though in young females the abdomen is a deep, honey-yellow colour. There is only one generation each year and the adults are on the wing from before white bud until the open-flower stage of pear.

The tiny, white eggs are laid by the female in batches of 10–30 within the unopened blossoms. They hatch in 4–6 days and the larvae make their way into the newly set fruitlets and start feeding on the young tissues, causing abnormal development.

The larvae are fully grown in 5–6 weeks. Just before this stage is reached, the fruits usually fall to the ground and the larvae escape from the rotting tissue. In those fruitlets which temporarily remain on the tree, cracks develop through which the larvae escape and fall to the ground. The larvae are able to make a jumping movement which helps them to escape from the fruit. On reaching the soil they burrow to a depth of 4–8 cm, sometimes a little deeper. Each larva then spins a delicate, silken cocoon about 2.5 mm long in which it hibernates, either as it is or after changing into a pupa. Development is completed in the following spring.

Red bud borer

When attacked by this species the buds on budded stocks in the nursery die or fail to grow properly. Larvae ranging in colour from pink to red are found feeding between the cambium layers of the bud and the stock.

There are three generations a year, with flights of midges occurring from late May to late June, from early July to early August, and from late August to early September.

The females lay eggs in fresh wounds in the bark, taking advantage of cuts made during budding, and the larvae feed on the cambium layers. Most damage is caused by larvae of the third generation in the latter part of August. As many as 18 have been found under one bud. When fully grown the larvae drop to the ground to pupate in the soil at a depth of about 5 cm, but the larvae of the third generation hibernate in the soil before pupating. The whole life cycle of each summer generation takes about 40–50 days.

Control measures

Pear leaf midge and apple leaf midge

In orchards In established orchards, unless infestations are severe, damage by apple or pear leaf midge is unlikely to be of economic significance. Spraying pears with broad-spectrum insecticides in summer should be avoided if possible because of toxicity to anthocorid bugs. These predatory bugs naturally regulate populations of pear sucker (see page 14), the main pest of pear.

If typical signs of leaf midge attack are discovered, the pest can, if necessary, be controlled by an insecticide spray applied during the summer. But where an infestation is known to exist, it is best to spray in the spring against the first generation. Ideally the spray should be applied after egg laying by emerging midges has ended but before the oldest larvae begin to drop from the leaves for pupation. In practice this is not easy to achieve unless leaf samples are examined regularly. It is probably best to spray as soon as leaf symptoms are found, i.e. rolled leaf margins showing the typically irregular puffiness that indicates the presence of larvae. The time at which this occurs varies from about mid-May to early June on pear and is probably similar on apple. If necessary, further sprays may be applied, timing them by the appearance of damaged leaves caused by larvae of later generations.

In gardens a recommended insecticide may be used as above. On small trees infestations can be reduced by picking off and destroying affected leaves before the larvae have left them.

Pear midge

Thorough spraying for one or two seasons should eliminate pear midge. Annual sprays should not be needed as a routine but, when infestations are known to be present, a spray is recommended at white bud.

In gardens infestations can also be reduced by picking and destroying infested fruitlets before the larvae have left them. Thorough and repeated cultivation of the soil beneath the trees will destroy many of the cocoons.

Red bud borer

Control measures are designed to prevent egg laying because damage occurs as soon as larvae start feeding under the buds; control is not then possible.

With chip buds the tie gives adequate protection, providing it completely covers the bud and shield and it extends slightly above and below the stock cut. If budding tape is used, it is necessary to ensure a good overlap each time it is wrapped round the stock. Where the older method of shield budding is practised, infestation can be prevented by smearing a film of petroleum jelly over the tie on the side of the stock containing the T-cut. It is important to apply the petroleum jelly *immediately after tying* as the midges will sometimes lay eggs very soon after the bud has been inserted.

45
Bean seed flies

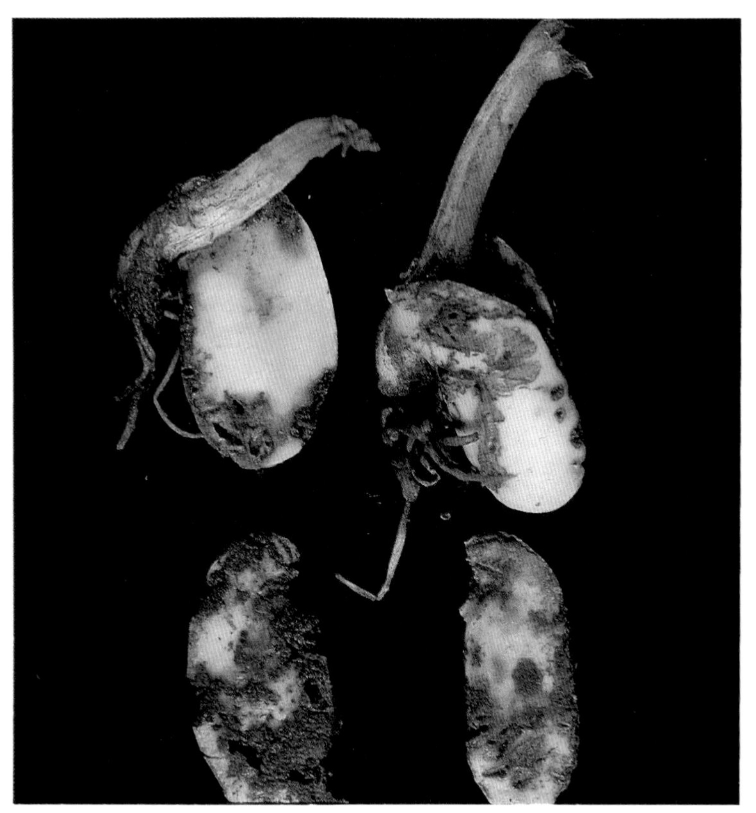

Fig.45.1 Two runner bean seeds cut open to show surface feeding and tunnelling by maggots of bean seed fly

Two species of bean seed fly (*Delia platura* (Meigen) and *Delia florilega* (Zetterstedt)) occur in Britain. Their maggots can cause severe damage to dwarf French and runner beans (*Phaseolus* species) (Fig. 45.1), onions, particularly those sown in late summer, and field-grown cucurbits. Other crops affected include broad bean, pea, cauliflower and other brassicas, lettuce, spinach, radish, beet, cereals (maize, rye, oats and wheat), clover, freesia and gladiolus. Onion is also attacked by maggots of the closely related onion fly (see page 294). Most species of *Delia*, including the bean seed flies, are general scavengers on organic matter in the soil; they are often numerous in the debris of cereal crops.

Nature of damage

Although the flies are common and widespread, damage is localized and sporadic; it is often worst in market garden soils which have a high content of organic matter. Damage is caused by the maggots mining in cotyledons (Fig. 45.1), stems, underground shoots and occasionally the roots of plants. The spotted millepede (see page 379) sometimes extends bean seed fly damage, especially in germinating runner beans.

In an infested field of runner beans, weak seedlings emerge with irregular holes in the leaves and a damaged growing point. With dwarf French beans the damaged cotyledons are carried above soil level and then wither and die. This damage arrests growth, allows secondary organisms to enter the young seedlings and sometimes results in death of the plants. When runner or dwarf French beans are very severely attacked, the growing point may be lost; seedlings then emerge in a twisted condition known as 'snake head' (Fig. 45.2). These distorted plants are unable to develop further and soon die.

On autumn-sown onions, damage is usually worst where a lot of organic matter or plant debris, often from the previous crop, was ploughed in immediately before drilling. Damage to onions may appear merely as poor emergence, since the maggots, unlike those of onion fly, often attack seedlings between germination and emergence. Plants are often killed at the 'loop' or 'crook' stage. Damage to emerged onions is indistinguishable from that caused by onion fly: plants wilt suddenly and collapse. Where very small plants are attacked, the maggots move from plant to plant and discrete lengths of row may be killed.

The most serious damage to cucurbits occurs on plants raised in peat blocks, soon after planting out. The plants collapse completely, often within seven days of planting out. Later attacks result in the wilting of plants during dry weather; the plants may die, often because of secondary organisms.

Description and life history

The adult is a small, greyish-brown fly about 5 mm long; expert examination is necessary to confirm identification. The eggs are brilliant white and elongate, measuring about 0.3 × 1 mm. The white legless larvae (maggots) are 6–8 mm long when fully grown; they have an indistinct head, curved mouth-hooks and there are 5–8 processes on the anterior spiracles, which are fan-like.

The adults lay eggs in the soil during May, preferring freshly disturbed soil. Hatching occurs after about 2–4 days and the maggots enter a seed or the cotyledon, hypocotyl or stem of a seedling. Here they feed for one to three weeks before pupating in the soil. The pupal stage lasts 12–21 days, after which the new generation of flies emerges. The life cycle from egg to adult lasts 3–5 weeks depending on soil temperature, weather and availability of food. First-generation adult females may continue to lay eggs for up to six weeks. There may be three or four overlapping generations during the summer.

It is usually only the first generation arising from overwintered pupae which causes economic damage to beans and cucurbits. Later generations feed on a wide range of host plants or on dead organic matter.

Control measures

Cultural control

Sowing beans when rapid germination is probable reduces the likelihood of a severe attack by bean seed flies at the critical stage when the cotyledons are swelling and before they appear

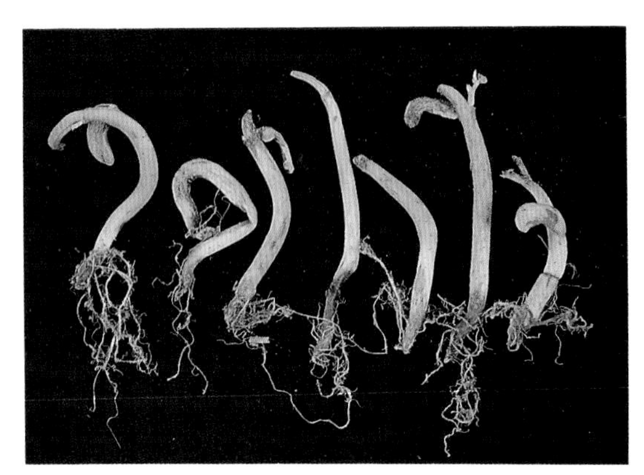

Fig.45.2 Dwarf French bean seedlings showing 'snake-head' effect following attack by maggots of bean seed fly

above soil level. Rapid germination is encouraged by early ploughing, adequate moisture, good firm friable seedbeds, freedom from compaction and proper burial of decayed organic debris. The sowing of seed into stale seedbeds may also help to prevent severe attacks by this pest. Once an attack has developed there are no effective cultural control measures.

Chemical control

On beans, chemical control is ineffective once an attack has started. Therefore, where damage is expected, control measures need to be applied as a precaution. Seed treatment with a recommended insecticide usually gives good control of the pest on beans and onions. On cucurbits, seed treatment is unlikely to be effective when plants are raised under protection and subsequently planted out. In these circumstances, a pre-planting application of a recommended insecticidal drench is likely to be the most successful control measure.

46
Beet leaf miner

Fig.46.1 Sugar beet plant with leaves damaged by feeding of beet leaf miner. After mining in cotyledons, maggots moved into leaves on left causing blister mines. Leaf on right shows blister mine developing from narrow mine made by very young maggots

The beet leaf miner is the maggot of the mangold (or mangel) fly (*Pegomya hyoscyami* (Panzer)) and is a well-known pest of mangel, sugar beet and spinach in northern Europe. The adult is a true fly and should not be confused with the mangold flea beetle (see page 174), which is also sometimes called 'fly'.

Damage is caused by the maggots, which mine extensively within the leaves and produce characteristic blister mines (Fig. 46.1). The extent of damage is most serious if the plants are small, since a few maggots can lead to complete defoliation, especially if weather and soil conditions prevent the rapid growth that compensates for loss of leaf area. The very low seed rates now being used, i.e. 'planting to a stand' with monogerm seed, lead to greater numbers of eggs per seedling and greater risk of damage should an infestation occur.

Attacks by beet leaf miner used to vary considerably from year to year, possibly because of control by natural enemies (see page 235), and were more prevalent in the north of England than in the south. Nowadays, despite wide seed-spacing, damage is less common generally; this is probably because of control measures applied against soil-inhabiting pests (see page 235).

Nature of damage

The discoloured patches caused by the maggots feeding within the cotyledons and leaves become brown and withered, giving a scorched appearance (Fig. 46.1). The growth of small plants is retarded and, when the attack is severe and the weather unfavourable for rapid growth, they may be killed outright. Fortunately, both mangels and sugar beet have remarkable powers of withstanding attack: plants may recover completely, even from a severe infestation, provided the central shoots survive.

Fig.46.2 Mangold fly (× 5)

Complete defoliation (artificially, using scissors or a knife) in different months caused the following loss of sugar yield at harvest in mid-November.

Month of defoliation	% loss of yield
mid-May	10
mid-June	20
mid-July	30
mid-August	45
mid-September	30
mid-October	15

Defoliation by beet leaf miner is most likely to be extensive, and occasionally even complete, in May and June. Later in the season, mining is usually confined to the outer leaves and only a very small proportion of the total leaf area is affected; yield decrease is thus unlikely.

Description and life history

The flies (Fig. 46.2) emerge in April and the females lay their eggs from the end of April until early June, normally on the lower surface of the leaves in batches of two or three but sometimes up to 20. In East Anglia the greatest number of eggs per plant is reached at about the end of May. The eggs are pure white and elongate, being about 1 mm long (Fig. 46.3).

Creamy-white larvae (maggots), which are legless and lack a distinct head (Fig. 46.4), hatch in 4–5 days and immediately bore into the leaf tissue, where their mining produces pale blisters on the leaves (Fig. 46.1). The maggots are fully fed in 10–15 days. They are then about 8 mm long and vary from green to cream. After dropping from the leaves and burrowing about 5 cm down in the soil, they turn into brown puparia from which the adults emerge 14–20 days later.

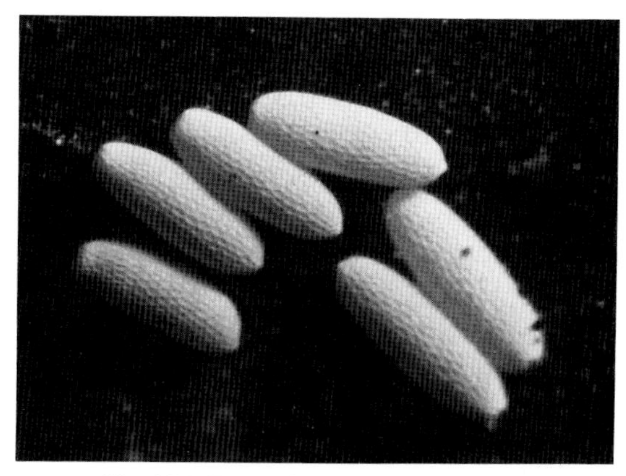

Fig.46.3 Eggs of mangold fly (× 25)

During the year there are two or three generations, which tend to overlap as the season progresses. The winter is passed in the pupal stage in the soil.

Natural enemies

Numerous parasitic insects have been reared in England from eggs, maggots and puparia of the mangold fly. These parasites decrease the numbers of the pest considerably. Hot, dry weather favours the parasites and prevents the rapid increase of the pest.

Various predators and fungus diseases also attack all stages of the mangold fly.

Control measures

Cultural control

Yield loss is best avoided by early drilling on soil adequately fertilized and in good tilth; this provides plants that are large and vigorous at the time when the first generation of this pest attacks. The crop rarely suffers once it has passed the eight-leaf stage. If a severe attack develops on seedlings that are to be singled, singling should be delayed slightly.

Chemical control

An attack can be controlled effectively with an insecticide spray. However, it is difficult to define the level of attack required to make

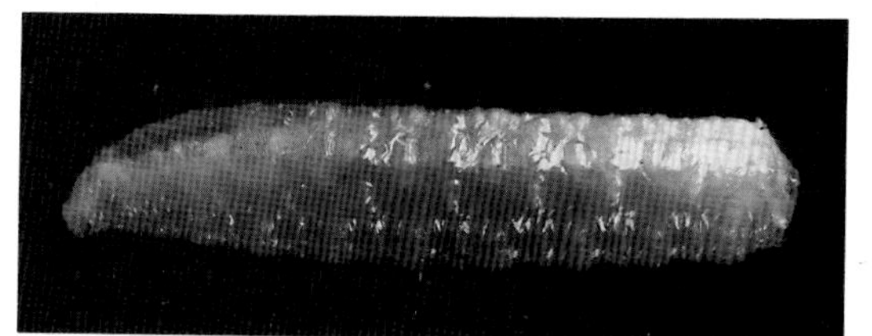

Fig.46.4 Beet leaf miner (maggot of mangold fly) (× 8)

spraying economic because so much depends on the size and vigour of the crop and on prevailing weather conditions. As an approximate guide, spraying is worth while when the number of fresh eggs plus living maggots per plant exceeds the square of the number of rough leaves. Thus, plants with four rough leaves need more than a total of 16 eggs plus maggots to be economically worth spraying. Such a heavy infestation is rare nowadays.

A recommended systemic insecticide should be applied at high volume when egg laying seems complete and the first mines are appearing. In years when aphids invade the crop early and need to be controlled, the two pests can be controlled simultaneously by the addition of a recommended aphicide.

Granular pesticides applied at sowing to protect the crop against soil-inhabiting pests will control mining of first-generation beet leaf miner.

Late attacks

Attacks by the later generations of beet leaf miner during July or September are never severe. It is preferable to leave these attacks untreated, so that the natural enemies of the pest can multiply and increase the natural control.

On spinach

Complete control of beet leaf miner on spinach is essential to satisfy the pre-pack and processing markets. Control measures must be applied at the very first sign of damage in order to avoid blistering of the leaves. Crops sown either early or very late in a continuity programme are likely to miss the main period of pest activity and, in most seasons, should not require treatment.

There are no recommendations for the use of seed-furrow granular treatments to control leaf miner on spinach. A spray of a recommended systemic insecticide can be applied at the same rate as for sugar beet. Care must be taken to observe the pre-harvest intervals.

47
Cabbage root fly

Fig.47.1 Cauliflower plants attacked by maggots of cabbage root fly

The cabbage root fly (*Delia radicum* (L.)) is a destructive pest of cruciferous crops throughout Europe and North America. In Britain it is most serious on cauliflowers and cabbage, particularly summer and autumn cauliflowers and summer cabbage, but it also damages Brussels sprouts, calabrese, Chinese cabbage, radish, swede, turnip, garden stocks and wallflower. The pest can also live on many cruciferous weeds.

Damage is caused by the maggots. Plants can be attacked at any stage of growth, but the most serious damage is usually done to young plants, which can be killed if severely attacked in the seedbed or shortly after transplanting. Heaviest attacks by maggots derived from the first generation of flies occur during late April and May. Brassica plant beds and transplants may be attacked in July by maggots of the second generation. Winter cauliflower and spring cabbage rarely suffer economic damage in the field, but attacks to Brussels sprout buttons in the autumn may lead to rejection by processors.

The turnip root fly (*Delia floralis* (Fallén)), a closely related species, occurs in some years in

Fig.47.2 Left: undamaged cabbage plant with fibrous roots removed to show long tap root. Right: plant attacked by maggots of cabbage root fly, showing tap root mostly destroyed and new roots growing from base of stem

the north of England and in Scotland. Maggots of this species cause damage within the bulbs of turnip and swede from late September onwards.

Nature of damage

Plants attacked in the seedling stage, or soon after transplanting, often wilt and die (Fig. 47.1). Pulling up a dying or dead plant usually reveals a badly damaged tap root, devoid of any fibrous roots. Subsequently, the damaged roots rot or wither; if the plant, though small and stunted, remains alive, some new roots may eventually grow from the base of the stem (Fig. 47.2). Less severe attacks will stunt plants to varying degrees and, for example, reduce the size of cauliflower curds. The outer (older) leaves of attacked plants tend to turn bluish or reddish and to wilt readily. More mature plants, or vigorously growing crops, can support quite large populations of maggots without showing obvious signs of attack, but yield may be reduced.

The roots, bulb or crown of kale, white turnip, swede turnip and radish are also frequently damaged by this pest. Even slight superficial damage (Fig. 47.4) which may not affect yield will impair the quality of crops for human consumption, especially those for prepackaging and processing. Early damage to the developing tap root of young fodder swedes in May and June can reduce their yield even without damage symptoms being visible, but only when mature plants are severely damaged later in the season is the yield of crops reduced for stockfeed.

The pest sometimes lays a small proportion of its eggs in or on parts of the plant above the soil, such as the folds of the enveloping leaflets of Brussels sprout buttons, particularly those towards the bottom of the stems. The maggots then burrow into the sprout (Fig. 47.3) and

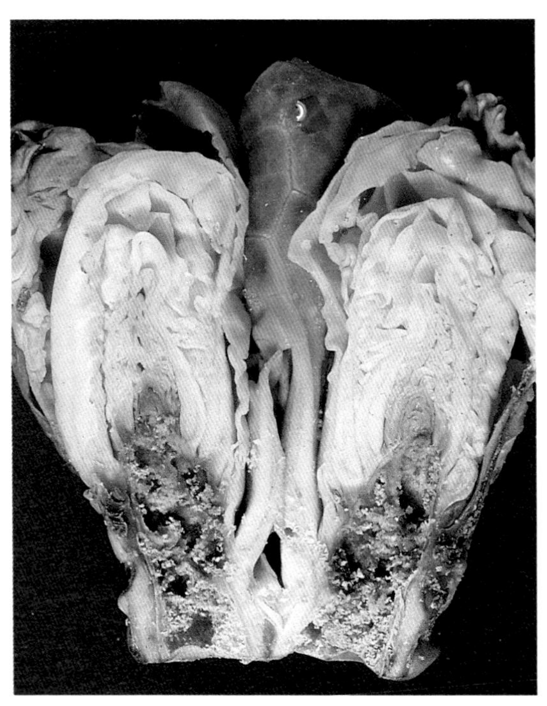

Fig.47.3 Brussels sprout cut open to show damage by maggots of cabbage root fly

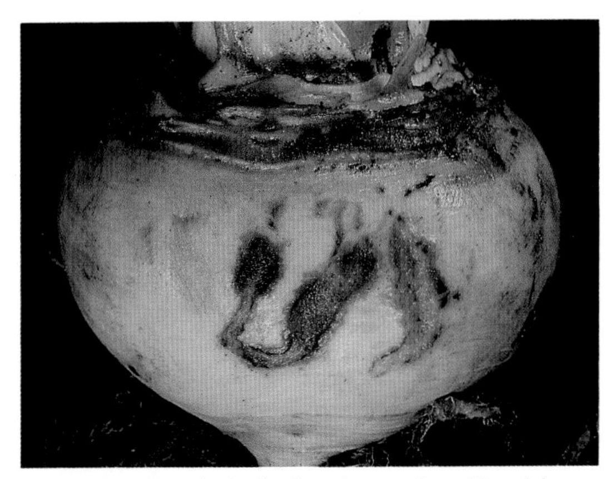

Fig.47.4 Turnip bulb showing surface blemishes and feeding tracks of maggots of cabbage root fly

remain undetectable until they leave to pupate in the soil three to four weeks later. Some maggots can thus be harvested with the sprouts; in crops grown for quick freezing, this is a grave problem even when only a few per cent of the buttons are affected.

Occasionally the growing points of swede or turnip are attacked, resulting in a multiple crown (Fig. 47.5); this symptom is sometimes called 'many neck' or 'many head'. Similar damage occurs in Chinese cabbage, as well as a scarring of the stems which is usually not noticed until the heads are cut open.

Description and life history

The adult (Fig. 47.6) is a dark grey fly, about 6 mm long and similar in general appearance to the house fly.

There are always two generations of the fly each year and a third generation in certain areas in some years. The first generation of flies emerges during the second half of April or early in May. The second generation usually appears in late June and July, and the third generation from about mid-August onwards. Because the later generations usually overlap, flies are found in the major brassica-growing areas almost continuously during July, August and September.

The females generally lay eggs on or just

Fig.47.5 Swede with multiple crown following attack by maggots of cabbage root fly

below the soil surface in cracks or between soil particles close to the main stem of a host plant. They may also lay a few eggs on the foliage or other parts of the plant such as Brussels sprout buttons. The first eggs can be found in late April or early May in the Midlands and southern England, coinciding with the opening of the first flowers of cow parsley (*Anthriscus sylvestris*), and about mid-May in northern England and in Scotland. The best conditions for egg laying seem to occur in most, but not all, years in the spring, so that infestations arising from the first generation of flies tend to be more concentrated and severe than those occurring later in the year.

The eggs, which are visible without magnification, are white, oval, finely ribbed and about 1 mm long. Hatching occurs after three to seven days depending on the temperature.

The larvae (Fig. 47.7) are known as maggots; they are legless, lack a distinct head and are white or cream-coloured. They feed on the root tissues, mostly staying close to the tap root. Occasionally a few are found in the stems, growing points and midribs of the leaves as well as in sprout buttons. After about three weeks, when about 8 mm long and fully grown, the maggot leaves the plant and moves several

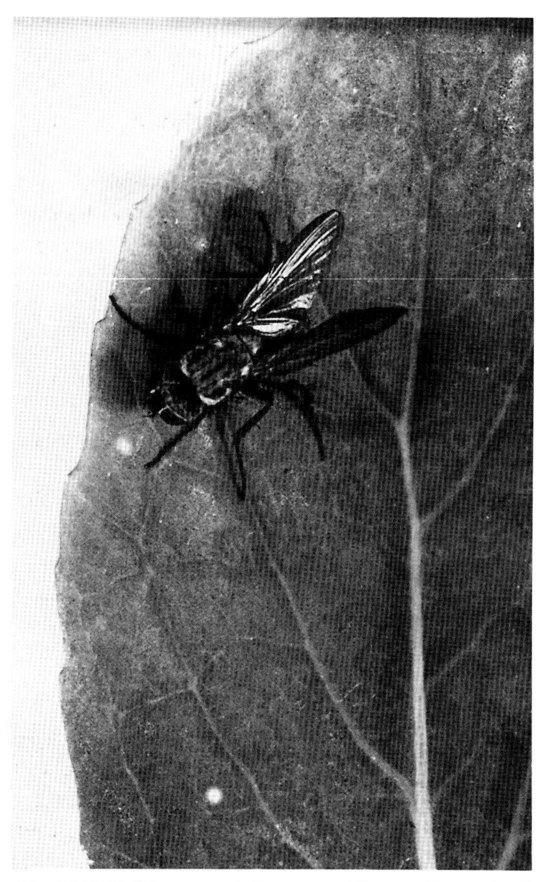

Fig.47.6 Cabbage root fly on seedling cabbage leaf (× 4)

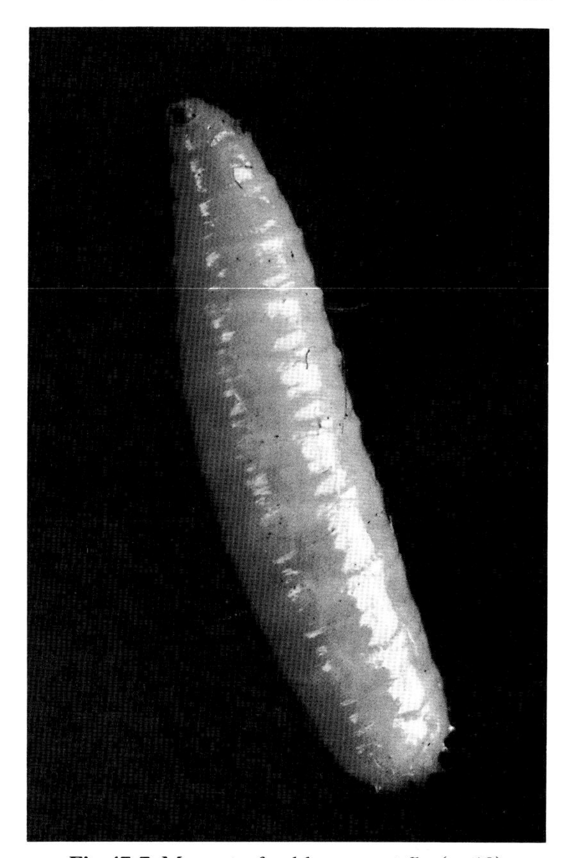

Fig.47.7 Maggot of cabbage root fly (× 12)

centimetres through the soil before changing into a pupa inside a reddish-brown pupal case (puparium), from which the fly eventually emerges. Pupae from the later generations overwinter in the soil and give rise to flies in the following spring.

Cabbage leaf miner (*Phytomyza rufipes* Meigen). The adult of this species is a small grey fly, about 4 mm long. Eggs are laid on cruciferous plants and the young maggots tunnel into the stems of seedlings or into the midribs and leaf petioles of older plants. The maggots are sometimes mistaken for those of cabbage root fly as both pests are often found attacking the above-ground parts of the same plant, e.g. Chinese cabbage. The smaller size of the maggots and pupae, and the absence of papillae at the hind end of the maggots, distinguish leaf miner from root fly.

Damage by cabbage leaf miner is seldom serious and does not justify control measures (see also page 187).

Natural enemies

All the immature stages of the cabbage root fly have natural enemies in the soil. Certain small beetles whose diet includes root fly eggs, maggots or pupae are known to be important in the natural control of the pest. In field experiments more than 90% of all root fly eggs laid in a brassica crop were eaten by beetles, of which *Bembidion lampros* (Herbst) and *Trechus quadristriatus* (Schrank) were important. The root flies surviving even this high degree of predation were, however, still sufficiently numerous to cause economic damage. The pupae are also destroyed by predatory staphylinid beetles, larvae of the parasitic staphylinid beetles *Aleochara* spp., and the larva of the parasitic cynipid wasp *Trybliographa rapae* (Westwood).

Effect of insecticides

Injudicious use of insecticides in the soil can cause heavy losses of these beneficial insects. Then, if the insecticide fails to give a high level of root fly control, the pest may survive in larger numbers than normal and so cause more damage to the crop than would have occurred if no treatment had been applied. Some insecticides tend to be more selective than others, only affecting beetle populations when used at very high rates. The effects of insecticides on populations of natural enemies of the cabbage root fly can be kept to a minimum by applying chemicals to the soil close to the plants rather than broadcasting them over the soil surface.

Control on field crops

Cultural control

Some cruciferous weeds, such as wild radish, are alternative hosts for the cabbage root fly and their presence acts as a reservoir for the pest. Hedgerows and tall weeds also provide shelter for the flies. Hedges should, therefore, be kept well trimmed and headlands not allowed to become weedy.

Crops planted or sown out of phase with the expected egg-laying periods of the two or three generations of female flies that emerge each year will be less liable to attack at critical stages of growth.

Low-level crop covers used to encourage earlier crop maturity can also reduce cabbage root fly damage by excluding flies and thereby minimizing egg laying.The time of removal of crop covers in relation to cabbage root fly activity and crop maturity may influence the effectiveness of this method of control.

Chemical control

Control measures must be aimed at preventing damage because little can be done to stop an attack once the maggots are down at the plant roots. Consequently, insecticides have to be placed accurately around the base of the plants, or with the seed at sowing, to kill either the eggs or, more usually, the young maggots as they leave the eggs. Most serious damage results from flies emerging in late April and May, so all sowings and plantings made from mid-April should receive routine treatment with a recommended insecticide. For earlier sowings and plantings, treatment should be delayed until late April.

Strains of cabbage root fly resistant to the organochlorine insecticides are widespread, so the effective control of this pest now depends on the use of chemical insecticides belonging to other groups.

On leaf brassicas

Seedbeds Treatment is necessary only where plants are to remain in the seedbed after the end of April. It will protect the plants only while in the bed; transplants need further treatment after planting out. Seedlings raised under glass or in frames will also need protection when root flies are active.

Before drilling, overall applications of certain insecticide granules can be incorporated to a depth of 25 mm. Alternatively, overall sprays of some insecticides can be used or granules can be applied by the bow-wave method (see below) when drilling the seedbed.

Band treatment Bands of granular formulations of some insecticides are effective when used on leaf brassicas. The insecticide may be applied (i) at drilling (bow-wave method), (ii) subsurface at transplanting, or (iii) over the rows after seedling emergence or after transplanting. These treatments are highly effective when applied correctly but, for maximum efficiency, treatment (iii) must be applied before the maggots hatch from eggs laid in the crop.

Granules applied to the soil surface well before egg laying may be less effective owing to loss of persistence. Also, when applied during very dry spells, granular treatments may remain inactive and result in poor control.

To ensure accuracy of treatment, the rotors and pulleys on the applicators must be correctly set and the calibrations checked periodically. Special rotors are needed for granules which are particularly abrasive.

Drilled crops Where sowings are made before mid-April, treatment should be delayed until late April, then a 7.5–15 cm wide band of granules applied over the rows. Alternatively, individual plants can be spot treated with granules using a hand applicator. Granules must be directed accurately so that they fall around the stems of the plants.

As an alternative to these granular treatments, spot drenches can be applied. The liquid should be placed accurately on the soil to flow around the stem, avoiding spillage which may scorch the foliage.

For crops sown after mid-April, it is advisable to apply a band of granules at drilling, using the bow-wave technique. The seed is sown through a 7.5–15 cm wide band of granules placed on the soil in front of the seed coulter. To ensure accuracy, bow-wave applications should be made as a single, combined operation at drilling and not as two separate operations. As an alternative, a band of granules can be applied along the rows after emergence when the plants have reached the two true-leaf stage. The granules should be applied when the crop is dry and, if possible, directed on to the row from the side.

Transplanted crops For crops planted before mid-April, treatment should be delayed until late April, then a surface band of granules, a spot treatment of granules or a spot drench applied as for drilled crops.

For crops transplanted mechanically after mid-April, consistently good control of root fly can be achieved by planting through a sub-surface band of granules. This is achieved by mounting a separate hollow coulter (Leeds Coulter) 30–40 cm in front of, and directly in line with, the planter share to deliver the granules about 5 cm below the soil surface. Feeding the granules directly into the planting furrow can result in poor control, with some risk of phytotoxicity in light soils during dry conditions. Where facilities are not available for applying granules sub-surface or where crops are hand planted, a surface band of granules, a spot treatment of granules or a spot drench should be applied within two days of planting out.

Combined cabbage root fly and aphid control Dual-purpose granules are available for the combined control of cabbage root fly and cabbage aphid on leaf brassicas. The granules are recommended for application at sowing or planting by the methods already described.

Incorporation of insecticide in peat blocks and loose-filled cells Good control of root fly can be achieved by thoroughly mixing certain granular insecticides into the peat compost before making blocks in which brassica plants are raised. For optimum protection the interval between insecticide incorporation and planting the treated blocks should be kept to a minimum. The layer of untreated soil over the planted blocks should be no more than 2.5 cm. This method is not satisfactory for loose-filled cells as the volume of peat used in these is much less than that used in blocks.

Blocks and cell-raised transplants can be treated by drenching with a recommended insecticide immediately before planting. Care must be taken to avoid uneven application and phytotoxic reactions.

Spraying Brussels sprout buttons Where attacks by late-generation root fly frequently result in maggots entering and feeding inside the sprout buttons (Fig. 47.3), spray applications of a recommended insecticide can be made at appropriate intervals before the expected date of harvest. A good spray cover, particularly on the lower buttons, is essential, requiring the use of inter-row pendant lance equipment.

On root brassicas

Bands of granular formulations of certain insecticides can be used on swede and turnip at drilling. Alternatively, a band spray of a recommended insecticide can be applied along the rows in front of the drill coulter at sowing. This treatment is not recommended for use after seedling emergence because of the risk of phytotoxicity.

Although a band treatment at drilling controls early root damage, it does not control crown damage satisfactorily or persist long enough to protect the crop against late root fly attack which results in damage to the bulb (Fig. 47.4). Damage to late-harvested roots can seriously affect the quality and marketability of crops grown for human consumption. To minimize this late-season damage, mid-season sprays of a recommended insecticide can be applied. The first application, in July or August, should be based on information on root fly activity. This can be measured in the field by trapping flies or extracting eggs from the soil. A mathematical model also exists which can predict peaks of activity.

Control in gardens and allotments

Cultural control

More labour-intensive methods of plant protection which are not possible in large-scale systems of production may often be feasible in gardens. Black felt discs, 15 cm in diameter, are available to fit around the stems of brassica plants after transplanting. These discs deter egg laying by cabbage root flies, provide shelter for predatory beetles and conserve soil moisture. The stress imposed by root fly maggots on all brassica plants is increased considerably in dry soil conditions, especially with cauliflowers. Frequent watering of plants showing symptoms of attack (wilting etc.) helps them survive root damage.

The sowing or planting of crops out of phase with the expected times of egg laying by the flies will also minimize damage, as will raising plants in pots or blocks rather than in seedbeds; such plants can be transplanted without much disturbance of the roots and they are more tolerant of damage. Gardens should be kept free from cruciferous weeds to eliminate additional host plants.

Chemical control

The insecticides available for use in gardens are generally not as efficient as those approved for farm use, but when used according to the manufacturer's recommendations they give some protection to young plants. Certain fungicides applied for the control of club root may also decrease cabbage root fly damage.

48
Carrot fly

Fig.48.1 Carrots damaged by maggots of carrot fly

The carrot fly (*Psila rosae* (Fabricius)) is widespread in Britain and its maggots frequently cause serious damage to carrot crops both on farms and in gardens. In some districts parsnips, celery and parsley are seriously damaged. Many members of the carrot family are known to be hosts of carrot fly.

Nature of damage

On carrot

Carrot is the most important crop damaged by this pest. The maggots feed on the root systems of young and mature plants. In early June, the first generation of maggots feeds mainly on the side roots near the base of the tap root; this is often the only obvious damage showing on roots in early summer. (In Scotland this damage is usually seen in late June to early July.) The foliage of severely attacked plants becomes reddish, wilts and dies, particularly in dry weather. Plants less severely attacked may become stunted, and damage to the tap root can cause 'fanging'. Carrot seedlings can be killed by early carrot fly attack; this contributes to the 'natural thinning' process which tends to reduce plant stand progressively during the season. If many seedlings are killed by carrot fly, the plant stand may become 'gappy' (Fig. 48.2) and the carrots will ultimately be very variable in size. Surviving roots can become extensively mined by the maggots.

During August and early September, many of the more superficial mines caused by the first generation of maggots will heal and little damage will be apparent at this time. Damage by the second generation of maggots becomes progressively worse during the autumn, leaving the brown and rusty-looking tunnels characteristic of this pest (Fig. 48.1) and from which the creamy-white maggots may sometimes be seen protruding.

Damage to carrot plants may also be caused by maggots of another fly, namely *Napomyza carotae* Spencer, which was first observed in England at Cambridge in 1974. The adults of this species cause characteristic feeding punctures and lay their eggs in the leaves. When the maggots hatch from the eggs they mine downwards through the leaf stalks into the roots. By late summer they have formed shallow tunnels near the crown of the carrots; these tunnels become open galleries as the roots continue to grow.

The foliage symptoms caused by carrot fly attack are similar to those occurring on carrots infested with willow–carrot aphid (see page 73) or infected with motley dwarf virus complex.

On parsnip

Carrot fly damage on parsnips is very similar to that on carrots. The maggots feed in the upper 15 cm of the soil so parsnips are damaged around and just below the shoulder of the root. The first generation of maggots may kill small parsnip plants by tunnelling through the main root and can mine extensively in the leaf petioles.

On celery

Where celery is grown intensively, carrot fly can be a serious pest. The maggots bore into the roots, crown, and bases of the leaf stalks (Fig. 48.3), causing a yellowing of the leaves

Fig.48.2 Carrot crop damaged by first-generation maggots of carrot fly

Fig.48.3 Celery cut open to show damage by maggots of carrot fly

and poor, uneven growth.

In the Fens of eastern England, most of the damage is done to small plants soon after they have been transplanted in the field, seriously reducing the size and quality of the celery. The earliest planted celery is covered with mulch and is little damaged, whereas the first planting that is unprotected suffers the most damage. Because the numbers of first-generation flies are usually decreasing by early June, crops transplanted after the first week of June do not usually require insecticide treatment, except after a very late spring. The second generation of carrot fly rarely causes economic damage to celery in eastern England.

In Lancashire the first flies appear later in the season than in eastern England and they are usually present in large numbers in June and most of July. Damage by the first generation of maggots is often more severe than in eastern England and crop failure more frequent. Flies of the second generation appear in August and September, but in most years in Lancashire they are not as numerous as those of the first generation.

Self-blanching or dwarf white varieties of celery can be severely affected if they are planted out before early June and are thus attacked by the first generation of the pest. Later varieties that make good growth in the autumn can partly recover if the attack is not severe. In Lancashire all celery crops are likely to be attacked to a greater or lesser degree and should be protected against carrot fly.

On parsley

The maggots mine in the surface of the tap root and also feed on the fibrous lateral roots. Similar damage can be found on the roots of hemlock.

Description and life history

The adult is a small, shining black fly with a reddish-brown head, yellowish legs and iridescent wings (Fig. 48.4). It is about 8 mm long with a wing-span of about 12 mm.

The eggs are white, ovoid, about 1.0 × 0.4 mm and have pronounced longitudinal ribbing. They are laid singly or in small groups in cracks in the soil surface near the host plant. Depending on temperature, they usually hatch in about seven days and the tiny, colourless larvae, known as maggots, burrow into the soil, feeding first on the side rootlets or externally on the young tap root. Later they burrow into the root forming mines; typically, the maggots found in the mines are third stage although a few are second stage. The older maggots are creamy-white (Fig. 48.5) and when fully grown are about 8–10 mm long. After the final (third) moult, they form puparia in the soil close to the tap root. These puparia are yellowish-brown, about 4.5–6.0 mm long and about 1.2 mm broad.

Although there are usually two generations of flies during the year, there is a partial third generation during the autumn in some years in the Fens. Thus flies may be found in some areas continuously from May to October or November. The first generation of flies is most abundant in the Fens during the latter half of May and early June, whereas the second generation of flies appears towards the end of July, being most abundant during August and September. There may be considerable overlap between the second and partial third generations of flies and sometimes there is only a very extended second generation. The dates of peak abundance of the flies vary from year to year and between different parts of the country.

Control on field crops

Cultural control

Cultural methods alone do not give complete protection against carrot fly, but attacks can be reduced by certain cultural practices and, conversely, they are likely to be enhanced by others.

Crop rotation and spacing A damaging level of the pest is quickly built up where susceptible crops (carrots, parsnips, celery or parsley) are grown close to previous crops in short rotations. The numbers of the pest can be limited by widely separating susceptible crops from one year to the next, by growing them less frequently in the rotation and by avoiding growing early and late crops close together. When carrots are first grown on land remote from areas where they have been grown previously, the pest is not likely to be troublesome for a few years, particularly if late and early crops are well separated to break the sequence of host plants in the pest's life cycle. The pest may become more numerous when crops are grown at high densities, even though the relative level of damage to individual carrots is usually less.

Reduction of shelter The flies spend much of their time in shelter round the edges of carrot fields. Ditches, hedges and nettle beds should therefore be kept well trimmed to reduce the shelter available to the flies.

Time of sowing Liability to attack varies with the sowing date of the crop. First and second earlies drilled in the autumn or in January/February are unlikely to be attacked provided the crops are harvested in June or July. In the main carrot-growing areas of eastern England, crops sown from February to mid-May are liable to attack by the first generation of flies. Those early and main crops still in the ground after the end of August are also liable to be damaged by the second generation of maggots. Crops sown after mid-May are unlikely to be severely attacked by the first generation of the pest, except in the north where the adults emerge later in the spring than in south-east and eastern England.

Varietal susceptibility Some carrot varieties are more severely damaged by carrot fly than others, but even the least susceptible varieties are sufficiently damaged by carrot fly maggots to require protection by an insecticide.

Crop covers Experimental work has shown that non-woven covers can be used to prevent carrot fly attack. The covers can be applied at drilling or later, but before the flies become active.

Disposal of infested crops Badly infested crops should be lifted by November and the damaged carrots fed to livestock or disposed of in such a way that the maggots cannot complete their development. Infested carrots should not merely be ploughed-in or dumped nearby.

Chemical control

The need to use insecticides should be considered carefully in the light of local conditions and the past history of attacks. Applications of insecticides may kill predators and/or parasites of the carrot fly as well as, or even instead of, the pest itself, especially if they are applied inefficiently. This may encourage the pest to build up rapidly and make it more difficult to control in future years. Insecticides should not be applied in excess of recommended doses. Increasing the dose does not significantly improve the level of control and causes adverse side-effects.

Yellow sticky traps are used by some growers to monitor numbers of adult carrot flies. These traps have proved useful in determining the optimum timing of the first spray in the mid-season spray programme. A mathematical model of carrot fly development is also available to aid the timing of insecticide treatments.

Phytotoxicity Even under normal conditions, and especially in light sandy soils, insecticides may affect germination or plant survival. Uniformity of application is very important. Most types of granule should not be sown with the seed.

Insecticide residues It is important to adhere to recommended methods and rates of insecticide application, and to check that applications are uniform, so as to ensure that residues in the carrots will be minimal at harvest, especially with crops grown in mineral soils. If the recommended doses are exceeded, insecticide residues may accumulate and reach undesirable levels in some soils; also the development of resistant populations of the pest may be accelerated.

Type of treatment The type of treatment differs for early, maincrop and late sowings. An early crop exposed to attack by only the first generation of flies needs protection for a much shorter period than a maincrop or a later sowing; a crop to be harvested after November will require the longest protection of all. Very early drilled and harvested crops (those drilled during the winter and harvested in June or July) are unlikely to require treatment.

The type of treatment also differs for soils containing high organic matter (peat) and low organic matter (mineral). Soil type affects the efficacy of insecticides, so that some treatments effective on mineral soils give less protection on peat soils.

Where heavy attacks occur, or where a long period of protection is needed, a soil treatment should be applied followed by supplementary sprays in mid-season. Even so, the level of control on peat soils is unlikely to exceed 90% for more than about 16 weeks. Therefore, on these soils, crops harvested after October may still be appreciably damaged.

Insecticide degradation Microbial activity in the soil can accelerate the degradation of pesticides, thus decreasing the efficacy of treatments. This has occurred on some intensive vegetable-growing farms where carbamates have been used regularly every one or two years and has led to a few cases of crop failure. There are also cases of microbial-accelerated degradation of organophosphorus insecticides on some farms in the Fens of eastern England.

Where there is a known problem of accelerated degradation, growers are advised to ring the changes by using pesticides belonging to different chemical groups and to widen their cropping rotations.

Fig.48.4 Carrot fly (× 5)

On carrot

Treatment at sowing In general, insecticides are more effective against carrot fly when they are placed 5–10 cm deep in the soil than when they are left on the soil surface. The method of application determines the depth to which the insecticide will be mixed into the soil and consequently its performance. Deep side-placement of granules at drilling has been shown to be advantageous.

When an insecticide is broadcast over the soil surface, whether as granules or as a spray, it should be mixed into the soil quickly to minimize loss by vaporization. An insecticide will be mixed to a depth of about 2.5 cm by harrowing and to about two-thirds of the depth of soil disturbance by rotary cultivation. Rotary powered harrows have proved to be highly satisfactory for incorporating chemicals in the upper few centimetres of soil. To obtain a firm seedbed after this cultivation, the soil should be reconsolidated by light rolling as necessary.

Some insecticides are most effective when placed as 2.5 cm wide bands under the seed, about 2.5–5 cm below the soil surface. This is not easy to achieve and can create undesirably loose seedbeds. There is also a risk that some insecticides may be phytotoxic when applied in this way.

If the insecticide is to be applied as discrete bands along the seed rows, this is best done as a combined operation when drilling. The insecticide should be delivered on to the soil as a wide band in front of the seed coulter, which them mixes it shallowly into the soil (the bow-wave method). If this is done as two operations, i.e. first distributing the insecticide in bands and then drilling the seed, the seed coulters will not always be travelling along the centres of the bands and the result is likely to be less satisfactory.

Mineral soils. Several granular systemic insecticides are recommended and give good control of carrot fly when applied by the bow-wave method (see above). Granules that are liable to be phytotoxic can also be applied to the soil as it moves into the closing seed furrow; direct contact between granule and seed is thus avoided. Alternatively, some types of granule can be broadcast and worked into the soil before drilling. If there is no risk of phytotoxicity, certain granules can be applied direct to the seed furrow with the seed. Where severe attacks occur, protection is unlikely to be adequate after October. It may be desirable to apply supplementary sprays of a recommended insecticide in mid-season, but taking care to avoid undesirably large residues accumulating in the crop and ensuring that specifications for the type of crop being grown are not contravened.

Fig.48.5 Maggot of carrot fly on damaged carrot (× 5)

The systemic insecticide granules applied by the bow-wave method will help to control early infestations of willow–carrot aphid (see page 76), but one or more sprays of a systemic insecticide to the foliage will usually also be necessary to prevent damage by the aphid and to minimize the spread of virus in the crop.

Some contact insecticides are recommended and will also control carrot fly (but not the aphid). These can be incorporated in the soil as granules before sowing, or they may be drilled under the seed (see cautionary note above).

Peat soils. Not all the granular insecticides recommended for use on mineral soils are effective on peat soils. Therefore, the choice of chemicals recommended for use on these soils is more limited. Also, most of them need to be applied at a higher rate than on mineral soils.

The application of a granular systemic insecticide helps to limit populations of willow–carrot aphid, but one or more sprays of a systemic or translaminar aphicide may be needed in addition for good aphid control.

Contact insecticides are generally less effective than systemics for control of first-generation carrot fly and will not control aphids.

Treatment at sowing time is unlikely to give adequate protection to carrots left in the ground after October in areas where carrot fly attacks are probable.

Application rates. Carrot crops are grown in many different row-spacing arrangements and plant densities. When calculating insecticide delivery rates to suit particular systems, several factors have to be considered. First, the total amount of active ingredient per hectare must not exceed the permitted maximum. Secondly, the concentration of insecticide close to the seed should never exceed that given by the bow-wave method when applying an 8 cm wide band to rows 38 cm apart — the standard system on which recommendations are usually based. Higher concentrations may reduce germination and cause excessive residues to be present in the crop at harvest. Thirdly, insecticide further than about 8 cm from the carrots is relatively ineffective against this pest.

The choice of method and the rate of application will often depend on the row-system being used. Broadcast applications are most suitable for crops sown in close rows. For close-row bed-systems with rows 13 cm or less apart in the beds, a significant saving of insecticide is possible by treating only the effective bed-width and not the whole of the wheelings. (The effective bed-width is the distance between its outermost rows plus 8 cm on either side.) Band applications by the bow-wave method are most suitable for wide-spaced (38 cm) single or scatter-band rows.

Treatments after sowing Crops at risk from the second generation of carrot fly, i.e. lifted after the end of October, may require supplementary mid-season spray treatments. One mid-season spray may be adequate for crops on mineral soils, but two or three may be required where there is a history of damage, e.g. central Norfolk. Two to four sprays may be required on peat soils. The number of applications depends on the lifting date, the history of damage and the rate of application. In Scotland supplementary treatments are rarely required for crops lifted before November; a single spray is adequate for crops lifted later.

There are several insecticides recommended for mid-season control. Each should be applied in at least 1000 litres of water per hectare and, for maximum effect, the spray should be aimed at the carrot crowns rather than just over the foliage.

On parsnip

Experience indicates that it is more difficult to reduce carrot fly damage on parsnips than on carrots. Several systemic insecticides formulated as granules are recommended for carrot fly control on parsnips.

On celery

Control measures on this crop are usually needed only against the first generation of carrot fly maggots.

For celery grown in rows spaced 122 cm apart (35 000 plants/ha), certain granular insecticides have given good control when incorporated in the soil around the seedling roots by means of an applicator mounted in front of a mechanical planter. Alternatively, the granules may be applied as a band before planting out.

Where self-blanching celery is grown in close rows at about 110 000 plants/ha, the application rate should be double that for the wide-row system. The amount applied will be about the maximum permitted per season for each chemical in mineral soils.

An increasing number of celery growers no longer use insecticide granules before planting out. Instead they apply recommended insecticide sprays timed against the first-generation carrot fly maggots.

Control on carrots in gardens and allotments

Cultural control

Certain recommendations given above for cultural control of carrot fly in fields apply also to carrots grown on a small scale.

Late sowing In the southern and eastern counties of England, sowing main crops at the end of May or early in June will minimize damage by first-generation maggots. Seeds should be sown thinly to reduce the need for thinning and so avoid attracting large numbers of flies. Where thinning is necessary, the soil should be reconsolidated along the rows by treading and watering, and the thinnings should be removed from the bed.

Sowing in exposed positions The flies like shelter. Where possible, sowing close to hedges or other sheltered places should be avoided. Sowing next to potatoes should also be avoided and neighbouring weeds, particularly nettles, should be kept down.

Crop covers Non-woven covers can be used to prevent carrot fly attack; these are available commercially. It should be noted that most netting is unsuitable for this purpose as the mesh size is usually too large to keep out carrot flies.

Time of lifting Carrots should not be allowed to remain in the soil throughout the winter, unless an insecticide has been used. They should be lifted and stored during the autumn. Heavily infested crops should also be lifted during the autumn and destroyed or fed to animals: they should not be dug in.

Chemical control

Contact insecticides recommended for carrot fly control and available to home gardeners can be applied *below* the soil surface at sowing. This treatment should be followed by a spray of a recommended insecticide in August or September if the carrots are to be lifted after October.

49
Celery fly

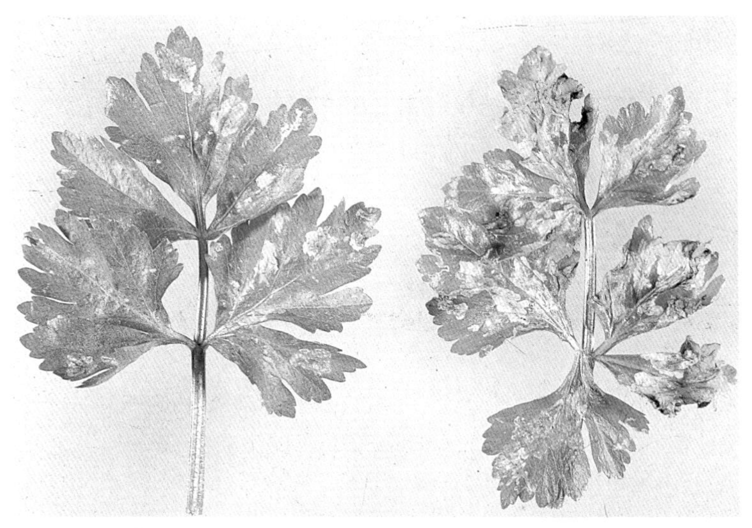

Fig.49.1 Celery leaves with blisters caused by mining of maggots of celery fly

The celery fly (*Euleia heraclei* (L.)) is a well-known pest in the British Isles and on the continent of Europe. The maggot, which is sometimes called the celery leaf miner, mines inside and destroys the leaves of celery and parsnip, thus indirectly affecting the yield and quality of the produce. Severe mining of the leaves of parsley has also been reported, though rarely. Attacks begin in April or early May, and in mild seasons may continue until early December.

Nature of damage

The maggots eat the soft tissues within the leaves, making blotch-like blisters which are pale at first but later turn brown (Fig. 49.1).The leaf soon contracts and then shrivels. When the loss of tissue from celery leaves is extensive, the stems cannot grow and fill out properly: they remain small and green and have a bitter flavour. When parsnips are badly attacked the roots are small.

Fig.49.2 Celery fly (× 7)

Description and life history

The adult (Fig. 49.2) is a fly about 5 mm long with a wing-span of about 10 mm. It is tawny-brown with mottled, iridescent wings and dark metallic-green eyes.

The larva, known as a maggot, is legless, lacks a distinct head and is white or very light green, with the dark line of the gut visible through the skin. The fully grown maggot is about 8 mm long.

The puparium, which contains the pupa, is oval, pale yellow, brittle, much wrinkled and about 3 mm long.

The celery fly has two main generations per year and possibly a third. As the generations overlap considerably, flies, maggots and pupae can occur simultaneously. The first generation of flies appears from April to early June, according to the season, and the second generation from July onwards.

Each female fly may lay about 100 eggs, usually inserting them singly into the leaf tissue but occasionally placing them on the surface. The eggs hatch after 6–14 days and the young maggots burrow within the leaf (Fig. 49.3). After a further 14–19 days the maggots are fully fed and pupate, either in the leaf or in the soil.

In the summer, flies emerge from the puparia after 3–4 weeks but, unless there is a third generation, the pupae of the second generation overwinter and develop into the first generation of flies in the following year.

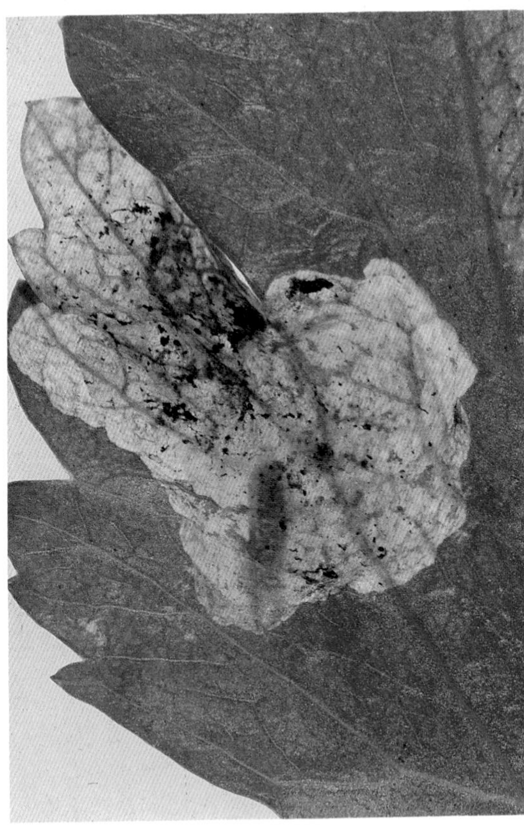

Fig.49.3 Close-up of blister in celery leaf showing a maggot of celery fly inside

Control measures

On field crops

Cultural control Celery fly attacks are more likely when celery, parsnip or parsley is grown near to the site of any of these crops in the previous year. This practice should, anyway, be avoided to reduce the risk of carrot fly attack (see page 244).

Chemical control The plants should be examined frequently for mines during May and June. On celery or parsnip seedlings, if there is more than one mine per five leaves, a spray may be worth while; spraying larger plants will rarely be economic but will prevent the development of a further generation of the fly. Parsley should be sprayed if there are enough mines to spoil the appearance of the marketable produce.

Several insecticides are recommended for control of celery fly and should be applied as sprays. Some granular treatments applied for control of carrot fly keep populations of celery fly at a low level.

In gardens and allotments

Unfortunately, allotment and garden crops cannot usually be isolated from sources of infestation.

Hand-picking Pinching blistered leaves as soon as they are seen will kill the maggots and reduce the next generation of flies. It is useless to postpone hand-picking until the autumn when the main attack has developed.

Burning Immediately after heavily infested celery or parsnips have been dug, the foliage should be collected and burnt. If it is dug in or put on compost heaps not actively fermenting, puparia will probably be formed. These puparia may be carried away with compost for celery or parsnips, or other crops grown nearby, and in due course the flies will emerge.

Spraying Where attacks occur regularly, a recommended insecticide spray may be applied as soon as the damage is seen.

50
Chrysanthemum gall midge

Fig.50.1 Chrysanthemum plant with galls on leaves and stem, and distorted flower, caused by feeding of larvae of chrysanthemum gall midge

The chrysanthemum gall midge (*Rhopalomyia chrysanthemi* (Ahlberg)) was first noticed in 1915 in North America as a serious pest of glasshouse chrysanthemum. The first two outbreaks in England were in 1927 and 1936, and were traced to chrysanthemums imported from the United States. There have been occasional outbreaks — all under glass — since then, but the pest remains uncommon in Britain and has caused little damage since the early 1950s.

Nature of damage

Small but conspicuous cone-shaped galls are produced as a result of the midge larvae feeding within the tissues of the plant. Where the infestation is slight, the galls are found only on the leaves, mainly on the upper surface but also occasionally on the underside. In severe attacks, however, the stems, buds and developing flowers are also affected (Fig. 50.1), resulting in severe distortion and stunting of the plants. There are fewer flowers and they are of poor quality, while little or no growth for cuttings is produced. Galled cuttings do not root readily and make weakly plants.

Description and life history

The adult (Fig. 50.2) is a small, delicate fly, not more than 2.5 mm long, with long legs and antennae. It is brownish, although the female has a red abdomen. Most of the adults emerge from the pupae in the early hours of the morning and mate almost immediately. In warm weather, egg laying is usually completed on the same day, but in colder conditions it may be prolonged for three days. The males die soon after mating and the females shortly after egg laying: the adult life thus occupies only one to five days.

The eggs (Fig. 50.3) are oval, bright red and exceedingly small, being practically invisible without magnification. Each female lays 5–150 eggs in groups in the buds and among the folds of very young leaves and the sepals of the flower buds.

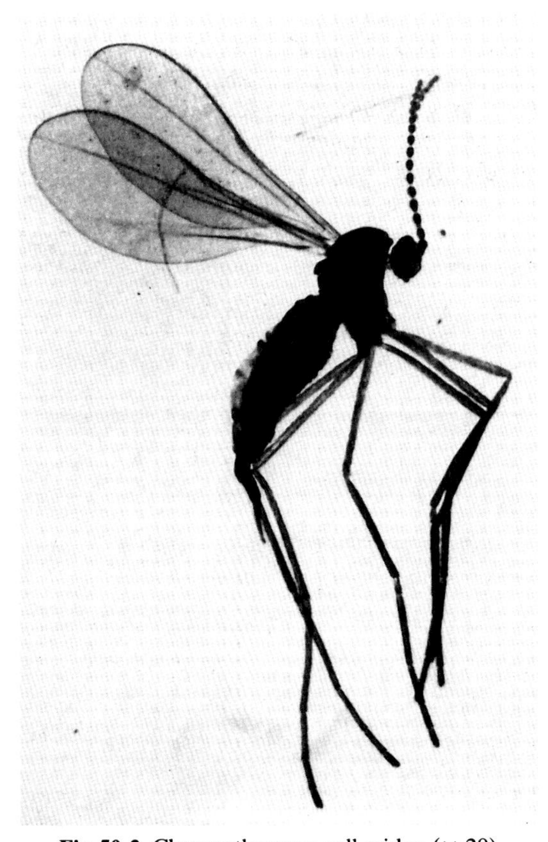

Fig.50.2 Chrysanthemum gall midge (× 20)

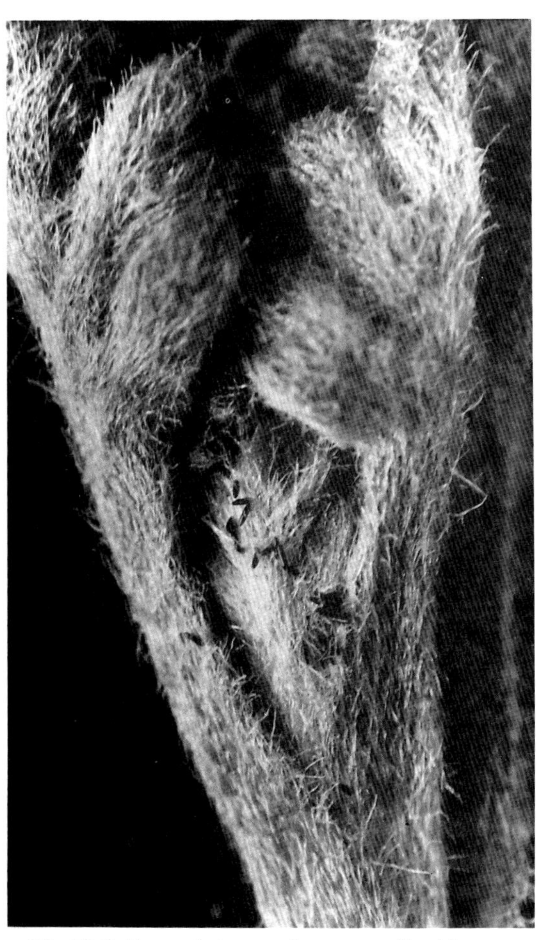

Fig.50.3 Eggs of chrysanthemum gall midge on young shoot (enlarged)

The eggs hatch in 3–14 days, according to the temperature. At first the tiny larvae are almost colourless but later become a distinct reddish-orange. After moving about on the surface for up to a few hours they burrow into the plant tissues on which they feed; a definite gall develops later. In June and July galls become readily visible in 16–25 days, whereas from mid-November to early March they may take as long as 14 weeks to appear.

When fully fed the larvae change to the pupal stage, which lasts from six days to several weeks. At the end of this period the pupa pushes itself half way out of the gall and the adult emerges. The time taken from egg to adult can vary from 26 days in summer to 21 weeks in winter.

Breeding continues throughout the year and, unless it is checked, the infestation will increase very rapidly in late summer, causing a serious loss of marketable blooms and cuttings.

Means of spread

The usual way in which the chrysanthemum gall midge is spread from nursery to nursery is by the distribution of infested stools, young plants and cuttings. However, introduction of this pest into a nursery should not occur if propagating material is brought in from reliable, specialist producers. Material brought in on a small scale should be carefully examined for galls and treated thoroughly to eradicate the pest if any signs of it are found.

Another means of spreading chrysanthemum gall midge used to be the return to the nursery of debris in returnable containers after marketing infested blooms. However, with the almost universal use of non-returnables, this source of infestation is now relatively unimportant.

Control measures

Hygiene is important: all infested material should be removed and burnt.

Hot-water treatment of stools for control of chrysanthemum nematode (see page 407) should help to control the midge.

Chemical control

Outbreaks of the midge can be controlled by spraying with a recommended insecticide as soon as galls are seen. Repeated application will protect uninfested plant material from attack. Three or four applications should be made at weekly intervals.

The use of a granular systemic insecticide for the control of other pests will probably help to control chrysanthemum gall midge.

51
Frit fly

Fig.51.1 Crop of forage maize severely damaged by maggots of frit fly

Frit fly (*Oscinella frit* (L.)) is a widespread species whose maggots cause occasional damage to cereals and certain grasses. In England and Wales there are normally three generations a year. The spring generation is most injurious to late-sown spring oats, ryegrasses and meadow fescue, sweet corn and both grain and forage maize (Fig. 51.1). The summer generation attacks the grain of oats, whereas the autumn generation impedes the proper establishment of ryegrass and meadow fescue in years when large numbers of frit fly are active. Economic damage to autumn-sown oats, wheat, rye and (very occasionally) barley can also occur when these crops are drilled after grass leys or cereal stubbles infested with grass weeds.

Grass and cereal flies sometimes cause damage similar to that caused by frit fly, particularly in winter wheat. However, attacks by the most common of these — the yellow cereal fly (*Opomyza florum* (Fabricius)) — are not noticed until well into the spring, long after frit fly attack has ceased.

Description

The adult (Fig. 51.2) is a small, shiny black fly about 1.5 mm long. It is active in fine weather but remains resting when it is cold and wet.

The larva, which is known as a maggot, is small, white, legless and lacks a distinct head (Fig. 51.4); it is usually found inside a damaged shoot. When fully grown (about 3 mm long) the maggot changes into a puparium (Fig. 51.3); this is a small reddish-brown case containing the pupa, which later turns into the adult.

Life history and crop damage

Spring generation

Maggots overwintering in grass or cereal shoots pupate in March or April and give rise to the spring generation of flies. These emerge in May and are usually abundant towards the end of this month in the south of England, in Wales and in the Midlands, but a little later further north. The date of their appearance may vary from year to year by up to three weeks.

Eggs are laid on, and at the base of, young oat plants, maize and grasses. On hatching, the maggots burrow into the young shoots and destroy the growing points. Very young plants are killed outright. In older plants, death of the main shoot is followed by the growth of several tillers which may be invaded in their turn. Such plants eventually have an abnormally large number of weak shoots which, in oats, produce a poor yield and later-ripening grain.

Maggots of the spring generation are fully grown in the second half of June, when they turn into puparia within the damaged shoot, in a lower leaf sheath or in the soil.

Summer generation

Flies emerging from these puparia during July are the summer generation. The female flies lay eggs on, or just beneath, the oat husks and the maggots feed on the developing kernels, which become shrivelled and blackened. Eggs may also be laid in the ears of wheat but not, apparently, in the ears of other cereals or grasses. The maggots turn into puparia within the husks.

Occasionally, oats are attacked while the ears are still within the sheath; then the stem of the ear is often injured causing blind spikelets ('bells'). This damage may be due either to late-emerging flies of the spring generation or early flies of the summer generation. Other forms of blindness, which may be confused with frit fly damage, occur in oats: one of the most common is 'blast' (blanched spikelets).

In the absence of a suitable cereal host,

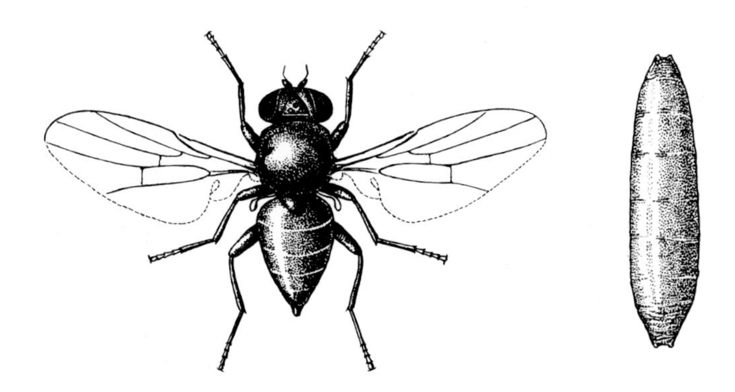

Fig.51.2 Frit fly (× 20)

Fig.51.3 Puparium of frit fly (× 12)

summer generation flies probably breed within the shoots of grasses.

Autumn generation

Adults are active during August and September. Eggs are laid mainly on grasses, especially ryegrasses, but also on volunteer cereals and grass growing in cereal stubbles. Rye sown in August for autumn or spring grazing is also vulnerable and, in a warm autumn, eggs may be laid on other germinating cereal crops.

The maggots feed slowly throughout the autumn and winter. Their feeding on established grasses does little obvious harm, but they may prevent successful establishment of some strains of ryegrass and meadow fescue in young leys, particularly those sown in August. The maggots become fully grown in the spring, when they pupate and give rise to the flies that emerge in May.

When a field of an infested ryegrass ley, or infested self-sown cereal, is ploughed and winter cereal, ryegrass or meadow fescue is sown soon afterwards, the maggots leave the grass or cereal and attack the seedlings, sometimes causing severe thinning.

Fig.51.4 Maggot of frit fly on oat stem (× 20)

Symptoms of attack

On cereals

Most serious shoot damage is seen in young plants up to the four-leaf stage; this is the stage of development in which there are three expanded leaves in addition to the unexpanded central leaf. The centre leaf of an injured plant turns yellow (Fig. 51.5), withers and dies ('deadheart') and then often falls out. 'Deadhearts' seen in wheat or rye from November to mid-February are often wrongly attributed to wheat bulb fly attack (see page 317).

Frit fly injury to young oat plants is sometimes confused with the injury caused by stem nematode (see page 441). Damage by the fly may be recognized by the fact that only the *central* leaf of the injured shoot turns yellow; the rest of the plant remains green (Fig. 51.5) or may assume a rusty brown colour. If the shoot is pulled apart carefully, the little yellowish-white maggot will be found in the centre. In oats attacked by stem nematode, the base of the plant is swollen and the leaves are often twisted.

In a bad frit fly attack in the spring, the

Fig.51.5 Close-up of spring oat plant showing typical damage caused by maggot of frit fly

continual production of tillers gives the crop a characteristic tufted, grassy appearance, and frequently there are numerous gaps caused by the death of plants that have been invaded when very young.

Unless the ears are examined closely, the only indication of the summer generation attack will be light or damaged grain at harvest.

On grasses

'Deadhearts' occur in grass seedlings and tillers in established pasture.

On sweet corn and maize

Severe attacks on sweet corn and maize cause stunting and distortion of the plants. Leaves are ragged and often appear to have been torn into strips; many leaves fail to emerge properly and are badly twisted. Where attack is less severe, there may be little effect on growth, but neat rows of small holes will be seen across the leaves when these expand. These holes are made by the maggots boring through the tightly rolled leaves. Maize usually has more than one maggot per plant.

Natural enemies

Frit fly is subject to attack by various insect and nematode parasites, but it is not known how much these affect the numbers of the pest in different years.

Cultural control

Time of sowing

Spring oats Plants that have developed more than four leaves (including the unexpanded central leaf) are very resistant to attack. Early sowing, provided it can be done on a good tilth, is the best method of avoiding attack (Fig. 51.6). As a general guide, the latest time for sowing oats in the spring so as to avoid frit damage is mid-March in the south of England, the end of March in Wales and mid-April in the north.

Oats sown early in cold soils are specially liable to damage by the seed-borne disease leaf spot (*Pyrenophora avenae*). Most strains of this disease are controlled by seed treatments.

When sown on a poor tilth or in unfavourable weather, oats often grow very slowly and may then be even less forward than oats sown later under better conditions. The aim should be to encourage rapid early growth so that the plants reach the four-leaf stage before mid-May.

Winter cereals If winter cereals are to be sown after a ryegrass ley, the land should be ploughed as early as possible (at least four weeks before sowing) to allow time for the frit fly maggots to die before the new crop is established. Grass-infested stubbles should be ploughed as soon as possible after harvest. However, in some areas the risk of wheat bulb

Fig.51.6 Oats sown on different dates in spring, showing damage by maggots of frit fly

fly atttack will be increased if fields are bare during August. Every effort should be made to prevent the growth of volunteer cereals, especially when winter cereal follows spring oats. When direct drilling or reduced cultivation techniques are adopted, it is necessary to obtain complete kill of volunteer cereals and weed grasses, with a suitable desiccant. This should be done well ahead of sowing winter wheat after oats or leys, or in stubbles infested with grass weeds.

Sweet corn With small-scale cropping of sweet corn, attack can be prevented by getting the plant to the five- or six-leaf stage before the time of attack. This can be done either by sowing early under cloches and removing the cloches in the second half of May or by sowing in large pots or soil blocks (*not* in boxes) in a glasshouse about 1st May and hardening off the plants before planting out at the end of May.

Some varieties of sweet corn mature quickly enough for it to be practicable to sow late. Plants sown out of doors at the end of May are much less likely to be attacked by frit fly than those sown in early or mid-May.

Forage and grain maize Early-sown crops shallowly drilled into well prepared seedbeds often grow away from pest attack. Whenever possible, sheltered fields with southerly aspects should be chosen in order to encourage speedy growth.

Grassland The risk of economic damage is usually greatest when crops have been drilled in August and eggs have been laid directly on the plants. Transfer of maggots from the previous sward can also occur but is not as important as was first thought. The risk of this transfer will be minimized if an interval of at least four weeks is left between sward destruction and drilling. Crops established in the spring are less vulnerable to attack because the plants normally grow rapidly through the susceptible stage before the peak emergence of flies occurs in May. Continued close-grazing of established grass or cutting of the sward is considered to increase the risk of damage from frit fly, as these practices ensure the production of new tillers which are susceptible to attack.

Resistant varieties

All the existing commercial cereal varieties are susceptible to frit fly attack. Ryegrass varieties, however, vary greatly in their susceptibility. Field trials and glasshouse screening tests have shown that some Italian and hybrid varieties are particularly prone to attack, whereas perennial ryegrasses are generally less susceptible.

Chemical control

On oats Current chemical control measures do not compensate fully for early sowing. Damage by the spring generation of frit fly to late-sown spring oats can be reduced by spraying, at crop emergence or at the first sign of damage, with a recommended insecticide.

On winter cereals The risk of damage is not usually great enough to justify a routine preventive treatment in crops sown soon after ploughing grass or grass-infested stubbles, unless regular damage has occurred previously. In these situations, the best control is achieved by spraying the grass, before ploughing, with a recommended insecticide.

Alternatively, crops at risk may be examined soon after full emergence before 'deadhearts' are apparent. A random sample of shoots should be split open and, if more than 10% are damaged, a recommended insecticide spray should be applied immediately. Killing the maggots will not always save shoots that have been attacked, but treatment with insecticide will prevent further invasion from the old grass and movement of maggots from damaged into undamaged shoots.

On forage maize Where conditions are less than optimum, insecticide application is often worth while. This applies particularly in the north of England. Treatment should be applied at drilling or soon after crop emergence; treatment after damage is seen will not prevent yield loss.

Granular insecticides should be applied, strictly according to the manufacturer's recommendations, at drilling or early crop emergence. Alternatively, recommended sprays may be applied at full crop emergence.

On sweet corn and grain maize Sweet corn and grain maize crops are more valuable and are more likely to suffer damage than forage maize as frit fly can seriously interfere with cob formation. Routine insecticide treatment is therefore justified.

On autumn-sown grass leys Sporadic damage occurs in some years, particularly to direct reseeds and in lowland areas. The risk of damage is greatest when seed is sown early and plants are vulnerable to egg laying. Additionally, direct reseeds may be damaged by maggots migrating from the old sward into the newly emerging seedlings. In certain years insecticide treatment will be justified.

52
Gout fly

Fig.52.1 Winter barley plant with swollen shoot indicating attack by maggot of gout fly

Gout fly (*Chlorops pumilionis* (Bjerkander)) is an occasional, minor pest of cereals throughout Britain. The damage is caused by the maggots. Barley, wheat and rye may be attacked but not oats or maize. The pest is also able to live on a wide range of weed grasses, especially couch-grass (*Elymus repens*).

During the 1950s damage by gout fly on barley declined to a negligible level, but in the late 1970s and the 1980s infestations increased slightly on late-sown spring barley and early-sown winter barley and wheat. Since then localized outbreaks have occurred, particularly on September-sown winter barley and winter wheat.

Nature of damage

Spring-sown cereals

The injury differs according to the stage of growth of the plant when attacked. Usually the damage is done during the two or three weeks before the ear appears and is caused by the maggot burrowing down the side of the ear to the first joint. As a result, growth is checked and the plant becomes swollen. Each maggot attacks only one tiller.

If the crop is backward, the attack may occur before the plants have begun to elongate; the attacked tillers then remain short, are gouty and swollen and never produce an ear. When the crop is well forward, the check to the plant is less severe: the ear emerges from the sheath but is poorly developed and the immature grains on one side are spoiled (Fig. 52.2). The yield of affected ears is reduced by about a half.

Fig.52.2 Spring barley ears damaged by maggots of gout fly

Autumn-sown cereals

The usual attack is by maggots which feed in the young plant during the winter and produce the same symptoms as in backward spring barley, i.e. the attacked tillers become swollen and gouty and remain short (Fig. 52.1). The plant may die altogether or other tillers may elongate in spring and ultimately produce ears that are usually weak. Winter cereals appear to recover well from attacks affecting up to 50% of plants.

Fig.52.3 Gout fly on barley leaf (× 16)

Description and life history

The adults are small flies, 4–5 mm long, and yellow with black markings (Fig. 52.3). They appear in May and June and the females lay minute, creamy-white eggs on the leaves close to the central shoots. Generally there is only one egg per shoot.

Hatching occurs after about 10 days and on emergence the larva, which is legless, lacks a distinct head and is known as a maggot, burrows into the centre of the shoot. Here it feeds, working down the ear and the stem, and forming a groove or furrow until it reaches the first joint. After about a month's feeding, the maggot is fully grown; it is then yellowish-white and about 5.0–6.5 mm long (Fig. 52.4). It moves upward along the feeding groove for a short distance and changes into a brownish, somewhat flattened puparium (Fig. 52.5) lying near the base of the leaves encircling the spoilt ear. The pupa forms inside the puparium and the adult emerges in about five weeks.

The flies developing from the summer generation of maggots are present during August and September, but when the weather continues mild they may remain alive and active well into October. Eggs are laid on couch and other weed grasses, volunteer cereals in stubble fields, or on early-sown winter cereals. The young maggots hatching from these eggs pass the winter in the centre of a plant, close to the root. Infested plants become very bulbous or gouty in winter or early spring. The maggots become fully fed and pupate in spring, giving rise to the flies that emerge in May and June.

Natural enemies

Two kinds of parasitic wasp, both ichneumon species, play an important part in checking the increase of the gout fly. Although many apparently sound puparia may be found in the straw or 'cavings' from combine-harvested barley, the majority of these contain one of the parasites. Most of the gout flies will have emerged already and flown away leaving empty puparia.

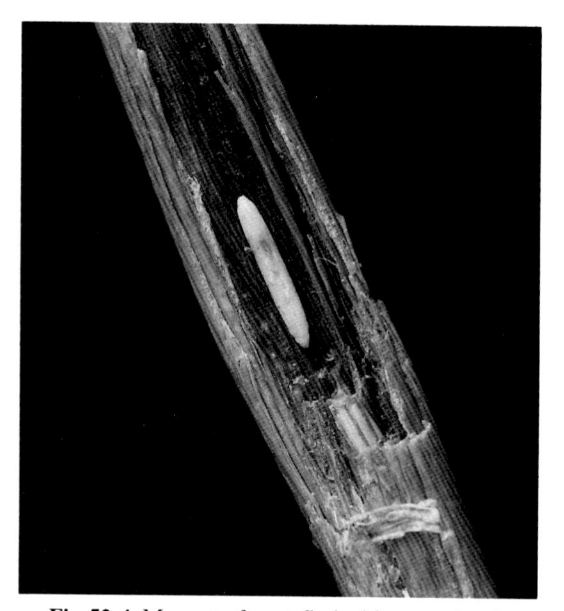

Fig.52.4 Maggot of gout fly inside stem (× 4)

Fig.52.5 Stem opened to show puparium of gout fly (× 2½)

Prevention and control

Control measures directed specifically against gout fly are not usually required. Good husbandry of the cereal crop will do much to prevent damaging attacks. Good weed control also helps to prevent gout fly attacks.

Early sowing of spring cereals
Early sowing is very important as a means of avoiding damage by gout fly, since in a forward crop most of the ears emerge before the maggots have had time to cause injury. Really serious summer attacks are likely only if sowing is late, say in late April, or if the crop is backward because of poor growing conditions. Early sowing is particularly important following exceptionally mild winters, when flies may emerge in late April.

Sowing of winter cereals
If winter cereals are sown very early, usually before mid-September, the plants are at the right stage of growth for egg laying when the autumn generation of flies is present. If sowing is delayed a little, the flies will lay their eggs on grasses and will have died before the crop emerges.

Clean cultivation
Clean cultivation of the previous crop or stubble is important. Crops are specially liable to attack on land which contains couch or other weed grasses or in fields where there is a regrowth of stubble.

Chemical control on winter cereals
In areas where gout fly occurs regularly, winter cereal crops emerging in September should be monitored during October for the presence of gout fly eggs. If eggs are found on more than 50% of plants, some control may be obtained by applying a recommended systemic insecticide immediately after the eggs have hatched, usually in late October.

53
Leaf miners of chrysanthemum

Fig.53.1 Chrysanthemum plant showing mines made by chrysanthemum leaf miner

Leaf miners are often found on chrysanthemums and related plants. Normally, these leaf miners are native insects, belonging to the species known as the chrysanthemum leaf miner (*Phytomyza syngenesiae* (Hardy)), which is a common pest widely distributed throughout the British Isles. However, alien leaf miners are sometimes introduced with imported cuttings and these species are described on pages 271–273.

The chrysanthemum leaf miner has a wide host range and occurs on cultivated plants (e.g. chrysanthemum, cineraria, calendula and lettuce) and on weeds (e.g. sowthistle and groundsel).

Fig.53.2 Chrysanthemum leaf showing bleached spots caused by adults of chrysanthemum leaf miner

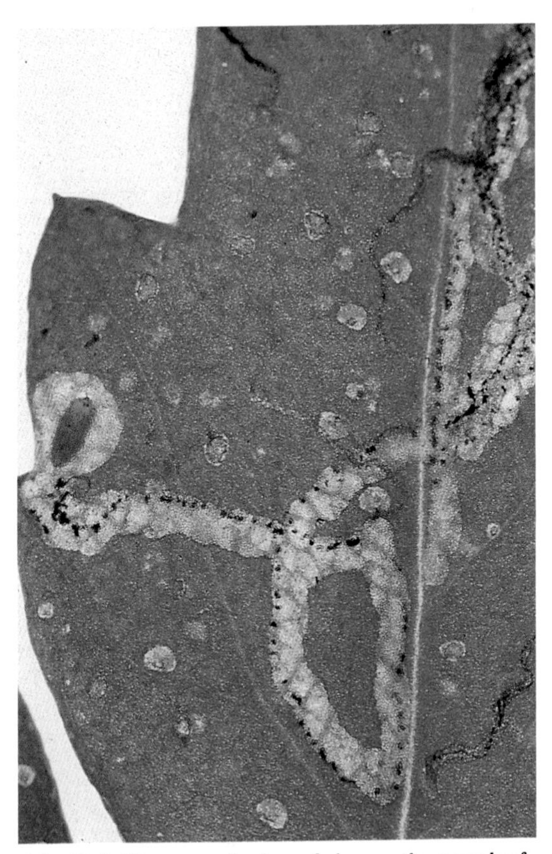

Fig.53.3 Close-up of mine of chrysanthemum leaf miner in chrysanthemum leaf. Note discontinuous frass trail and pupa at end of mine (left)

Nature of damage

The female fly makes feeding and egg-laying punctures in the leaves. These punctures are clearly seen from the upper surface of the leaf as very small bleached spots (Fig. 53.2) which later, in some varieties, develop small wart-like growths. These spots are particularly noticeable on young growth and have been confused with the early stages of white rust (*Puccinia horiana*).

The larvae feed inside the leaf, making narrow, serpentine mines that appear to be whitish when seen from the upper surface (Fig. 53.1). Small, dark, regular blobs inside the mines, usually along one side, are pellets of larval excrement (frass) (Fig. 53.3). Several mines may occur in the same leaf, crossing and recrossing, often giving the effect of a large blotch mine. Severely infested leaves shrivel and die, weakening the growth of the plant.

Although leaf miner will attack all chrysanthemum varieties, some are less susceptible than others.

Larvae of another leaf-mining fly — the chrysanthemum blotch miner (*Trypeta zoe* (Meigen)) — may also infest chrysanthemums, producing large irregular, blister-like mines in the leaves. It is seldom seen on commercial nurseries.

Description and life history

The adult is a small fly, about 2 mm long, with dark brown eyes tinged with red, a greyish-black body and distinctive pale yellow markings on the sides and under-surface (Fig. 53.4). It flies rather slowly making short, hopping flights of about a metre at a time.

The length of the life cycle is greatly influenced by temperature. The adults live for two to three weeks and each female lays about 75 microscopic eggs; these are laid singly in small incisions in the leaves. Relatively few incisions contain eggs: the majority are used as feeding points by the adult flies.

The eggs hatch in 3–6 days and the young larvae begin to feed in the leaf tissue, making the mines shown in Fig. 53.3. When mature the larva is about 3.5 mm long, legless and greenish-white. This stage is reached after feeding for seven to 10 days. Pupation occurs in the mine (Fig. 53.3). The pupa is oval and yellow to dark brown; it can readily be seen when the leaf is held up to the light. The pupal period is very variable, lasting about nine days at 21 °C, 12 days at 15.5 °C and 55 days at 4.5 °C. The adult finally emerges through a circular hole at one end of the pupal case.

Breeding is continuous under glass and infestations may occur at any time on year-round crops. Chrysanthemum cuttings may become infested at an early stage. The first signs of infestation are usually seen on areas of crop near doorways and paths. Outdoor crops can be attacked throughout the summer and reinfestation occurs from wild hosts. The incidence of damage tends to decrease in early autumn.

Fig.53.4 Adult of chrysanthemum leaf miner (× 20)

Control measures

On glasshouse crops, infestations may develop from several sources, including infested cuttings, flies and pupae surviving from previous crops, and flies entering the houses in spring and summer.

Hygiene is an essential aid to the control of this pest: all dead or unwanted plant material, such as debris from de-budding, should be bagged up and burnt. Weeds such as sowthistle are alternative hosts and should be eliminated from the glasshouse.

Monitoring the crop

Adult leaf miners are attracted to yellow sticky traps and these can be used as an aid to monitoring the crop for the presence of the pest. Such traps should be hung just above the crop at the rate of about one per 100 m^2. Traps should be inspected frequently and, if there is any doubt as to the identity of the flies, a sample should be sent for expert identification. Note that alien leaf miners (see page 271) may be present.

Chemical control

Treatment with a recommended insecticide can give effective control of leaf miner, but to obtain the best results it must be applied when the first feeding punctures or small mines are seen. To prevent an attack from developing, a spray should be applied when feeding punctures are first seen and again seven days later. When mines have already developed and pupae are present, at least two and preferably three applications are needed to control the infestation. The interval between treatments depends on the temperature. At an average temperature of 21 °C the interval should be 10 days and at 15.5 °C 14 days. High-volume applications give good control of larvae if the plants are thoroughly sprayed; this is often difficult on mature year-round beds.

As a general rule, pesticides should not be applied to plants which are dry at the roots or during periods of bright sunshine because damage may occur. Whenever possible, chemical treatment should be applied before the buds show colour as pesticides are more likely to cause damage after this stage has been reached. Chrysanthemums, particularly certain year-round varieties, are very susceptible to damage by some chemicals. Therefore, before using a pesticide on chrysanthemums, it is important to consult the manufacturer's recommendations for information on phytotoxicity to the varieties being grown.

Integrated control

Where integrated pest management systems are being used and western flower thrips (*Frankliniella occidentalis* (Pergande)) and other major pests are being controlled biologically, chemical control of leaf miners cannot be practised because the predators will be killed. In these circumstances it is necessary to use a fully integrated control programme in which parasitic wasps are introduced to control leaf miners. Some of these wasps, e.g. *Dacnusa* and *Opius* spp., attack mature larvae, while others, e.g. *Diglyphus* spp., feed as ectoparasites on the newly hatched lavae. Both types are available to growers. Although they are capable of maintaining leaf miner populations at insignificant levels during the summer, they are unreliable in the winter.

It should be noted that statutory eradication of alien leaf miners would disrupt any integrated control programme.

ALIEN LEAF MINERS

Several alien species of leaf miner may be found on imported chrysanthemums and other imported plants and produce. These species include: the South American leaf miner (*Liriomyza huidobrensis* (Blanchard)), the American serpentine leaf miner (*Liriomyza trifolii* (Burgess in Comstock)) and the South American vegetable leaf miner (*Liriomyza sativae* (Blanchard)). The mines made by each species can easily be confused with those made by our native chrysanthemum leaf miner. The North American cabbage leaf miner (*Liriomyza brassicae* (Riley)) also produces similar mines and may be found on imported plants, but this species has not been recorded on chrysanthemums.

Each of these alien species has a very wide and overlapping host range.They can damage lettuce, chrysanthemums, celery, beans, brassicas, tomato, *Gerbera* and a wide range of bedding plants. They can also infest many common weeds including groundsel, nightshade and ragwort.

In recent years the South American and American serpentine leaf miners have been found on imports from many countries and numerous outbreaks have occurred in protected crops. Statutory eradication measures continue, but there is a possibility that these species may become established in the future. The Plant Health (Great Britain) Order 1987 requires that the Ministry of Agriculture, Fisheries and Food is notified whenever the presence of an alien pest is suspected.

Nature of damage

The feeding and egg-laying punctures of these alien leaf miners are smaller than those of most native species and tend to be near the leaf margin. The larval feeding mines are generally more contorted and a narrow continuous trail of frass is usually noticeable in the mine (Fig. 53.5).

Description and life history

The adults of these leaf miners are flies that are 1.5–2.0 mm long and black or greyish-black with yellow markings (Fig. 53.6). Accurate identification depends on microscopic

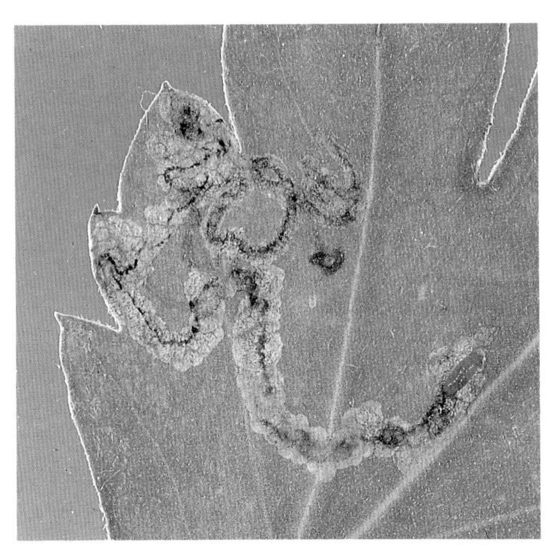

Fig.53.5 Close-up of mines of American serpentine leaf miner in chrysanthemum leaf. Note continuous frass trail and larva at end of mine (right centre)

Fig.53.6 Adult of American serpentine leaf miner (× 24)

examination. A native species, the tomato leaf miner (*Liriomyza bryoniae* (Kaltenbach)), is extremely similar both in appearance and the damage that it causes, but it has only occasionally been found infesting chrysanthemums.

On year-round chrysanthemums growing at 20 ± 6 °C, adult American serpentine leaf miners live from two to four weeks. They can fly over 100 metres in a few hours but, left undisturbed, they tend to aggregate on individual leaves. Each female may lay up to 600 eggs; these are laid singly in incisions in the leaves. Although individual leaves commonly have more than 100 punctures, only a few of these contain eggs. The eggs are translucent and 0.2 × 0.1 mm in size; they hatch in 1½–5 days depending on temperature.

The length of larval development is greatly influenced by temperature and the host plant. Larvae of the American serpentine leaf miner take 4–7 days to develop at temperatures over 24 °C. The mature larva is about 2 mm long, legless and orange-yellow. It leaves the mine and crawls or drops to the soil, where it pupates. The pupal stage may last 7–13 days at 20–30 °C. On average, 60 adults emerge successfully from the progeny of one female. On celery, the total development time is 54 days at 15 °C, 26 days at 20 °C and 12 days at 35 °C.

Breeding is continuous under glass and infestation may occur at any time on year-round crops. During the winter the level of pest activity is reduced by the low light intensities and low temperatures. With the return of suitable conditions in early summer, the pest, if uncontrolled, is able to increase rapidly.

Although unable to survive severe winters, there is a substantial risk that the American serpentine and the South American leaf miners can overwinter outdoors in Britain in sheltered positions, especially in mild winters, and reinfest the crop. During the summer many outdoor plants may be infested, including potatoes and ornamentals, and the pest may complete 4–6 generations. At the end of the summer some of the adults may enter glasshouses.

Control measures

Growers who use imported propagating material should look for leaf miners on receipt and during establishment or rooting.

Growers who find unusual leaf miner symptoms should send samples of these to the local office of the Ministry of Agriculture, Fisheries and Food as above, mentioning that they are suspected alien species.

If alien leaf miners are discovered, eradication is required by statutory notice. This notice, after consultation with the grower, will describe the methods and insecticides to be used.

54
Leatherjackets

Fig.54.1 Field of barley showing patches of plants damaged by leatherjackets

Leatherjackets are the grubs (larvae) of crane flies (*Tipula* spp. in particular), which are also known as daddy longlegs. They are important soil pests troublesome to both the farmer and the gardener. They are usually most numerous after prolonged damp weather in late summer and early autumn. The largest populations occur in grassland, particularly leys, lawns and golf courses.

Arable crops following infested grassland may suffer serious injury from these pests. Cereals, sugar beet and other root crops, vegetables such as brassicas and courgettes, and many herbaceous garden plants are frequently damaged. The worst attacks usually occur in spring, young plants being most affected. Closely sown crops are sometimes infested with many leatherjackets without showing serious loss but, when these crops are thinned, the remaining plants may be damaged severely. Similarly, thinly sown or transplanted crops may suffer severely. Some stubbles, particularly in western Britain, can become very grassy in the early autumn, especially during a wet period; leatherjacket trouble often follows such conditions.

In gardens even a few leatherjackets can cause serious losses, especially to seedlings. Unfortunately, the grubs are easily overlooked in potting soil owing to their earthy colour.

Nature of damage

Leatherjackets usually feed just below the surface of the soil, destroying the roots and more particularly the underground stems of plants; they may also eat into the seed of cereals. On warm, damp nights they also feed on the surface, making ragged holes in leaves and cutting off the plants at soil level like cutworms (see page 113). The presence of leatherjackets on a cultivated field is often indicated by birds, particularly rooks, constantly turning over small sods that have been left lying on the surface of the ground.

On heavily infested cereal fields or grassland, bare patches appear where the plants have been destroyed (Fig. 54.1). Winter cereals, particularly late-sown crops, may be damaged during late autumn and mild periods of the winter, but spring cereals at the seedling stage in April and May, when leatherjackets are large and feeding voraciously, are usually most affected. Once tillered, a crop is less likely to suffer economic loss.

Reseeded grass is particularly vulnerable and may be destroyed quickly when plants are at the single-shoot stage. Established *lowland* grass is unlikely to be noticeably damaged unless leatherjacket populations exceed 400/m^2, although production in spring and early summer is reduced by smaller populations. Clover is the most susceptible species in a sward.

In other crops, such as brassicas or strawberries, the wilting of the plants following root destruction is usually the first sign of attack.

Identification should be confirmed by finding the leatherjackets (Fig. 54.2).

Description and life history

Leatherjackets are greyish-black or brown, legless grubs without a distinct head (Fig. 54.3). They are plump and soft, yet with a tough skin. Other larvae bearing some resemblance to them are cutworms and bibionid larvae. Cutworms, however, are caterpillars with legs and an obvious head and are usually curled up in the soil when found. Bibionid larvae are legless but have a distinct black head and small processes on the sides of some segments. These larvae are sometimes mistaken for leatherjackets and are frequently found in large numbers in grassland and sometimes other crops, particularly if the land has been dunged. They also feed on cereals above and below ground but are usually of minor importance unless present in large numbers, e.g. 1000/m^2.

The grubs of several kinds of crane fly are destructive, but for practical purposes the habits of the most common species (*Tipula paludosa* Meigen) may be taken as typical of the group.

Crane flies are large flies with slim bodies, long ungainly legs and narrow wings (Fig. 54.4). They emerge mostly in late August and early September when they often enter houses in the evening; emergence is rather later in the

Fig.54.2 Leatherjackets exposed along a drill row of barley showing damage caused

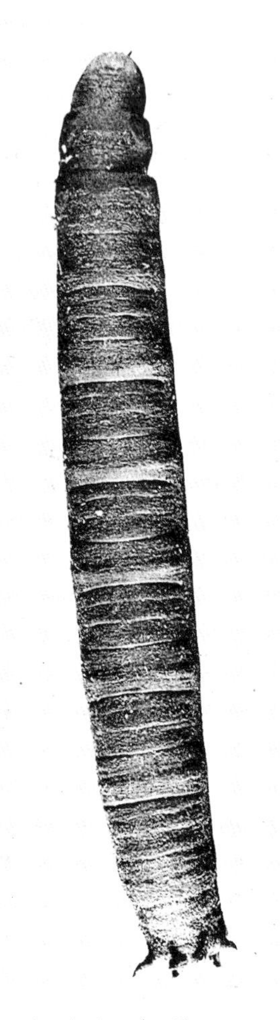

Fig.54.3 Leatherjacket (× 3)

south of England than in the north.

Eggs are laid in the soil; one female may lay as many as 300. The eggs are small, oval and black, like minute seeds. The small, legless grubs hatch in about 14 days. They are very susceptible to sunlight and drought: large numbers perish unless the weather is cool and moist at hatching time. By November the grubs are usually about 10 mm long; they grow slowly during the winter but, when fully grown after the main feeding period in the spring, they are about 40 mm long. During the summer the grubs change into pupae in the soil; these push partly above the surface just before the adults emerge.

Natural enemies

Leatherjackets are attacked by viruses, parasites and birds. The *Tipula* iridescent virus is widespread, but it is unusual for a virus or a parasite to affect a large proportion of the leatherjackets in a population. Starling is the main bird species feeding on leatherjackets. Cultivation helps predation by exposing a large number of the insects, but the birds also feed in grassland. In well-grazed pastures starlings can decrease leatherjacket populations considerably between late autumn and early spring.

Control measures

On field crops

Cultural control Attacks are frequent on crops following grass or clover leys but may be largely prevented if the land is ploughed in July

Fig.54.4 Male crane fly (about life size)

or early August (before the main egg-laying period) and the herbage is left well covered. If ploughing is later, thorough consolidation and a good tilth are important to enable the crops to grow away. A well grown crop is less likely to suffer significant damage than a crop drilled too deep and too early in a poor seedbed.

Cultivations decrease leatherjacket populations and, even after an attack has started, rolling with a Cambridge roller will give some control of damage to cereals and grass. If crop thinning is uniformly distributed, there may be no yield reduction providing adequate nitrogen is available to the plants.

Chemical control Recommended insecticides can be applied to infested land to prevent damage and halt attacks. In several regions forecasts are made of likely leatherjacket damage to crops in the following spring. These forecasts are based on annual surveys of leatherjacket populations in permanent pasture during the winter. Cereal crops following grass (permanent or temporary) or grassy stubbles should be inspected immediately after brairding. Thirty-cm lengths of drill should be examined in a diagonal across the field and, if 15 or more leatherjackets are found per 10 lengths of drill, treatment is probably justified.

Poison baits or sprays usually give good control. They are most effective if applied under mild, humid conditions when leatherjackets are feeding on the surface at night. Baits are prepared by mixing certain insecticides with moistened bran, although ready-prepared bran pellets are available. Such poison baits should be broadcast thinly and evenly during the afternoon or evening; they are particularly useful for treating isolated patches of damage. Insecticide sprays are more convenient for treating large areas. Generally, sprays should be applied following recognition of crop damage. However, when leatherjackets are known to be abundant in a field, sprays applied before ploughing, drilling or crop emergence may be more effective.

On grassland

In some situations, particularly on lowland pasture, leatherjacket damage to grass may be offset by increased application of nitrogen if optimum rates are not already being used. It is difficult for farmers to make an accurate assessment of leatherjacket numbers in grassland; however, unless fairly large populations are present (more than $300/m^2$) or damage is obvious, extra nitrogen may give greater yield response than an insecticide.

Upland pasture is more susceptible to leatherjacket damage but less responsive to nitrogen. The application of an insecticide rather than additional nitrogen will therefore be justified more often on this type of land. Where total spring nitrogen applications are already in the range of 100–150 kg/ha, the cost of an insecticide treatment is repaid when about 130 leatherjackets per m^2 are present. Loss of yield is best prevented by applying recommended insecticides between October and February before severe damage occurs.

On garden crops

Grass or waste land which is to be made into gardens should be broken up in spring or early summer and a careful watch kept for leatherjackets. If they are numerous, the ground should be kept free from weeds and no crops planted before the summer, i.e. until the grubs have pupated. If they are not numerous, potatoes may be planted. Crops normally planted after midsummer should be safe from attack and, if the ground is kept clean, there should be no trouble in subsequent years.

Occasionally leatherjackets move on to borders from lawns and damage plants. When this occurs the leatherjackets should be dealt with by hand-picking.

On lawns and golf courses

Some success has been obtained by trapping leatherjackets under tarpaulins. Lawns or greens should be thoroughly soaked with water in the evening and then covered with tarpaulins. On the following morning, the large numbers of leatherjackets that will be found under the tarpaulins can be swept up and destroyed.

Insecticide treatment is most successful if applied during mild, humid weather in spring or autumn.

55
Mushroom pests

Fig.55.1 Pinheads damaged by larvae of sciarid fly

Cultivated mushrooms grown in modern, purpose-built houses are affected by only a few pests. Most of these live within the compost but rarely survive the pasteurization, or peak-heat process, so they colonize the crop only at, or soon after, spawning. Casing material, whether subsoil or peat, is unlikely to be an important carrier unless stored in dirty, wet conditions, when cecid flies and nematodes may also become a problem. Many insects and mites associated with manure heaps or stacks are harmless, living on decaying organic matter or preying on other insects and mites. Pest species may be recognized by the characteristic damage they cause, but identification should be left to the specialist.

The pests affecting cultivated mushrooms are described and specific control measures are suggested below. Not all these measures may be necessary, particularly where strict hygiene and efficient peak-heating are practised. Serious pest damage occurs spasmodically and, although a series of crops may be affected, attacks may cease suddenly and unexpectedly.

FLIES

Sciarid, phorid and cecid flies are the most common pests of mushroom. Other flies resembling house flies may also be seen, but their larvae breed harmlessly in the compost.

Sciarid flies

About a dozen species have been identified from mushroom houses, but *Lycoriella auripila* (Winnertz) and *Lycoriella solani* (Winnertz) are the most important. The former appears to be the more common and potentially injurious species. *Bradysia* spp. may also be found.

The adult *L. auripila* is a small, delicate fly about 2 mm long with a black head and thorax, dark brown abdomen and long, thread-like antennae. It does not fly readily, except near lights, but runs rapidly over the trays and makes short, jumping flights. It is active throughout the year except in midsummer. Each female lays about 150 blunt, ovoid, white eggs, singly or in groups in the casing soil or at the base of a mushroom stalk. The eggs hatch within three or four days.

The larvae (Fig. 55.2) are white with a shining black head and have no legs; they feed for 3–4 weeks and are 5 mm long when fully grown. Younger larvae feed on the compost straws. Older larvae tunnel into the mushroom or, more often, feed in the casing. Both the stalk and cap of a mature mushroom may be tunnelled and as few as three larvae can spoil the market value of a mushroom. Damage to the caps occurs about 25 days after eggs are laid. More commonly, larvae burrow into pinheads (Fig. 55.1) and small buttons, reducing the whole to a sponge-like mass. Another type of damage is the severance of the mycelial attachments at the base of the stalk, causing buttons to die and turn brown.

The fully grown larvae make a silken cocoon in which they pupate; the adults emerge one to two weeks later. The adults normally mate on emergence and the females begin to lay eggs two or three days later. At first they prefer to lay in newly pasteurized compost, though later generations concentrate in the casing.

Sciarids can occur in the manure heap, but the larvae are usually eliminated by the high temperatures between successive turns. Some modern composting techniques, especially the use of brewers' grains, induce fermentation odours which appear to attract the adults: vast numbers fly to newly peak-heated compost during the 'cool-down' before spawning.

In addition to direct damage, sciarids often carry mites, nematodes and pathogens. They normally feed on decaying plant material and are associated with tyroglyphid mites (see page

Fig.55.2 Larva of sciarid fly (*Lycoriella auripila*) (× 12)

285) and bacteria-feeding nematodes (*Rhabditis* spp.). They readily transmit the sticky spores of *Verticillium* spp.

Control

The high temperature of the peak-heating process kills sciarids in the compost, so most infestations occurring during cropping are caused by flies entering the building and laying eggs either during 'cool-down' or after spawning. Yellow or white sticky traps, especially when used near light sources, are excellent for monitoring fly activity.

It is most important to ensure that stalks, spent manure and organic waste on which sciarids might feed are sterilized or removed from the farm. Ventilators should be screened with a sufficiently fine mesh: a gauze with apertures of 0.3 mm is recommended.

Chemical control may be directed against adults or larvae. Smokes, fogs, low-volume sprays and aerosols are widely used against adults. The frequency of application depends on the number of flies and the weather. Treatments are generally most effective when applied daily for 10 days before first flush pinning.

As adult sciarids spend much time crawling on exposed surfaces within the building, insecticide treatment of walls, boxes and bed-boards decreases the number of flies. A commercial product consisting of polybutene with added insecticide can also be sprayed on to plastic sheeting covering walls.

A sciarid population will almost certainly become established if control measures are directed against the adults only. Ideally, the compost or casing (or both) should be made lethal to the larvae.

Treatment of the compost with certain recommended insecticides will kill newly hatched larvae, but it does not prevent adults from laying eggs.

Thorough incorporation of certain insecticides during casing preparation is an effective control measure. Larvae in the casing can also be killed by drenching with a recommended insecticide. Beds should be picked hard before treatment. Where resistance to organophosphorus compounds is evident, an insecticide belonging to a different chemical group should be used.

Some nematodes are parasites of various insects, including sciarids, that spend part of their life cycle in the soil. These nematodes include *Steinernema* spp. and *Heterorhabditis* spp., which are being commercially produced for use against soil pests of other crops and should soon be available for use in mushroom crops.

Phorid flies

Adult phorids are small, dark flies with a slightly humped body which is stouter than that of sciarids and has shorter wings. The bristle-like antennae are inconspicuous. The larvae, known as maggots, are white, lack a distinct head and are legless; they have a narrow head end which is not black (cf. sciarid larvae) and a blunt hind end with prominences.

Until 1953 the only phorid known as a mushroom pest in Britain was *Megaselia nigra* (Meigen). This species lays its eggs in daylight only, is restricted to outdoor mushrooms and those growing near the doors of cropping houses. The eggs are laid on the gills, or at the base of the mushroom, and the larvae tunnel within the tissues (Fig. 55.3). Development is rapid, the eggs hatching in three days and the

Fig.55.3 Mushrooms damaged by larvae of phorid fly (*Megaselia nigra*)

larvae pupating five days later. Pupae are formed in the casing and the flies mature in five days. No chemical control measures are necessary for this species.

In 1953, *Megaselia halterata* (Wood), known as the Worthing phorid, was responsible for epidemic attacks in the Worthing area, as well as in nurseries in East Sussex, Kent, Lancashire and Norfolk, and became established as a major pest.

Adults of this phorid shelter in inclement weather, flight being prevented by even moderate winds, although crawling still occurs. As the temperature rises above 15 °C, flight increases. The females, when fertilized, are attracted to odour from growing mycelium and may respond over several kilometres.

Each female lays about 50 elongate, ovoid, white eggs close to hyphae growing in the compost. The subsequent rate of population increase is influenced by temperature, there being a considerable difference between cropping and spawn-running. At 24 °C the eggs hatch within two days, the larvae mature in five days and the pupae produce flies after a further seven days. At 15 °C the respective periods are 4, 14 and 28 days. The larvae feed entirely on the growing mycelium and damage is restricted to the destruction of newly implanted spawn. This is unlikely to occur now that through-spawning with grain is usual, though damage may occur where more than 100 larvae are present in each handful of compost.

Phorids are regarded mainly as a nuisance to pickers, some people developing allergies. However, recent work has shown that, where *Verticillium fungicola* is present, as few as seven flies per 0.1 m^2 of bed surface may be a disease threat to the crop because spores of the pathogen are readily spread by the flies.

Control

A mushroom crop is attractive to adults of the Worthing phorid for about three weeks after spawning, the maximum danger occurring in the second week. It is essential to prevent access of the females to the spawn-room. Particular attention must be paid to the sealing around doors and ventilators. Insecticides should be directed against the flies entering the compost to lay their eggs during the critical spawn-running phase. Protection must be provided whenever the weather is suitable for flight. Spraying with a recommended insecticide can provide the required protection, but application must be frequent enough to maintain a toxic aerial concentration during daylight hours whenever the outside temperature is 15 °C or above. Flight is considerably reduced at twilight, irrespective of temperature. Slow-release formulations are not suitable. Aerosols, fogs or smokes can be used but are effective for only a short time and may not prevent infestations during weather favourable to adult activity.

Some chemicals can reduce egg laying when applied to the bed. However, as flies can enter the compost through cracks and crevices in the boxes, it is difficult to achieve an effective distribution of chemical where the boxes are closely packed within the spawn-room. A more suitable treatment, provided that a reasonably uniform distribution can be obtained, is to incorporate in the compost chemicals that are toxic to the young larvae. Incorporation of a recommended insecticide should be made either at the end of composting or, more efficiently, during spawning. If the insecticide has to survive peak-heating, some breakdown may occur.

To control adults, a recommended insecticide can be used during cropping.

Nematodes that are parasitic on sciarids (see page 281) also attack phorids.

Cecid flies

The adults are minute, delicate midges which are seldom seen. The larvae are slender with an inconspicuous head and a pair of dark spots in the body just behind the head; they have no legs. Just before pupation a dark structure known as the 'anchor process' appears on the underside at the front end.

Normally each larva becomes a 'mother

larva' and at 24 °C gives birth to 12–20 daughters within a week of its own birth, without any adult being present. This rapid multiplication gives rise to enormous populations within the compost — as large as half a million per 0.1 m^2 of bed area — but there is little direct effect on yield. However, at certain times the larvae swarm from the compost and contaminate the growing mushrooms.

By far the commonest species is *Heteropeza pygmaea* Winnertz, whose larvae are white and have a spear-shaped anchor process. The mother larvae are about 3 mm long, whereas the daughter larvae, which are more commonly seen, are only half as long. Each female midge lays only one or two eggs and these hatch into larvae which produce daughter larvae until overcrowding occurs. Some of the larvae then swarm or migrate, while the rest each produce about five pupa-larvae. Within four or five days these latter develop anchor processes and pupate. Adults emerge about five days later. Although thousands of adults are produced they are so minute that growers are unaware of their presence. Their larvae usually swarm about 10 weeks after spawning. Unlike other cecids, larvae of this species are carriers of a bacterium which produces longitudinal, brown, discoloured stripes on the stipe as well as minute, black pustules of fluid on the gills. Where the beds become infested soon after spawning, as much as 20% of the crop may be spoiled, mostly in the last three weeks of cropping. Under dry conditions larvae tend to remain within the compost, but watering often brings large numbers to the surface, where they accumulate on the side boards and fall to the ground. On a dry floor they jump spasmodically and become tangled together in a writhing mass, sometimes as large as 20 mm across.

Both *Mycophila speyeri* (Barnes) and *Mycophila barnesi* Edwards have orange larvae with a trident-shaped anchor process. They are frequently reported when their larvae climb on the mushrooms of first or second flushes. They rarely appear later, but contamination of the early flushes may reduce the marketable yield by 50%.

Two other species have also been recorded on mushrooms in Britain: *Lestremia cinerea* Macquart is common and occasionally attacks mushrooms outdoors; *Henria psalliotae* Wyatt was described from a single outbreak on cellar-grown crops in Cornwall.

It is uncertain how cecid infestations arise, though all species breed in materials such as garden manure, rotting potatoes and decaying wood. Occasionally, larvae in fresh peat or peat stored in wet and unhygienic conditions provide a source of contamination. The adults are regarded as unimportant in starting fresh infestations.

Control

Cecid larvae are easily carried about the farm on cultivating equipment, so infested houses should be 'isolated' by disinfecting tools etc. with a recommended disinfectant. Separate protective overalls should be used in each house. Larvae may survive between crops and certain forms are specially adapted to survive several months in dry conditions. Peak-heating should kill all larvae, but it is essential to maintain 65 °C for 4–8 hours to ensure death of those hidden in crevices of trays. Treatment of the empty trays or bed-boards with a cold dip of a suitable disinfectant or with steam is a useful supplementary technique to kill superficial mycelium.

It is impossible to kill larvae in the compost without affecting the crop, but larvae will not contaminate as many mushrooms if a recommended liquid insecticide is thoroughly mixed with the casing material. The larvae in the compost will not be affected. Incorporation of recommended insecticide granules will effectively reduce population build-up.

Nematodes that are parasitic on sciarids (see page 281) also attack cecids.

MITES

Mites are very common but, owing to their small size, are not noticeable until they are present in large numbers. They are often numerous in the manure heap though the species present are usually beneficial. Most are yellowish-brown, very active and feed mainly on other mites and nematodes. The mite population of the manure heap generally dwindles as composting proceeds because most are killed by the fermentation heat, though some escape to the cooler, outer layers at each turn. The most important mites to the mushroom grower are tarsonemids and, to a lesser extent, tyroglyphids.

Tarsonemid mites

The most important mite pest of mushrooms is the mushroom mite (*Tarsonemus myceliophagus* Hussey). This is a pale brown, shining mite, less than 0.2 mm long, with a whitish transverse band across the body. Each of the fourth pair of legs terminates in two long hairs, one much longer than the other. This mite feeds entirely on the hyphae of mushrooms, causing a bright, reddish-brown discoloration at the base of the stipe, and may cut through the basal attachments of the stipes. The typical damage symptoms do not necessarily appear when many mites are present, whereas symptoms may develop in the presence of few mites. The discoloration may be caused by secondary decomposition that develops only under certain environmental conditions.

Where infestation occurs at, or soon after, spawning, vast populations can develop and large numbers of mites then swarm on the mushroom. At 24 °C development of the mite is completed in nine days but at cropping temperatures this extends to 12 days.

The long-legged mushroom mites (*Linopodes* spp.) and a closely related species of *Ereynetes* are also found on mushroom. They are small, whitish mites with extremely long forelegs (three times the length of the body) which are waved about like antennae. The mites move rapidly and hide in the compost when disturbed. They are said to sever the 'roots' of affected mushrooms, which tend to topple over, and cause a reddish-brown discoloration to the bases. However, it has been suggested that this damage is caused by the mushroom mite (see above) and that long-legged mushroom mites are predatory on it.

Control

Hygiene and thorough disinfection at the end of the crop are essential (see Crop and House Hygiene, page 288).

Chemical control of tarsonemids is rarely efficient as acaricides are often toxic to the mushrooms and cannot penetrate the compost. Growers incorporating a liquid insecticide in the casing for cecid control have found that fewer mushrooms are damaged by tarsonemids, but it is doubtful whether this measure can be used as a specific control for the mites.

Fig.55.4 Mushroom showing holes in cap and stipe caused by tyroglyphid mite

Tyroglyphid mites

These mites have soft, translucent, whitish bodies, often clothed with long fine hairs, and move slowly. They are not usually common in stable manure now that peak-heating is almost universally practised. The common species include several species of *Tyrophagus*, known as fungal mites, as well as species of *Caloglyphus* and *Histiostoma*, the latter often being found in over-wet composts.

The adults eat small, irregular pits in the stalks and caps (Fig. 55.4); buttons may be completely hollowed. The feeding is accompanied by bacterial decomposition. The feeding cavities should not be confused with those caused by bacterial pit disease, which begins at a point of breakdown just *beneath* the surface, the covering skin ultimately collapsing and leaving an open pit. Sometimes mites wander into these pits and may be mistaken as the primary cause of this damage. Tyroglyphid mites may also feed on mycelium and have been known to destroy grain spawn before any mycelial growth has occurred.

Tyroglyphids are generally introduced by insects, especially sciarids, carrying the migratory stage on their bodies. This specially adapted stage is frequently produced when colonies become overcrowded or the substrate becomes deficient. The migratory stage has suckers with which the mite clings on to passing insects; on reaching a suitable environment it transforms into an adult ready to breed. Under spawn-running conditions development is completed in 13 days, but at the lower, cropping temperatures this is extended to 36 days.

Control

The use of fresh manure, together with efficient composting and peak-heating, is probably the best means of preventing trouble from tyroglyphid mites. It is most important that organic debris in the buildings, especially in glasshouses, should be cleared away or destroyed.

Red pepper mites

These mites (*Pygmephorus* spp.) are minute (about 0.25 mm long), flattened, wedge-shaped and yellow-brown with a whitish, longitudinal band along the body. They swarm in vast numbers on the casing surface and the mushrooms, forming a jostling mass which makes the infested area reddish-brown and visible a few metres away.

The commonest two species feed on the weed moulds *Humicola*, *Monilia* and *Trichoderma*; they do not feed on mushroom or *Chaetomium*, but the association with this mould indicates that the compost is, simultaneously, a suitable substrate for other weed moulds.

When provided with abundant food, the females swell into a bubble-like sphere resembling a droplet of fluid. Eggs are produced every few minutes over two days until a heap of up to 150 lies beside each dead female. Development is completed in seven days. Their high fecundity, coupled with rapid development, provides an enormous reproductive potential. Thus, teeming thousands of individuals may swarm even a few days after casing, although they may disappear as rapidly as they appear.

Although it is claimed in the USA that these mites may cause mycelial damage, British species seem merely to indicate that the compost is unsatisfactory and that weed moulds are rampant.

Control

As red pepper mites only indicate poor composts, there is no point in considering direct control. Normally the swarms disappear within a few days.

SPRINGTAILS

Springtails are very small, wingless insects with an appendage on the underside of the abdomen which can be released violently downwards flicking them into the air. They are common in the soil and in decaying vegetable matter in damp, shady situations. They occur in manure but are important only where mushroom beds are made on the soil surface in glasshouses. Under favourable conditions vast numbers may develop and the soil surface appears coloured.

Several species of springtail, including *Hypogastrura armata* (Nicolet), *Proisotoma minuta* (Tullberg) and *Xenylla mucronata* Axelson, have been found in mushroom beds. *H. armata* is about 1.5 mm long and varies from silver-grey to dark bluish-black. It occurs widely on damp soil, among dead leaves, under bark, on pools of water and among fungi. Sometimes it occurs in great numbers and assembles in heaps on the beds and floor, hence its common name 'gunpowder-mite'. It excavates small pits in the stalk and cap of mushrooms. In contrast to the moist cavities eaten out by tyroglyphid mites, springtails make *dry* pits leading to tunnels in the stalk and cap. Springtails may be attracted to newly implanted spawn and the 'spawn run' may be inhibited temporarily.

Control

Springtails are eliminated by normal composting and pasteurization, but they can easily invade floor beds. The lowest trays should be kept off the floor; beds made on the ground should be protected by a sound sheet of polythene. Incorporation of a recommended liquid insecticide in the compost is the best preventive.

NEMATODES

Many nematode species are associated with decaying organic matter: some feed on bacteria, others on fungi, some prey on other nematodes, and a few are parasitic on insects. They are usually visible only with the aid of a microscope, being slender, colourless creatures about 1 mm long swimming in surface films of water. As large numbers are present in compost they are potentially important in mushroom culture.

Contamination with nematodes usually occurs before the end of cropping; the effect on mycelial growth depends on the species and number present. The majority of species found in mushroom beds feed on bacteria present in decaying plant material; they are therefore unlikely to cause direct injury but may indicate poor growing conditions. Sometimes vast numbers swarm on top of the compost casing, adhering to any objects which they touch: each larva stands on its hind end, waves its body in the air and twists round others to form glistening, writhing spires.

The fungus-feeding nematodes may make mushroom growing uneconomic, but in hygienic conditions and with short cropping periods, i.e. up to eight weeks, they rarely cause serious trouble. The mushroom spawn nematodes (*Aphelenchoides composticola* Franklin and *Ditylenchus myceliophagus* J.B. Goodey) are the most common cause of damage. They are responsible for the disappearance of mycelium in patches which gradually enlarge, becoming noticeable in the second or later flushes. In these patches the compost becomes dark and soggy, tends to sink and usually has a foul, pungent smell. Buttons may turn brown and die, and crop production falls. An infestation is difficult to eradicate and may be carried over from one crop to another by nematodes in the crevices in wooden trays or bed-boards.

D. myceliophagus is able to withstand slow drying and may survive desiccation for as long as three years; when the compost is re-wetted the nematodes revive. This emphasizes the danger of allowing infested organic debris to blow about the farm. As the desiccated nematodes are resistant to heat and chemicals, dry infested materials should be moistened before control treatments are applied. A grey mould over the surface of affected patches has been regarded as a primary cause of crop failure, but it is now known to be a predatory fungus which traps the nematodes.

A. composticola has been reported from

many countries on a wide range of fungi in addition to mushrooms. It breeds rapidly, often increasing at rates of 100 000 times in six weeks; the nematodes become so numerous that they cluster together in curds, and in water they stick together in whitish clumps.

Control

Affected patches should be destroyed and a generous margin of apparently healthy bed around them should be removed. Effective and uniform pasteurization is essential. If peak-heating is impossible, 'finished' compost can be fumigated, but this must be done only by contract operators.

'Cooking-out' at the end of cropping (see Crop and House Hygiene, page 288) is the best way to control nematodes. The resting stages are destroyed by maintaining a temperature of 55 °C for at least three hours, so the normal practice should be completely effective.

Nematodes can be introduced in casing material and, where they are troublesome, it should be pasteurized before use — with steam at 60 °C for at least 10 minutes. It is most important not to cause recontamination by storing casing material on infested surfaces or by using dirty barrows or tools.

MINOR PESTS

Woodlice (see also page 387)

Woodlice normally inhabit damp situations and may infest outdoor beds, glasshouses or makeshift buildings near suitable breeding places. Several species have been recorded eating irregular holes in the caps. Control can be obtained by lightly applying a recommended insecticide dust to the beds and blowing it into crevices where woodlice hide.

Millepedes (see also page 377)

Three species have been known to damage mushrooms: the spotted millepede (*Blaniulus guttulatus* (Fabricius)), which is small and pale yellowish-white with a row of reddish spots along each side, the palm millepede (*Choneiulus palmatus* (Nemec)) and the lesser glasshouse millepede (*Cylindroiulus britannicus* (Verhoeff)), which are larger and pale fawn with lateral rows of dark spots. All eat holes in the stalks at the casing level. Attacks are most likely to occur on floor beds in glasshouses and are easily controlled with a light dusting of a recommended insecticide.

Slugs (see also page 467)

Slugs are unknown in modern mushroom houses but may cause trouble on outdoor beds and in damp cellars. They eat large, irregular cavities in the stalk and cap, sometimes destroying most of the mushroom. Their rasp-like tongues make marks similar to those made by mice, but their identity is betrayed by glistening slime trails.

Control

The most satisfactory control measure is to distribute small heaps (half a handful) of molluscicide bait, 2–3 m apart, around the beds. The bait should comprise one part of powdered molluscicide mixed with 25–30 parts of dry bran. It should be covered by a piece of tile or board so that the slugs will crawl underneath. Affected slugs should be collected and destroyed. Alternatively, recommended molluscicide pellets can be distributed around the beds.

ROUTINE USE OF PESTICIDES

Because of the risk of encouraging the selection of pesticide-resistant strains of pests, the greatest care is required in choosing routine control programmes. With sciarids and phorids, the crop is most vulnerable in the first three or four weeks after spawning. Incorporation of a recommended insecticide in the compost at spawning is a valuable control technique but, as the larvae are exposed to the chemicals continuously, the high selection pressure may lead to resistance. These treatments should *not*, therefore, be continued for very long and should be restricted to the period of greatest risk, i.e. between June and November.

At other times recommended space sprays or smokes of chemicals with different modes of action should be used in rotation. Applications should be made twice weekly between spawning and pinning, economizing where possible by omitting routine applications when the weather precludes insect flight.

Insecticides need to be used regularly where efficient peak-heating is impossible or where pests have easy access to the crop, such as in makeshift structures, outdoor beds or floor beds in glasshouses.

CROP AND HOUSE HYGIENE

General precautions

Hygiene is one of the most important aspects of mushroom culture. Elimination of pests and diseases is of prime importance for, once pests or diseases gain entry, their spread is rapid and elimination difficult and costly. Every effort must be made to avoid carry-over from one crop to the next and no contaminated equipment, compost or other raw material should be taken into newly cleaned houses.

The growing crop should be protected by removing old mushrooms and cut stumps from the bed. As workers may spread disease from one bed to another, the removal of debris should be a separate operation from picking; debris should never be left about but removed in covered bins and destroyed.

Entry of insects, such as sciarid and phorid flies, can be checked by screening ventilators with copper gauze; this is more durable than other materials though more expensive. Screening impedes ventilation and, unless special equipment is used, the mesh cannot be reduced to exclude all insects. Insect traps do not generally decrease the numbers of flies sufficiently to achieve control, but they are valuable indicators of pest activity.

The danger of introducing pests and diseases on hands, clothing and boots, as well as on tools and transport equipment, should be kept in mind and strict rules observed to protect crops until the appearance of the first flush. Many insects persist in rubbish, which should be removed from the vicinity of the farm. Rank vegetation also harbours insects and should be replaced by concrete or short grass near mushroom houses.

Sterilization and disinfection

Prevention of trouble is better than control and it is essential to ensure that every new crop has a clean start. Laboratory examination of compost or casing samples at various stages can help to check the efficacy of the various treatments.

Cooking-out

Where possible all houses should be 'cooked-out' for about 12 hours between crops by raising the temperature to 71 °C to destroy all organisms. Some buildings, however, can suffer from repeated 'cooking-out' and fumigation may be more convenient, but must be done only by specialist contractors.

Steam sterilization

Empty structures can be sterilized with a high-pressure steam jet, care being taken to reach all cracks and crevices. All movable trays, bed-boards and boxes should be removed and treated separately.

Disinfectants

Several chemicals are suitable for use in empty structures and for spraying paths and roadways connecting the houses. Some are suitable for use as a dip for trays, boxes, tools, etc.

The interior of a building should first be washed down and then sprayed with a solution of a recommended disinfectant. The wetter the interior, the stronger the solution required. A high-pressure sprayer should be used. The whole of the inside fabric of the building including floors, wall and roof should be treated thoroughly and particular attention paid to all cracks and crevices. Portable shelving, trays and boxes should be treated by dipping in an appropriate disinfectant.

Disposal of spent compost

The greatest care should be taken in disposing of spent compost, whether 'cooked-out' or not, as sterilization may not have been completely effective. Spent compost should never be stacked near mushroom houses nor used as manure on cultivated land nearby. Ideally, it should be sold off the farm and loaded direct from the house into the lorry. All tools, equipment and clothing contaminated with the old compost should be cleaned thoroughly.

56 Narcissus flies

Fig.56.1 Narcissus bulb cut open to show maggot of large narcissus fly inside

Fig.56.2 Rotting narcissus bulb cut open to show secondary infestation of maggots of small narcissus fly

There are three species of narcissus fly or bulb fly in Britain: the large narcissus fly (*Merodon equestris* (Fabricius)) and two small narcissus flies (*Eumerus strigatus* (Fallén) and *Eumerus tuberculatus* (Rondani)). The two latter, from the practical point of view, may be treated as the same. All three species are important to growers of narcissi, but the large narcissus fly is the most important as it is a major pest of narcissus, especially in south-west England, and causes severe losses throughout much of Europe and North America. The small narcissus flies normally attack only diseased or damaged bulbs. Freedom from these pests in imported bulbs is a requirement of several overseas countries and each year some of our consignments fail pre-export inspections because of infestation by maggots of narcissus fly.

The large narcissus fly is mainly a pest of narcissus, but occasionally it attacks snowdrops and other bulbous plants. The small flies are found in a wide range of plants, including narcissus, hyacinth, iris, lily, carrot, parsnip, potato and onion.

Nature of damage

It is important to distinguish between infestations of large fly and those of small flies because the large fly is a primary pest and therefore attacks healthy bulbs, whereas the small flies usually attack only damaged bulbs. The two types of fly damage are easy to distinguish: in an attack by the large fly there is a single maggot feeding in the centre of an otherwise undamaged bulb (Fig. 56.1), while the maggots of the small flies are usually found in groups in rotting bulbs (Fig. 56.2). The maggots of another closely related fly (*Syritta pipiens* (L.)) also occur in decaying bulbs. They are very similar to maggots of the small narcissus flies and are also found in groups.

Large narcissus fly

The symptoms of attack by large fly vary according to the time of year. After lifting, during storage in July and early August, the infestation is difficult to detect because the maggot is still small. The bulb still feels firm and the only sign of attack which can be found is the entry hole in the base plate of the bulb. If the surface of the base plate is carefully scraped with a sharp knife, the entry hole can be seen as a small rust-coloured spot surrounding a hole the size of a pin-prick.

By late September and October the large narcissus fly maggot will have grown and will have tunnelled extensively in the bulb. The centre of the bulb will have been eaten and the single, large maggot will be found surrounded by a large amount of mud-like frass (excrement) (Fig. 56.1). These bulbs are then softer than unattacked bulbs, particularly in the neck region.

In the spring, bulbs attacked by the large fly can usually be detected by the difference in their growth. A small bulb may have been killed completely so that no growth will appear above the ground. A large bulb will not produce a flower, as the flower bud will have been eaten, and the axillary buds will grow out producing a clump of abnormally narrow leaves, aptly called 'grass' by many growers. Damage by the large fly sometimes causes the bulb to produce weak, yellowish and distorted foliage; these symptoms can be confused with those produced by other pests, e.g. bulb scale mite or stem nematode (see pages 335 and 449).

Fig.56.3 Large narcissus fly (× 3)

Fig.56.4 Small narcissus fly (× 7)

Small narcissus flies

The small flies are not regarded as primary pests; their maggots are often found in narcissus bulbs (Fig. 56.2) but only after these have been damaged by fungi, large fly or nematodes. Other plants are probably only attacked after they too have been weakened by other causes. Infested bulbs are softer at all stages of growth, with obvious rotting of the bulb tissues.

Description and life history

Large narcissus fly

The adult (Fig. 56.3) is about 13 mm long, has a hairy body and looks like a small bumble bee. The colour of the body varies considerably: it may be black, banded with yellow, buff or orange, or one of these colours may predominate. Hover flies, which belong to the same family as the narcissus flies but are beneficial, have glossy bodies with few hairs.

The large narcissus flies may be on the wing from early May until July; they are active on bright sunny days and have a characteristic whining hum when in flight. Each female lays about 40 eggs, one to a bulb, on the foliage below the soil surface.

On hatching, the maggot (larva) crawls down the side of the bulb and burrows into the base of the bulb through one of the root canals. After tunnelling in and around the base plate for a while, it works its way upwards to the fleshy scale leaves surrounding the growing point. By the time it is fully grown it has eaten out a large cavity. In the early spring the fully grown maggot, which may be as long as 18 mm, moves either to the neck of the bulb or into the soil and then pupates. The pupal stage lasts about five or six weeks and the adult then emerges. There is only one generation a year.

Small narcissus flies

The adults are about 6 mm long, shining black with white crescent-shaped marks (Fig. 56.4). The first adults emerge at about the same time as the large narcissus flies, in late April or early May, and the females lay their eggs in groups of 10 or more, on or near diseased or damaged bulbs.

When the maggots hatch, they enter a bulb in the region of the 'nose' or at any point where it is damaged. They work through the bulb and finally destroy it completely. In July most of the maggots pupate and produce another generation which infests new bulbs, but some may spend the whole summer feeding, then remain in or near the bulbs during the winter and pupate in the following spring.

Prevention and control

The large narcissus fly is a serious and constant pest in south-west England, and routine preventive treatment is required. Although this species occurs in eastern England, precautionary measures are usually unnecessary there. Small narcissus flies attack only unsound bulbs. An infestation of small flies therefore indicates some other problem.

Treatment of growing crop

In trials, good control has sometimes been obtained by applying systemic granular insecticides to the crop during the early stages of larval invasion, before there has been any visible injury to the bulbs. The optimum time of treatment may vary from year to year.

Physical methods of discouraging egg laying, such as crop defoliation, surface cultivation, covering the rows with fine netting or non-woven fleece during May and June, or early lifting and windrowing, are partially effective but involve considerable labour and some risk of reduction in bulb yields.

When roguing for stem nematode or other disorders, plants attacked by narcissus fly should also be removed.

Precautions after lifting

After lifting, bulbs are often left in the fields for a while to dry. At this stage, especially if there is much soil on them, the bulbs are very attractive to small narcissus flies; they may also be damaged by 'sun-scald'. Protection can be obtained by using some form of covering, such as empty trays, sacking or even dried bulb foliage.

Inspection of bulbs

During sorting and grading, all soft bulbs should be discarded. Bulbs should never be replanted without some form of treatment. Those attacked by small narcissus flies, in addition to harbouring other pests and diseases, will never grow satisfactorily and should be destroyed by burning. Bulbs of valuable varieties attacked only by the large narcissus fly may be saved by hot-water or chemical treatment, though in practice this is seldom worth while.

Control of maggots in bulbs

The maggots of any of the narcissus flies are easily killed by hot-water treatment. All bulbs should be treated to control stem nematode each time they are lifted (see page 451); if this is done by hot-water treatment, any maggots in the bulbs will also be killed. If a grower does not wish to treat his bulbs for the full three hours at 44.4 °C, fly maggots can be killed by one hour's treatment at 43.5 or 44.4 °C, but this will not kill nematodes. Formalin (40% formaldehyde) should always be added to the hot-water treatment tank.

57
Onion fly

Fig.57.1 Salad onions attacked by maggots of onion fly

The onion fly (*Delia antiqua* (Meigen)) is widely distributed in Britain and the maggots can cause serious damage to bulb and salad onions. Leeks and shallots are also attacked, but damage on these crops is less common and not so severe. Although commercial crops grown anywhere in the country may be damaged, the most severe and regular attacks tend to occur in certain areas, such as the Vale of Evesham, Bedfordshire and Essex. Damage is often serious in gardens and allotments.

Maggots of the closely related bean seed flies (*Delia platura* (Meigen) and *Delia florilega* (Zetterstedt)) (see page 230) also attack onions and cause similar damage to that caused by onion fly maggots.

Nature of damage

The worst damage usually occurs in June and early July, but damage can also occur in August and early September. Small plants are most seriously affected and rapidly wilt or collapse completely (Fig. 57.1) as a result of the maggots feeding in the shank just below the soil surface. When numerous, the maggots can kill a large proportion of young plants, resulting in a patchy crop. In closely planted crops such as salad onions, discrete lengths of row may be killed as the maggots move from one plant to another.

On larger plants, the maggots feed in the bulb of the onion or in the shank of the leek (Fig. 57.3). The first symptom of attack on larger plants is wilting of the younger leaves and yellowing (on onions) or whitening (on leeks) of the older leaves (Fig. 57.2). These old leaves can easily be pulled away from the stem. Where the attack is heavy, up to 30 maggots per bulb may be present and the whole bulb usually rots away. With very light attacks, symptoms may be absent on the aerial parts of the plant, so the damage is only found at harvest. In late attacks on bulb onions and leeks, the maggots may fail to gain entry to the bulb or stem base; damage is then concentrated on the basal plate with some superficial splitting of the bulb or stem.

Onion fly damage is often confused with damage by stem nematode (see page 459). The latter commonly causes rotting of the base of the plant, frequently in patches, and onion fly attack may be only secondary and incidental.

Fig.57.2 Leeks damaged by maggots of onion fly

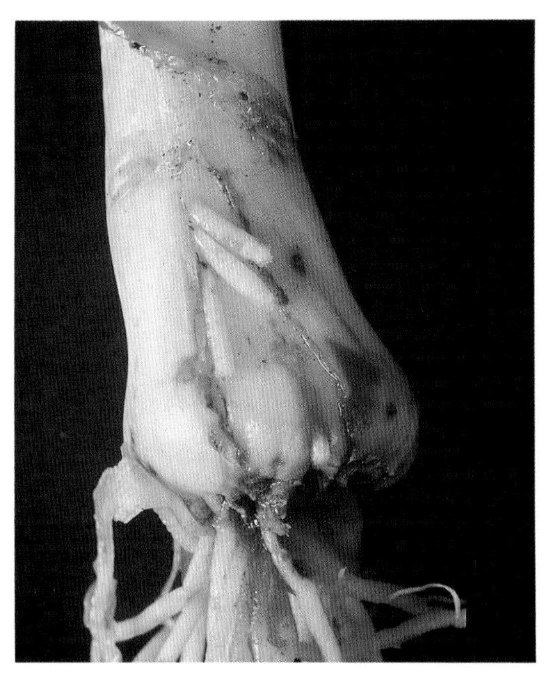

Fig.57.3 Maggots of onion fly feeding in base of leek

Damage caused by maggots of bean seed flies is often identical to that caused by onion fly maggots. However, bean seed flies are active earlier in the season and may therefore be responsible for damage that is occasionally found much earlier than is normal for onion fly. Because maggots of bean seed flies can be present in the soil when the crop emerges they are capable of killing plants in the 'loop' or 'crook' stage. Autumn-sown onions are particularly vulnerable to this type of damage. Onion fly eggs are not laid until the crop is present, so damage by this species cannot be seen until the plants reach a more advanced stage.

Description and life history

The adult (Fig. 57.4) resembles a house fly. It is grey, with four brownish stripes on the thorax and a dark stripe along the mid-line of the abdomen.

The adults emerge in May and after about a week the females start to lay their eggs on the young leaves or necks of the onions, or in the

Fig.57.4 Female onion fly on onion plant (× 6)

soil nearby. The eggs are white, oval and about 1.25 mm long. Larvae hatch after about three days and immediately burrow through the soil to an onion plant where they begin to feed, destroying the tissue as they probe into the plant (Fig. 57.3). The larvae, known as maggots, are dirty-white, legless and lack a distinct head. There are 11–13 processes on the anterior spiracles, which are shaped like an elongated hand. After about three weeks, when fully grown, the maggots are about 8 mm long. When mature they leave the onion, burrow into the soil to a depth of 5–8 cm and turn into ovoid, chestnut-brown puparia. From some of these puparia, adults emerge in about 17 days and give rise to a second generation. The majority of those pupating in the autumn remain in the pupal state until the following spring, but in some years a third generation occurs.

Bean seed flies are active much earlier than the onion fly, i.e. in April, and have up to four generations in a season. All stages of these species are very similar. However, eggs of bean seed flies are usually laid on recently disturbed soil, whereas onion fly eggs are only laid in the presence of a host crop. Maggots of bean seed flies can, therefore, be present in the soil before the emergence of the crop.

Control measures

In small gardens and allotments, attacked plants should be burnt to kill the maggots. When removing badly decayed plants, care is needed to avoid leaving maggots or puparia in the surrounding soil. Sub-standard onions left at the edge of a field or elsewhere on the farm should be buried to reduce the build-up of fly populations.

Where feasible, crop rotation should be practised and onion fields should be situated as far apart as possible. Onions grown in the year following an attack should be sown as far away from infested land as possible.

Chemical control

Once the maggots are feeding in the plant, chemical control is unlikely to be effective. Treatments should therefore be targeted at the maggot as it moves from the egg to the plant, or from one plant to another.

One of the most effective methods of control is to treat the onion or leek seed, before drilling, with an insecticide recommended for this use. Dry or slurry treatments have been widely used, but pelleting or film-coating the seed is becoming more common. Although all these treatments give effective control of onion fly, pelleted or film-coated seed is more likely to carry an even dose of insecticide, is safer to handle and usually less phytotoxic than either of the other treatments.

Treatment of the soil with a recommended insecticide can also be effective. Some treatments applied for control of other pests, e.g. stem nematode, will give incidental control of onion fly.

58
Pea midge

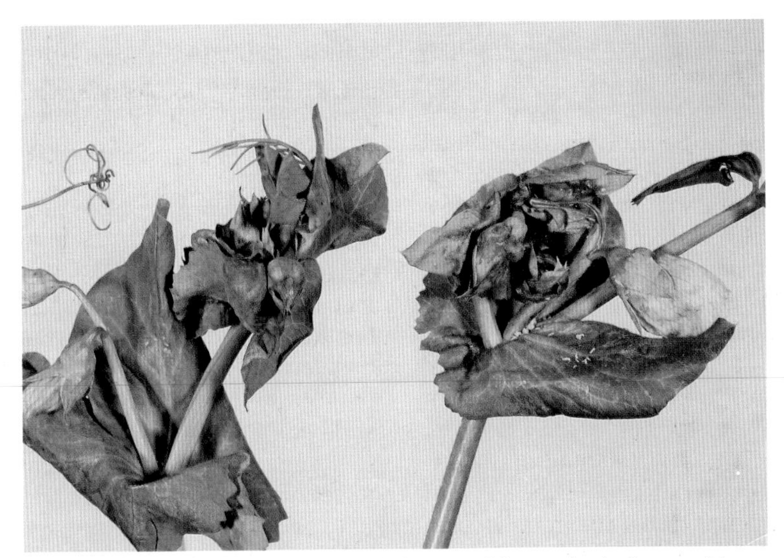

Fig.58.1 Pea plants with terminal shoot and flower buds damaged by larvae of pea midge. Left: knuckled calyces. Right: early stages of 'nettle-heading' and larvae on stipule

Pea midge (*Contarinia pisi* (Winnertz)) is a localized pest in Britain, large populations occurring in north and south Humberside, Lincolnshire, north Cambridgeshire and parts of Norfolk. Elsewhere in Britain pea midge incidence is currently slight and seldom important. Serious damage to peas and crop loss may follow heavy attacks, but the incidence of attack varies from year to year, from area to area and from crop to crop within a season.

Nature of damage

Leading shoots and flower buds of peas are deformed or killed by the feeding of numerous midge larvae. This can limit extension growth and result in the plant becoming 'nettle-headed' (Fig. 58.1, right). Infested buds and flowers swell giving a knuckled appearance to the calyx (Fig. 58.1, left), which becomes uneven and paler than normal; the petals do not open properly and remain balloon-shaped and yellowish. Attacked flowers may abort or produce only dwarf, misshapen pods. Infestations inside the developing pods are uncommon but have been seen in the absence of obvious shoot or flower damage, usually in later-sown crops.

Description and life history

Midge larvae from previously infested pea crops overwinter in cocoons in the soil. In late May or early June these larvae start to pupate and adults emerge about 14 days later, the time taken depending on soil temperature and moisture.

The adults are delicate, grey-brown midges which are about 2 mm long (Fig. 58.2). They have long legs and thin, many-segmented antennae which, in the males, are often held curved back over the head. On emergence the females quickly mate and disperse to the new season's crops. These dispersal flights frequently occur in the still air conditions of early evening and morning, or after rainstorms. The females immediately invade the leading shoots to lay their eggs on the stipules or on and in the flower buds. More than 100 eggs have been found in a single flower but 20–30 are more usual. The adults live for only about four days.

The eggs are translucent and jelly-like, oval with a tail-like tip, and about 0.25 mm long so are just visible without magnification; they are often laid overlapping one another. They hatch in about four days and the young larvae feed by scraping the plant tissue, causing the distorted growth described above. The damaged tissue may be invaded by secondary organisms which develop as rots, especially in wet weather.

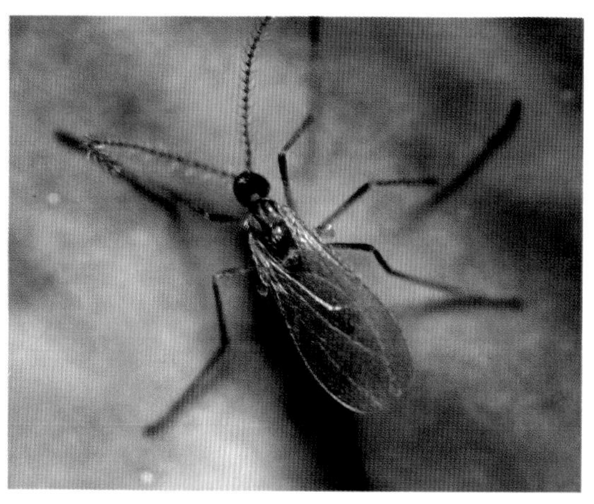

Fig.58.2 Pea midge (× 14)

After feeding for about 10 days the larvae, which are dirty-white and about 2–3 mm long (Fig. 58.3), descend to the soil. They may pupate immediately and within a fortnight give rise to a second generation of adults which invade later flowers of the same or other pea crops. Larvae from this second generation, together with those of the first that did not pupate, overwinter in the soil. Most of them pupate early in the following summer, but some persist as larvae in the soil for several years before completing their development to emerge as adults.

Crop damage and midge activity

Pea midges lay their eggs on pea crops throughout the bud and flowering period. Serious damage and loss of yield occur only when the flower trusses contributing to harvested yield are affected. The longer growing period of dry-harvesting peas frequently allows them to escape, or compensate for, midge damage. Late midge attack on dry-harvesting peas can even be beneficial by stopping their growth and allowing the pods to fill. In vining peas the first three trusses contribute most to yield. With programmed sowings the earliest crops usually escape significant attack, but throughout the period of midge activity one or other of the later crops will be at the most susceptible stage.

As the adults are short-lived, the periods of risk depend on the frequency and intensity of flights of adults emerging from the soil throughout June and July. The magnitude and occurrence of flights cannot be predicted accurately as they vary according to previous midge infestations and soil and weather conditions. Thus damage ranges from severe nettle-heading at an early stage of growth, reducing plant yield to well below an economic level, to

Fig.58.3 Infested pea flower with calyx and petal removed to show larvae of pea midge feeding on flower parts

mild nettle-heading at a later growth stage when sufficient pods have been set to give an acceptable yield. Damage of this second kind may not allow time for the larvae to complete their feeding and escape to the soil before the crop is vined, thus limiting the build-up of surviving populations in the soil. Intermittent, light attacks cause individual flower or pod loss throughout the flowering period without noticeable check to extension growth.

Inspection of crops

On receipt of warnings of the imminent emergence of adult pea midges, growers in areas previously affected by midge should inspect their pea crops regularly to determine those at the susceptible stage.

The earliest susceptible stage seems to be when the 'oldest' green bud is about 6 mm long, which is about a week before the crop shows its first white flowers. This stage can be determined only when the leaves and stipules around the growing point are folded back. If the leading shoots are gently squeezed before being opened to determine their growth stage, any midges present will be immobilized. Plants should be inspected to windward of any fields infested by midge in the previous two seasons and in areas sheltered from the prevailing wind where midges are likely to have settled.

Other insects found within pea shoots at the time of pea midge invasion and likely to be confused with them are pea thrips and the small parasitic wasp *Pirene graminea* Haliday. Pea thrips (see page 89) is yellow or brown, very slim and active with 'bristly' wings contrasting with the membranous wings of the midge. The parasitic wasp is shiny black and smaller than pea midge but looks more robust; its 'elbowed' antennae are usually held in front of the head as it searches for its host. The finding of these parasites indicates the presence of pea midge in the crop. Although there may be two or three parasites per flower shoot, they do not give sufficient control of the pea midge.

Chemical control measures must be applied when one or more midges are found within the leading shoots of one in five plants at the susceptible growth stage.

Shoots and buds already attacked by pea midge are easily detected by their distorted appearance and the presence of larvae. If the crop has not set sufficient pods, immediate application of chemical control measures may ensure continued extension growth to obtain sufficient pods, but their uneven maturation may still affect vined yield.

Growers in areas usually unaffected by pea midge should inspect their most susceptible crops when pea midge warnings have been issued, so that any new areas of infestation can be detected and the necessity for control measures determined.

Control measures

Cultural control

Crop isolation to avoid pea midge attack is not feasible in the main pea-growing areas in Britain, but the sowing of peas on land adjoining previously infested land should be avoided. For dry-harvesting peas, early drilling, so that the earliest pods are set before the period of midge attack, ensures that extension growth is maintained and losses minimized, but this is not possible with all sowings of a cropping programme for peas grown for processing.

Chemical control

The use of insecticides to control this pest in dry-harvesting peas should rarely be necessary. On vining peas insecticide sprays have given variable and often disappointing results. Kill of adult midges and larvae has been observed but has not always resulted in yield increase. This is due to the following factors.

(1) The difficulty of timing sprays to coincide with midge flight.
(2) The rapid growth of the crop so that unsprayed haulm is present when later invasions of midge occur.
(3) A single effective spray is likely to protect only two flower trusses capable of producing about 4–6 pods; their contribution to the final yield is determined by their position on the vine, later pest and disease attack and the growing conditions that affect the rate of maturation.

Provided the above limitations are appreciated, a routine spray programme should be used in areas where this insect is a persistent pest. Once warnings have been issued or adult midges are found in the crop, the first spray of a recommended insecticide should be applied to vining crops in sequence as they reach the earliest susceptible stage. If monitoring of midge activity indicates prolonged attack, or crop growth is slow owing to adverse weather, a second spray should be applied 6–10 days after the first. Such a spray programme must be considered as an attempt to secure a vinable crop: it offers the best possibility of successful control at present. Correct timing of sprays is essential for the control of midge. Too often, growers apply sprays at the wrong growth stage of the crop and obtain poor control.

The spraying of susceptible crops must not be delayed. Where necessary, land machines should be used to apply spray at medium to high volume (450–700 litres/ha).

59
Raspberry cane midge and midge blight

Fig.59.1 Second-generation larvae of raspberry cane midge feeding beneath rind of first-year raspberry cane

Raspberry cane midge (*Resseliella theobaldi* (Barnes)) is mainly a pest of cultivated raspberry but occasionally attacks loganberry. This midge was unknown in Britain until 1920 when it was reported in Kent and Surrey. It spread throughout England and by 1973 had reached

Scotland, where it became established in the raspberry-growing areas.

The feeding of the midge larvae damages infested canes but, more important, it renders them liable to infection by certain fungi which cause the disease known as 'midge blight'. Like 'cane blight', which is generally associated with mechanical wounds, midge blight can lead to death of the fruiting canes. However, midge blight is a disease of raspberry only, whereas cane blight is a disease of several types of cane fruit. Midge blight was not recognized in England until the 1950s and remained undetected in Scottish raspberries until the early 1970s when a widespread outbreak occurred. It is prevalent in many raspberry-growing areas in Europe, where serious outbreaks often occur in cycles, each lasting several years.

Description and life history of the midge

The adults are small and reddish-brown. They are difficult to identify in the field because they are similar to other midges. The adults of the overwintered generation emerge from late April until early June in England and from late May until mid-June in Scotland. Males usually emerge a few days before females. Mating occurs soon after emergence of the females. Many small, elongate, translucent eggs are laid in rows under the flaps of splits or wounds in young canes. Splits suitable for egg laying by the overwintered females usually develop in the lower 40 cm of the canes, often below the leaf stalks. An infestation reaches damaging proportions only in years when the peak of adult emergence of the overwintered generation coincides with the development of splits.

The larvae hatch from the eggs after 7–10 days and feed in the succulent outer tissues (cortex) of the cane. They are translucent at first but soon become bright pinkish-orange or yellow. After two to three weeks the fully grown larvae, which are about 3 mm long, fall to the ground and burrow into the soil to a depth of 1–4 cm where they form a cocoon and pupate. Adults start to emerge in late June, just before harvest, in England and in July and early August, during harvest, in Scotland. At this time susceptible raspberry varieties develop extensive splits as the cork matures and the rind peels. Numerous eggs are laid in this type of split at the base of the young canes and the larvae of this second generation feed on the newly exposed tissues (Fig. 59.1). When fully grown, these larvae drop to the ground, burrow into the soil and spin cocoons. Some larvae pupate and emerge to produce a third generation of larvae on the canes, particularly in the warmer areas of England such as Kent and the south-west, but most overwinter as larvae to emerge as adults in the following spring.

The adults fly weakly and the pest is probably spread when larval cocoons are accidentally distributed with soil adhering to new planting stocks.

Nature of damage

There is a marked difference in the damage caused by the first and by the second and subsequent generations of larvae. The first-generation larvae cause deeply penetrating lesions which become cankered by August (Fig. 59.2); they damage the tissues and allow fungi to destroy part of the vascular cylinder before the stem cork is mature. Wound cork is produced around the lesions, preventing further damage, and the undamaged vascular cambium continues to produce vascular tissues normally. However, these cankered canes are physically weak and liable to be broken by wind, machinery or pickers; if unbroken, they may survive to fruit in the following year.

The second and third generations of larvae damage the mature cork layers when feeding and allow fungi to penetrate the conducting tissues at a time when canes cannot repair the damaged tissues. The lesions (Fig. 59.3) again arise from infection of feeding areas by fungi, which in severe attacks girdle the canes. The lesions, which appear as irregular brown

Fig.59.2 Raspberry canes showing cankered lesions caused by feeding of first-generation larvae of raspberry cane midge

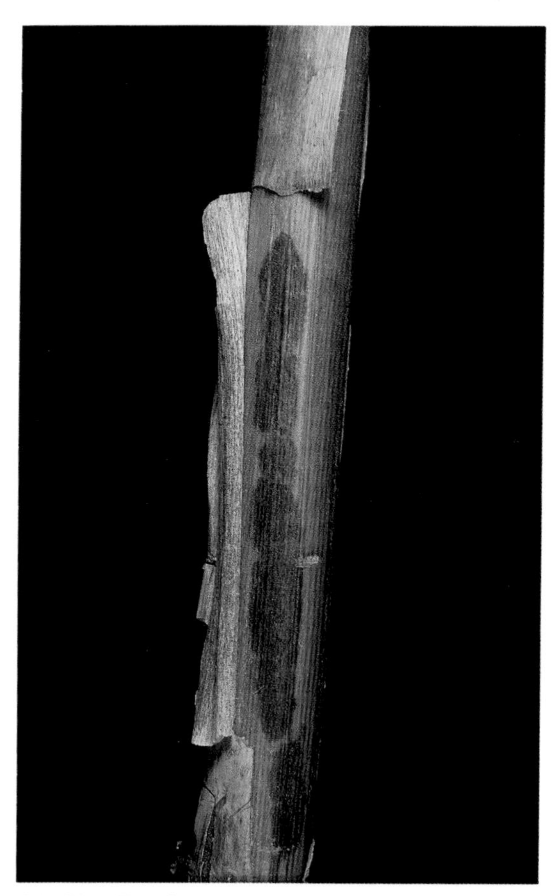

Fig.59.3 Raspberry cane showing lesion caused by fungal attack of tissues damaged by feeding of larvae of raspberry cane midge: surface view after pulling back split rind

patches (patch lesions), can best be seen by scraping the cane in winter to remove all the rind and cork; by contrast, the uninfected canes are green. In some older plantations, some of the scraped canes show spreading brown lesions caused by the cane blight fungus, *Leptosphaeria coniothyrium*, spreading from the irregular patch lesions caused by midge blight. This suggests that *L. coniothyrium* can infect midge-feeding sites on canes if inoculum is available on the old wood.

A cane damaged by midge blight may appear healthy during the winter and so be selected for fruiting the following season. However, if it has been attacked by many larvae, some fruiting lateral shoots may not grow or, if girdled by lesions, no laterals emerge and the cane may wilt and die.

Varietal susceptibility

Raspberry varieties in which growth splits occur freely in the young canes, such as Glen

Clova, Glen Moy and Malling Jewel, are more likely to be affected by cane midge and midge blight. Varieties in which growth splits are rare, such as Norfolk Giant, are less affected.

Natural enemies of the midge

In Britain several parasites of raspberry cane midge have been identified but only two are common: *Synopeas craterus* (Walker) (a platygastrid wasp) and a species of *Tetrastichus* (an eulophid wasp). *S. craterus* has been found in Scotland infesting second-generation larvae. The parasitized larvae complete their development on the cane, fall to the ground and spin cocoons as usual, but the parasite continues to develop within the cocoon and an adult wasp emerges in the following July. *Tetrastichus* spp. have been recorded from Kent. Larvae parasitized by one of these species remain on the cane while the parasite completes its development within them.

None of these parasites adequately controls midge populations in Britain, so other control measures are needed.

Control measures

Cultural control

The first flush of young canes of vigorous varieties such as Glen Clova may be removed, when 10–20 cm high, to reduce the risk of cane damage and subsequent infection by the cane blight fungus. A second flush of canes grows but is shorter and less likely to be damaged. These replacement canes escape serious midge infestation because their splits are small and develop after most of the overwintered female midges have emerged and died. Thus, where vigour control is practised, very few midges succeed in laying eggs, so the second generation is small or absent and the use of insecticides should be unnecessary.

Biennial cropping

In Britain there is no evidence that cane midge can lay eggs on the fruiting canes. Therefore, biennial cropping has potential for control of this pest, providing large blocks are grown in the same growth phase to minimize spread of adults from annually cropped raspberries in neighbouring plantations.

In the non-fruiting year, sprays timed to coincide with the second generation are easily applied in the absence of fruiting canes. The young canes should be examined when the rind begins to peel and sprayed if midge larvae are present. In the fruiting year, no sprays need be applied, providing all the primocane growth is removed early, either physically or chemically.

Chemical control

The most serious crop damage is caused by the second-generation larvae, which are difficult to control with insecticides because they emerge just before or during harvest. However, serious damage can usually be avoided by controlling the first generation. A recommended insecticide applied at high volume (1000 litres/ha) to the basal 40 cm of the cane when the overwintered midges begin to emerge, and again 10–14 days later, should give adequate control of adults and first-generation larvae. The addition of a non-ionic wetter to the spray will improve cover of the young canes.

The timing of sprays is important and a warning system is operated throughout Britain. Flight of adults of the overwintered generation depends on temperature and varies from season to season. It occurs 2–3 weeks later in Scotland than in south-east England, but even within one locality midges may emerge at different times, partly because of differences in altitude and aspect.

Routine spraying should be avoided because, if natural splits in the cane develop late, spraying is unnecessary. Scraping and examining samples of susceptible varieties during the winter can give some estimate of an overwintering midge population and the possible need for spraying.

In southern England sprays may be applied after harvest to control the third generation of larvae if sprayers can be used in the plantation without causing excessive damage.

60
Saddle gall midge

Fig.60.1 Barley stems showing symptoms of attack by larvae of saddle gall midge

The saddle gall midge (*Haplodiplosis marginata* (von Roser)) is associated with frequent cereal growing on heavy land. It is widely distributed in most European countries and was first recorded in Britain at Alford, Lincolnshire, in 1889. Sub-economic levels are now common throughout the Midlands and in eastern England from Yorkshire to Kent. Heavy localized attacks have resulted in yield losses of winter wheat and spring barley in most of these areas.

All cereals and most grasses can serve as hosts of the larvae, but there is a wide variation in the degree of susceptibility. Wheat and barley are high susceptible with spring-sown crops at greater risk than those sown in late autumn. Oats are poor hosts and rarely suffer economic damage; rye is intermediate. With the exception of couch-grass (*Elymus repens*), which can support and maintain large populations, grasses are also poor hosts.

Description and life history

The adults are red midges up to 5 mm long. They appear from late May onwards. The large egg-bearing females can easily be seen resting or laying eggs on leaves, but the relatively slim, very active males are rarely seen. After mating, the female lays groups of red eggs in a raft or chain-like pattern on the upper or lower surface of leaves (Fig. 60.2).

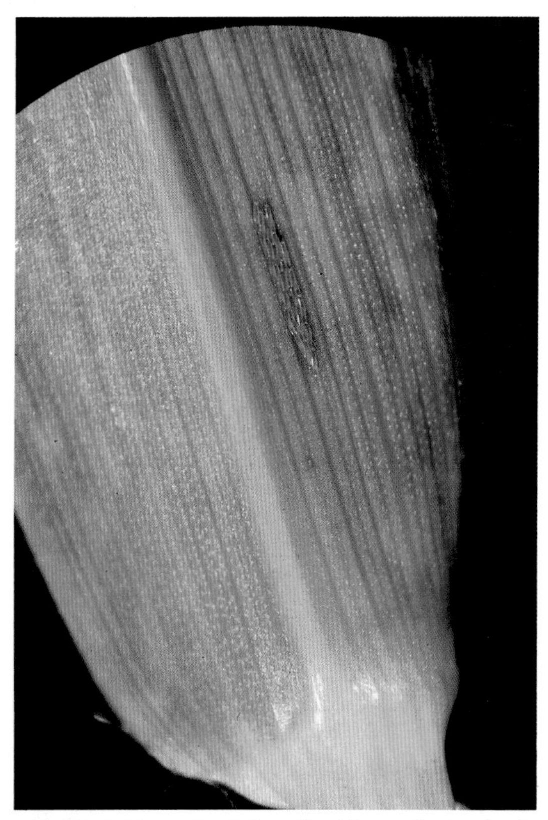

Fig.60.2 Eggs of saddle gall midge on barley leaf

The eggs hatch within 1–2 weeks and the newly hatched larvae move down the leaf to feed on the surface of the stem within the protection of the leaf sheath. At first the larvae are whitish-green but become orange-red by the time they are fully grown in mid-July (Fig. 60.3). They then fall to the ground, enter the soil and overwinter as larvae in tiny mud cells. Larvae which are going to become adults in that year pupate in May; the adults emerge 10–20 days later according to rainfall and temperature. Other larvae may remain in the soil for one or more years, particularly if the weather during May and early June is very dry and the soil very hard. There is only one generation a year.

Nature of damage

Larval feeding results in the formation of galls (Fig. 60.3), which appear as saddle-shaped depressions, swollen at either end; it is from these that the pest derives its common name. These symptoms occur mainly on the top three internodes but may be present on the lower internodes of backward crops. If larvae are numerous, the galls may fuse together and the stem may be completely destroyed. Visual symptoms may not be apparent during the growing season, except in prolonged wet periods when moulds may develop on the damaged areas. In the absence of visual symptoms, an attack can be diagnosed by the stem surface having uneven contours caused by the underlying galls. In dry seasons, damage does not become apparent until the straws ripen and harden. At this stage, attacked stems, particularly of barley, bend over (Fig. 60.1) and may even break off at the point of injury, causing the crop to look untidy and ragged.

Galls interfere with the flow of nutrients in the stem. Direct yield loss may therefore result

Fig.60.3 Galls and larvae of saddle gall midge on wheat stems

from incomplete filling of the ear or shrivelled grain. Indirect losses arise from unharvested heads which have become completely severed at the point of damage or hang below the level of the combine cutter-bar. Such losses are particularly severe in barley, but minimized in wheat by the stiffer nature of the straw.

Factors influencing attack

Although infestations occur on crops growing in a range of soil types, damaging populations are mostly restricted to heavy soils. With a pre-sowing soil population above the economic threshold of 6–12 million larvae per hectare, the severity of attack depends on weather conditions and crop growth.

High temperatures and showery weather in May and June, which create warm and damp soil conditions, encourage mass emergence of midges and abundant egg-laying in a short period, leading to maximum damage. By contrast, in drier, cooler conditions adult emergence is protracted but crops continue to grow and can withstand attack. In very dry soil a large proportion of the larvae delay pupation until more favourable conditions occur in succeeding years. Yield losses are minimal in hot, dry summers, when many eggs fail to hatch, or when heavy rain occurs in early summer, washing eggs off the foliage. Yield loss is increased by wet, humid conditions because there is secondary invasion of the damaged areas by bacteria and fungi.

Loss of yield is also related to the growth stage of the plant at the time of attack and hence to sowing date and crop vigour. Yield is little affected when egg hatch coincides with, or is near to, ear emergence, but backward crops or those in the process of stem extension at the time of attack suffer much heavier losses.

Control measures

Most infestations are below damaging levels and have little effect on yield. Population increases are limited by soil type and various husbandry factors, so control measures are rarely necessary. Where damaging infestations do occur, populations can be reduced by the following cultural or chemical means.

Cultural control

Crop rotation A break from wheat or barley by the substitution of a non-cereal crop for one, or preferably two, years will allow populations to decline to a relatively safe level. Oats can also be grown as a break crop but, to achieve maximum benefit, should be sown early as numerous eggs have occasionally been laid on young, late-sown oats.

Early sowing Yield loss can be reduced by early sowing and by husbandry practices designed to ensure quick germination and vigorous early growth.

Weed control Couch-grass can maintain midge populations in the absence of a host crop. A high standard of weed control is necessary to take full advantage of crop rotation.

Chemical control

Attempts to kill overwintering larval populations in the soil or larvae within the leaf sheath by systemic insecticides have been unsuccessful. Preventive treatment with a recommended insecticide spray is therefore necessary and must be directed against newly hatched larvae as they migrate from the egg-laying to the feeding sites. Timing is critical and application must coincide with peak egg hatch and larval migration, i.e. about 10 days after peak midge activity.

61
Wheat blossom midges

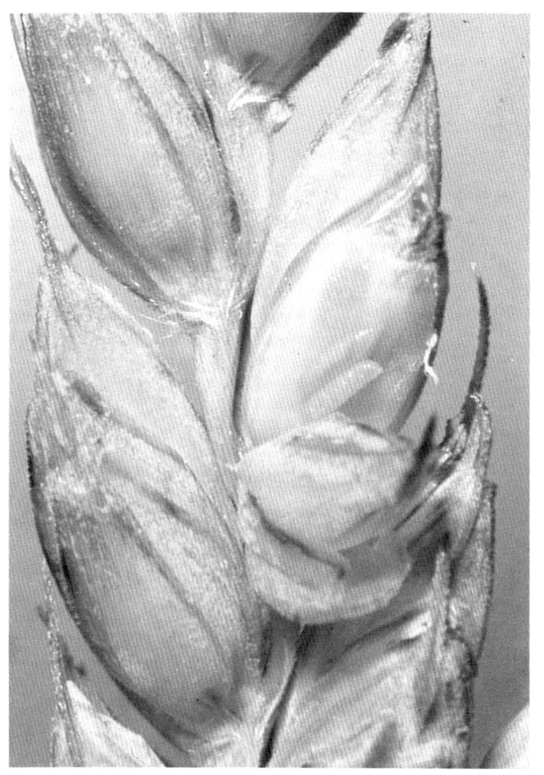

Fig.61.1 Two mature larvae of orange wheat blosom midge feeding on wheat grain. Parts of the ear have been removed

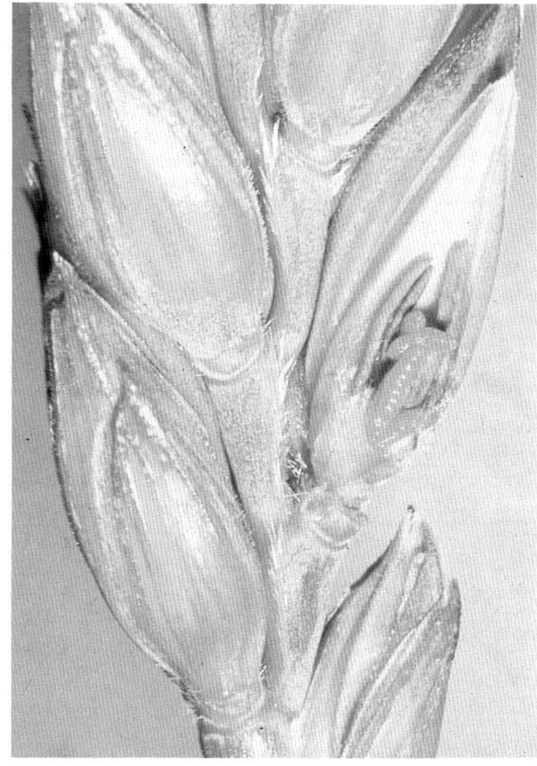

Fig.61.2 Larvae of yellow wheat blossom midge feeding on blackened anthers in wheat floret where no grain has formed. Floret has been exposed to show larvae

Two species of wheat blossom midge are pests of cereal crops: the orange wheat blossom midge (*Sitodiplosis mosellana* (Géhin)) and the yellow wheat blossom midge (*Contarinia tritici* (Kirby)). The larvae of both species feed within the floret and can be distinguished, except in the final stage, by their colour.

The two species occur sporadically as pests throughout Europe, Asia and North America, particularly where wheat is cropped intensively. Although both species attack wheat, rye, barley and oats, the only significant damage reported in Britain has been to crops of wheat and rye. The yellow wheat blossom midge sometimes has a second generation which attacks couch-grass (*Elymus repens*).

The larvae of grain thrips (*Limothrips cerealium* Haliday) may also be found feeding within cereal florets; these larvae are also orange but have dark heads and legs and are more active than larvae of wheat blossom midges.

Nature of damage

Orange wheat blossom midge

The larva feeds on the grain (Fig. 61.1) as it swells during July and early August, causing a depression and a loosening of the seed coat, which often cracks during ripening. The plant tissue is broken down by enzymes secreted by the larva, which then absorbs the fluid that is formed. This feeding results in smaller grains (Fig. 61.3) with reduced germination capacity and poorer milling and baking qualities. When more than two larvae feed on a grain it is reduced to an empty husk.

Yellow wheat blossom midge

The larva feeds on the flower (Fig. 61.2), killing the stigma and preventing pollination and development of the grain. It feeds to maturity on the anthers, which remain within the floret. Infested florets do not swell so the ear has a flattened appearance when heavily infested. If fewer than five larvae are present in a floret or if the larvae do not hatch before the flower is pollinated, the grain develops more or less normally although some damage, similar to that caused by the orange midge, can occur.

Description and life history

Orange wheat blossom midge

All stages are orange. The adult is a medium-sized midge about 3 mm long (Fig. 61.4). The male has long feathery antennae. The female is more robust in appearance and has a short ovipositor. The adults emerge throughout June. The males mate with the females at the emergence site and then die. The females, which live for two or three days, fly at dusk, seeking crops at the right growth stage on which to lay their eggs. They fly within or just above the crop and can travel about one kilometre (half a mile), but they do not fly at wind speeds above 10 km/h (6.25 mph) or at temperatures below 15 °C.

Eggs are laid in florets that are clear of the 'boot' but have not flowered (Growth Stages 55–65) (Fig. 61.4). Midges start laying on the lower ears in a crop at about six o'clock in the evening and work higher as the light fails; egg laying continues until dark. The eggs are cylindrical and most are laid within the floret, though the orange midge also lays on the outside of the floret and on the glumes. Each female lays many batches of 1–8 eggs and under ideal conditions in the laboratory can lay 100 eggs or more.

The time taken for the eggs to hatch depends on the mean daily temperature: more than 10 days at 15 °C and less than four days at 20 °C. The number of larvae in each floret varies from one to about 60 in the heaviest attacks, when several batches of eggs of either or both species may be laid in the same floret. Normally only one or two orange midge larvae survive from each egg batch. They feed for 2–4 weeks and grow to a length of about 3 mm. The mature larva (Fig. 61.1) retains its skin after the last moult and is therefore paler than the other stages.

When fully fed in late July or early August the larva leaves the ear, often after rain, and drops or crawls to the ground. It burrows a few centimetres down into the soil and spins a cocoon, which is later buried more deeply by cultivation.

Not all larvae become adult in the following summer; most do so within three years but orange midges have been recorded emerging after 13 years in the soil. The reactivated larva leaves its cocoon and moves up near the soil surface to pupate. Emergence of adults is

Fig.61.3 Wheat grains damaged by larvae of orange wheat blossom midge

Fig.61.4 Orange wheat blossom midges on wheat ears (Growth Stage 59)

controlled by fluctuations in temperature and varies according to local climatic differences, but it can occur within a week of pupation.

Yellow wheat blossom midge

All stages are pale yellow. The adults (Fig. 61.5) are similar in size to those of the orange wheat blossom midge but are thinner with longer antennae in the male and a longer ovipositor in the female. They emerge mainly in early June. The life cycle is generally similar to that of the orange midge but differs in several important respects.

The yellow midge starts to lay its eggs as soon as the 'boot' splits to reveal the ear (Fig. 61.5). It does not lay once the floret hardens, which occurs about a day after the floret passes out of the 'boot' (Growth Stages 51–59). Each female lays a few batches of 8–30 eggs (cf. orange midge).

As with the orange midge, egg hatch is affected by temperature, so that at normal daily temperatures the flower will be pollinated before the eggs hatch. This is important as the larvae must arrest development of the flower before pollination in order to retain the anthers in the floret. Where pollination succeeds, the grain develops more or less normally and damage is only slight.

Usually 4–15 yellow midge larvae survive from each egg batch and up to 60 per floret have been recorded in heavy attacks. The larvae tend to leave the ear before those of the orange midge and can jump into the air to facilitate their downward journey. Unlike orange midge larvae, a few larvae of the yellow midge may pupate without spinning a cocoon and develop without a resting period, giving rise to a second generation of midges in the same year — in August/September. These midges lay eggs in couch-grass and a second generation of larvae is produced. When fully fed, these larvae find their way to the soil where they spin cocoons.

Fig.61.5 Yellow wheat blossom midges on wheat ears (Growth Stage 51)

Most of the yellow midge larvae pupate in the following spring and adults emerge shortly afterwards. However, some larvae remain dormant in the soil for up to three years before producing adults; longer emergence periods have not been recorded for this species.

Natural enemies

During the emergence of the adults and the return to the soil by the slow-moving larvae, a large proportion may be taken by predatory beetles or spiders.

Several parasites lay their eggs within the egg of wheat blossom midges and, about a week after the main midge emergence, large numbers of these parasites may be found searching for midge eggs. The most common of these are the wasps *Leptacis tipulae* (Kirby) and *Pirene penetrans* (Kirby), which can exert useful control of orange wheat blossom midge populations but only rarely attack the yellow wheat blossom midge. The egg of the wasp is laid within the midge egg but remains dormant while the midge larva feeds and overwinters. It hatches when the midge larva becomes active in the spring. The parasite larva then rapidly consumes the midge larval tissues, pupating within the midge larval skin. The adult wasps are most active during the late flowering stages of the crop (Growth Stages 65–71). When there is a severe midge infestation, the application of aphicides at this time, or late midge sprays, should be avoided to protect the parasites.

Forecasting attacks

The overwintering larvae can be extracted from soil samples to obtain an estimate of the potential population on a farm during the following season. The amount of damage caused depends on the proportion of the total population that hatches during the susceptible stages of crop growth. When the soil temperature in May is low or the mean daily temperature during ear emergence is below 15 °C, damage will be slight. Most damage will occur in a season in which temperatures rise rapidly in May or June and remain high throughout the period of ear emergence.

Control measures

Cultural control

A regular rotation of wheat crops will do much to keep midge numbers down. Significant damage occurs only in first wheat crops if the adjoining field was severely damaged in the previous year. Second and subsequent wheat crops will have a resident source of infestation and will be at greater risk from attack. On a farm cropping a third or more of its land with wheat, there is likely to be a rapid increase in midge numbers if conditions favourable to midge reproduction persist for two or three years. This increase can happen even where wheat crops are rotated, as there is little chance of the adults dispersing.

Chemical control

Estimates of midge numbers The adult midges can be seen on the ears if the field is examined at dusk. At other times the midges rest at ground level but fly when the crop is disturbed. Some estimate of numbers can be made during the day by parting the crop and relating the number of midges that fly to the number of cereal heads disturbed. If more than one adult midge per ear (or per two ears in a seed crop) is seen in a traverse across the whole field, then an insecticide treatment is warranted.

Sprays Adult midges are killed by many insecticides, but the best control is achieved by chemicals which also permeate the floret and kill the eggs within. A spray of a recommended insecticide should be applied between the middle of ear emergence and the beginning of flowering (Growth Stages 55–61): spraying later than the middle of flowering (Growth Stage 65) will kill the parasites but few midge larvae. Yield responses of up to one tonne per hectare have been recorded following chemical control of midges.

62
Wheat bulb fly

Fig.62.1 Wheat damaged by maggots of wheat bulb fly: (left) two seedlings attacked before or about time of emergence; (right) two plants attacked after emergence

The wheat bulb fly (*Delia coarctata* (Fallén)) is probably the most serious insect pest of winter wheat in Britain. Damage is caused by the maggots feeding within the shoots. There is considerable annual variation in the intensity of the attacks but occasionally these are severe. Wheat bulb fly is particularly prevalent in the eastern half of Britain as far north as the Angus District of Tayside Region, but even within this area it is rather local.

Wheat, rye and barley are liable to be attacked but, because of the habits of the fly, susceptibility is greater when the crop is sown in autumn or winter. Spring wheat and barley are at risk if sown before late March but suffer little or no damage if sown later. Oats are immune.

Conditions for attack

Wheat bulb fly tends to occur under two different sets of conditions which influence the control measures that can be taken.

On heavy land, such as the boulder clays of East Anglia, serious attacks may occur where wheat follows a fallow or a bastard fallow. Attacks are not usually serious after potatoes, other root crops or field beans.

On light land, including peats and silts, serious attacks occur often after potatoes and, to a lesser extent, after other root crops such as beet, celery and onions. The most serious damage may occur after crops which are removed early.

On both heavy and light land, attacks may follow dwarf French beans or peas, particularly vining peas which are removed early and, to a lesser extent, dry-harvesting peas. Attacks may follow oilseed rape crops if the stubble is cultivated early, leaving bare soil in the field during early August.

Description and life history

The adult is similar to a house fly but slightly smaller. The males are dark brown and the females yellowish-grey. They may be seen from late June until August, often in calm weather, on the ears of wheat where they feed on fungal spores.

Eggs are laid from late July until early September. The females have the curious habit of depositing their eggs in bare soil or in soil under a root crop. The eggs are laid just beneath or on the surface of the soil and are unusual in that they hatch during the winter, starting about the middle of January and continuing throughout February. If a long spell of frost occurs during these months, hatching will be delayed until the ground thaws. The newly hatched maggots (legless larvae lacking a distinct head) die unless the field has been sown with wheat, barley or rye, or unless host grasses such as couch are present.

The young maggots are white, about 1.25 mm long and cylindrical, but pointed at the front end and blunt at the hind end. They bore into the plant below ground at the so-called 'bulb' at the base of the stem. Within a few days their feeding affects the central shoot of the plant, which withers or becomes yellow or stunted. Although the outer leaves remain green, attacked plants have a rather dull appearance. Without a careful examination the attack may not be noticed until February or March. The maggots feed slowly for the first few weeks, but in the latter part of their life they feed much more voraciously and move from shoot to shoot on the same plant and sometimes through the soil to another plant. Thus fields which at the end of March appear to have survived the attack may succumb in April.

When fully fed the maggots are about 10 mm long and creamy-white (Fig. 62.2). According to the season, this stage is reached between mid-April and early May, when the maggots leave the plants and pupate within a brown barrel-shaped puparium in the soil. The adults emerge 5-6 weeks later, usually during the first half of June. At first they remain in the wheat, barley or rye field but gradually disperse as the females seek egg-laying sites.

Fig.62.2 Maggot of wheat bulb fly in tissues of damaged plant (× 4)

Symptoms of attack

The presence of the pest can easily be overlooked in the early stages, when the farmer may think that unfavourable weather is affecting his crop. To look for wheat bulb fly, plants having yellow central shoots should be slit open. A careful search inside will reveal the small white maggots if present. The point at which the young maggot has entered the plant is readily seen as a brown discoloration if the outer sheath is stripped down (Fig. 62.1(D)).

The maggots of other flies also occur in wheat; damage due to frit fly (see page 260) may be seen from September to January and should not be confused with that caused by wheat bulb fly, which occurs later. From February to April, maggots of the yellow cereal fly (*Opomyza florum* (Fabricius)) may be found in shoots, particularly in wheat sown before mid-October. They are thinner than wheat bulb fly maggots and do not leave an entry hole in the plant because they invade it through the top of the shoot, spiralling down through the inner leaf in a characteristic fashion.

Crop damage

The amount of damage to the crop as a whole depends on several factors. The population of maggots and the stage of growth of the crop are of considerable importance, but factors affecting the vigour of the crop, such as weather, fertility and soil consolidation, are all important.

In exceptional circumstances, e.g. where late sowing and unfavourable weather have delayed germination until about the time when the eggs hatch, the seedlings can be attacked and killed before they have emerged (Fig. 62.1(A)). This may result in a very thin stand.

Control measures

Cultural control

Damage can be reduced effectively by cultural methods.

Theoretically, the pest could be controlled by a change in crop rotation, because it occurs only where wheat is taken after a fallow or a crop which provides suitable egg-laying conditions. A change in the sequence of cropping is worth considering where wheat bulb fly is regularly a serious pest. There are, however, practical limitations to this method because, in the heavy and light land areas where the pest is troublesome, wheat is not only the traditional but also the most suitable crop to sow after early-harvested crops.

On heavy land, if the ploughing of leys or seeds for bastard fallows is delayed, egg laying will be reduced, but this may be at the expense of weed control. Fewer eggs are laid in fallows with a smooth tilth than in those where it is rough and cloddy. On bare fallows, a crop of mustard sown to cover the soil by mid-July will also reduce egg laying. Land lying bare after harvest of other crops should not be worked in early August. A herbicide can be used pre-ploughing to kill weeds, especially grassy weeds in leys, since an interval of only 3–4 days is required before ploughing or cultivating and re-sowing. However, a short interval between destruction of the sward and sowing may increase the risk of infestation by aphids carrying barley yellow dwarf virus (see page 39) or by the maggots of frit fly if either are present in the ley.

On both light and heavy land early drilling with an increase in the seed rate is the best control measure. If the wheat can be sown in October, or at the latest the first week in November, in a normal season it will have started to tiller before the eggs have hatched. This should enable the plant to withstand attack by the maggots. The same population of maggots can do more damage to a crop which has not tillered, for a rapidly tillering crop may quickly replace shoots that are killed.

Deep sowing due to incorrect drill setting or to insufficient consolidation will give a weak and backward plant susceptible to attack.

Spring wheat planted in areas subject to wheat bulb fly attack should be drilled from late March onwards to avoid damage, although such late drilling will incur some loss of yield.

Chemical control

In spite of much careful experimental work, there is still no insecticidal treatment that is completely effective.

Seed treatments can give useful protection, particularly to later-sown crops, i.e. after mid-October, or to crops that are likely to be backward at the time of attack owing to unfavourable conditions. These treatments are more effective on shallow-sown crops than on crops sown more deeply. However, good soil cover should always be provided to minimize risks to wild and game birds.

Granules or sprays applied at, or soon after, drilling Recommended persistent insecticides may be incorporated in the top 25–50 cm of soil at, or immediately after, drilling. These treatments are more costly than seed treatment but are worth while on fields with large wheat bulb fly populations. Some treatments may be less effective on soils with a high organic matter content.

Sprays applied before, or at, egg hatch Sprays of recommended persistent insecticides may be applied at the early stage of egg hatch, which is normally during early January. The effectiveness of these egg-hatch sprays is reduced on soils with a high organic matter content; in these situations different application rates may be recommended.

Sprays applied at the first signs of damage Sprays of recommended systemic insecticides may be applied at the onset of plant damage. Such treatment cannot restore lost tillers but will reduce further damage. The effectiveness of these sprays decreases as the maggots mature; they are therefore of greatest value during the early stages of plant attack.

Re-drilling

As wheat has remarkable powers of recovery, it is always difficult to decide whether to re-drill, patch or leave a badly attacked crop. Given good weather, an apparent failure may develop into a good crop. If re-drilling is decided upon, a period of about 2–3 weeks should be allowed between ploughing the existing crop and re-drilling. If this interval is not allowed, there is a risk of maggots migrating to and attacking the newly sown crop before it emerges. Patched crops may prove difficult to handle at harvest time because of uneven ripening; re-drilling is generally preferable.

WASPS

Hymenoptera: Apocrita

63
Wasps

Fig.63.1 Wasps feeding on ripe apple (× 3)

Wasps are one of the most familiar and generally disliked groups of British insects. Their striking coloration makes them easily recognizable and their sting causes them to be feared. However, this fear is partly misplaced because wasps of most species rarely sting unless they are aroused or frightened. Contrary to popular belief, they are beneficial in spring and early summer when they feed their grubs mainly on insects. But from midsummer onwards the worker wasps feed on ripening fruit (Fig. 63.1) as well as on other sweet substances. In late autumn, flies and other protein foods are collected for feeding to the last brood.

Nature of damage

Wasps can cause serious damage to apples, pears, plums, autumn strawberries and certain varieties of grape. With apples and pears, however, it is unlikely that ripening fruits will be attacked unless they have first been damaged by other agents, e.g. birds. Dahlias may also be damaged by wasps gnawing the stems at ground level.

Considerable wastage of goods, as well as persistent worry to people, is caused by wasps in sugar warehouses, jam factories and other

Fig.63.2 Common wasp on flower (× 4)

Fig.63.3 Hornet on piece of bark (× 2½)

places containing sweet, aromatic substances which attract them. In houses they are a nuisance during the cooking and eating of meals.

Wasps are a serious pest of hive bees in some seasons. In the spring they pounce on foraging worker bees visiting flowers, bite off their wings and legs and carry away the bodies to feed to the wasp grubs. Later in the year they rob the hives of honey and take away the bee grubs and pupae; an attacked colony is seriously weakened and may ultimately die out. Wasp attack is particularly devastating in a queen-raising apiary where small nucleus colonies are too sparse to defend themselves.

Description

Wasps are fairly large insects with a very marked constriction (waist) in the middle of the body. They are brightly banded in yellow and black (Fig. 63.2) or, in the case of the hornet, yellow and brown (Fig. 63.3). The needle-like sting, possessed only by the females, is concealed near the tip of the abdomen. As with bumble-bees and hive bees, both queen and worker wasps are females and develop from fertilized eggs. Workers or queens are produced according to the diet of the grubs. Workers are smaller than queens and never lay fertilized eggs. Male wasps, which are also smaller than queens, develop from unfertilized eggs laid by queens or workers.

Wasps are related to bees and ants and, as with bees, there are both solitary and social kinds; only the latter need be considered as pests. Including the hornet, there are seven species of social wasp in Britain. Of these, by far the most abundant are the common wasp (*Vespula vulgaris* (L.)) (Fig. 63.2) and the German wasp (*Vespula germanica* (Fabricius)), both of which nest underground and in cavities in trees, walls and buildings, e.g. under roof tiles and in lofts. The tree wasp (*Dolichovespula sylvestris* (Scopoli)) is a locally common and very aggressive species, which builds its nests in trees and other aerial sites or underground. The Norwegian wasp (*Dolichovespula norwegica* (Fabricius)) is rare in southern and south-eastern England, quite abundant in the rest of England and all of Wales, and is the dominant wasp species in northern Scotland. It makes nests among twigs and branches; often a nest includes twigs which pass right through it giving extra support. The cuckoo wasp (*Vespula austriaca* (Panzer)) is a species lacking workers and living in the nests of the red wasp (*Vespula rufa* (L.)), which are

Fig.63.4 Honey bee on raspberry flower (× 2½)

underground. The hornet (*Vespa crabro* L.) (Fig. 63.3) is the largest species but is not very aggressive. It occurs locally in southern England, nesting in hollow trees and buildings.

Insects mistaken for wasps

Honey bees (*Apis mellifera* L.) (Fig. 63.4) can be distinguished from wasps (Fig. 63.2) by: (i) the less conspicuously banded abdomen which is orange-brown and brown (not yellow and black); (ii) the brown and furry thorax (not black and shiny); (iii) the brown, furry and strongly built hind legs (not yellow, shiny and slender), which often carry pollen (wasps do not carry pollen); (iv) the wings which at rest are not folded; and (v) the waist which is less obvious than that of wasps.

Fig.63.5 Currant hover fly (*Syrphus ribesii* (L.)) on flower (× 4)

Some insects that do not sting are also mistaken for wasps or hornets: hover flies for the former, the giant wood wasp (*Urocerus gigas* (L.)) for the latter. **Hover flies** can be distinguished by their alternate darting and hovering flight and the fact that they have only one pair of wings, unfolded at rest, whereas wasps have two pairs which they fold at rest. There are several species of hover fly which vary considerably in appearance: some (e.g. Fig. 63.5) closely resemble wasps, while others are more like bees. The **giant wood wasp** is distantly related to the social wasps but is larger and lacks the narrow waist of the latter. The female has a long, rigid, egg-laying structure which is often mistaken for a sting.

Life history

In spring the overwintered queens leave their hibernating quarters to seek nesting sites. Having selected a site, a queen starts to build her nest with a papery material that she makes by chewing small pieces of wood mixed with saliva (Fig. 63.6). The nest consists of a canopy fixed to the top of the chamber or cavity and, under the canopy, a stalk to which several cells (generally four) are attached. The queen continues to add further cells until she reaches the sides of the nesting cavity, when she suspends a new tier from the first. The openings of the cells face downwards and, as she lays an egg in each completed cell, she sticks it to the roof. When the grubs hatch, they do not completely emerge from the eggs and are thus prevented from falling out of the cells. The queen feeds the grubs until they are ready to pupate, when they seal their own cells with a silk-like substance.

By about the beginning of July, there are enough adult workers to take over the duties of nest building and foraging to feed the grubs.

Fig.63.6 Nest of wasp

The queen then stays in the nest and devotes herself entirely to egg laying. Nest building is continued until there are possibly seven or eight tiers, with enough room between each tier for the wasps to move about freely. A flourishing colony of the common or German wasp may eventually consist of several thousand workers.

During mid- and late summer, males and young queens are produced. The fertilized young queens fly off to hibernate in dry protected places, such as under bark and in sheds. Males and workers in dwindling numbers continue to be active into late autumn or even early winter.

The wasp community resembles that of the bumble-bee in being annual and never producing swarms. Nests are not recolonized the following year, but a specially favourable site may be used again and a new nest built.

Food

Whereas the grubs require a largely protein diet to maintain healthy growth, the intensely active workers need mainly energy foods, i.e. carbohydrates. The food of workers consists of the nectar of certain flowers, e.g. cotoneaster and ivy, and a variety of other sweet substances including fresh and processed fruits. This diet is also given to the very young grubs for a short period. The grubs are fed mainly on other insects, portions of which are first masticated by the workers, but fresh and decaying meat and fish are also used. There is also an exchange of foods between the grubs and the workers feeding them, the grubs secreting a sweet fluid which is imbibed by the workers. This exchange may help to ensure that the workers tend the brood adequately.

The queens, when confined to the nest, are fed by the workers on a liquid mixture of nectar, fruit and meat juices. Males in the nest also obtain food from the workers but, once outside, probably feed only on plant juices.

Wasp stings

The pain of a wasp sting is caused by a toxic fluid containing a complex protein, which is injected through the needle-like sting as it penetrates the victim. Individuals react differently to being stung: some are hardly affected, others suffer considerable pain and swelling, and a few can become seriously allergic to being stung, which in rare cases may result in sudden death due to anaphylactic shock. For normal reactions, a cold compress soaked in witch-hazel may give some relief, while creams or sprays containing certain anti-histamine drugs are beneficial if applied within 20 minutes of being stung. Localized swelling can be a severe problem, especially if the sting is around the mouth or on the neck. For this, antihistamines will help but emergency treatment is essential. In the event of any severe reaction, the patient should be taken at once to the casualty department of the nearest hospital.

Control measures

Nest destruction

Although it is worth while destroying wasps' nests in and around orchards and other places where wasps are a nuisance, no great decrease in numbers can be expected from isolated efforts because workers may forage a kilometre (half a mile) or more from their nests. The nests can be traced by looking at likely sites on a fine day and noting any signs that wasps are associated with a particular spot. Flight lines of wasps should also be watched to see whether they are converging (towards a nest) or diverging (away from a nest).

There are various chemical means of destroying nests.

Insecticide dusts Some insecticides formulated as dusts are effective. A liberal application should be made in and around the entrance of the nest during the evening. Returning workers stir up the dust with their wings and carry it into the nest, so that it reaches the queen and immature stages. A spoon tied to a cane is useful for applying the dust from a distance to avoid being stung. Some dusts may not kill all the immature stages, but there is no need to remove and burn a treated nest as the grubs and pupae are unlikely to survive long enough to reach the adult stage.

Dusts that are repellent to wasps are not recommended for nest treatment as they are likely to cause many hundreds of wasps to hover around the entrance of the nest unwilling to enter.

Contact insecticides When a nest is built inside a cavity wall, under the eaves, or in an attic with the entrance under a tile for example, the only accessible point is the area on which wasps are alighting. Such nests can be destroyed quickly by applying a small quantity of a recommended contact insecticide to the area around the entrance so that the wasps pick up the insecticide on their legs as they land. The product should be applied undiluted with a small paint brush, preferably tied to a cane. The brush must be thoroughly washed after use. Aerosols sold as 'crawling insect killer' may also be used to treat alighting areas. Treatment should be applied during the evening.

Some contact insecticides are extremely dangerous to honey bees; beekeepers should not allow such chemicals on their premises at any time.

Smoke generators Certain insecticide smokes are sometimes used for nests in cavities where there is no fire risk. A smoke pellet should be lit and placed in the nest entrance and the opening sealed.

Fumigants If the nest has been built in an accessible cavity, a slow-release insecticide strip can be used in the space and a polythene sheet placed over the entrance hole.

Certain liquid fumigant mixtures are effective in destroying nests but should be used only by experienced operators of servicing companies, local authorities or government departments and should not be used indoors.

Killing foraging wasps

There is no evidence to prove that killing queens in the spring is a worthwhile means of control, even when it is done intensively over a wide area.

Where worker wasps are a nuisance indoors, one of the following methods may be helpful in controlling them.

Bait-trapping Jam-jar baits may be effective in domestic buildings but are of doubtful value in the open and are ineffective in jam factories, sugar stores, etc. Jars should be filled with a mixture of jam, water and detergent. Sugar or honey should not be used because such baits are a danger to bees.

Spraying and fogging Aerosol fly sprays can be useful for dealing with the occasional wasp indoors, but food must be screened to prevent dying insects from falling into it. Additional care should be taken when spraying working areas because partially affected wasps will increase the risk of being stung.

Electrocution Insect electrocutors can give good control of wasps provided that the grid is sited correctly and alternative sources of light are not too strong.

Screening

Wasps can sometimes be excluded from factories and warehouses by screening window and door openings with mesh of maximum size 3 mm. Where this is impossible, those parts of the building most attractive to wasps should be screened off. This is particularly important where wasps could fall into food.

Bees

Some British species of solitary bee occasionally nest in walls of buildings, excavating holes in weak and crumbling mortar. These are incorrectly referred to as mason, masonry or mortar bees, but the true mason bees belong to a separate family which does not occur in Britain.

The use of insecticide on a wall occupied by solitary bees is unlikely to achieve eradication or prevent further weathering. Mortar joints in brickwork should be scraped out and repointed in late summer or autumn.

If a swarm of honey bees makes its home in a cavity wall or roof space, the local beekeepers' association should be asked to remove or destroy it. (The local police usually have the appropriate telephone number.)

MITES

Acari

64
Bryobia mites

Fig.64.1 Apple foliage in June showing typical bronzing caused by bryobia mites

Bryobia mites infest a range of plants including apple, pear, several *Prunus* species, walnut, gooseberry, hawthorn, ivy, clover, grass, herbaceous plants and, occasionally, cucumbers under glass. For a long time these mites were referred to as gooseberry red spider mite, but now they are regarded as a group of species and perhaps also biological races. Five species have been recognized in this complex. These are the apple and pear bryobia (*Bryobia rubrioculus* (Scheuten)), the grass–pear bryobia (*Bryobia cristata* (Dugès)), the gooseberry bryobia (*Bryobia ribis* Thomas), the ivy bryobia (*Bryobia kissophila* van Eyndhoven) and the clover bryobia (*Bryobia praetiosa* Koch).

In contrast to the fruit tree red spider mite (see page 341) and the two-spotted spider mite (see page 369), bryobia mites are rarely important pests of commercial horticultural crops. They are, however, common on neglected fruit trees and bushes and in gardens. They occasionally cause nuisance by invading dwelling houses.

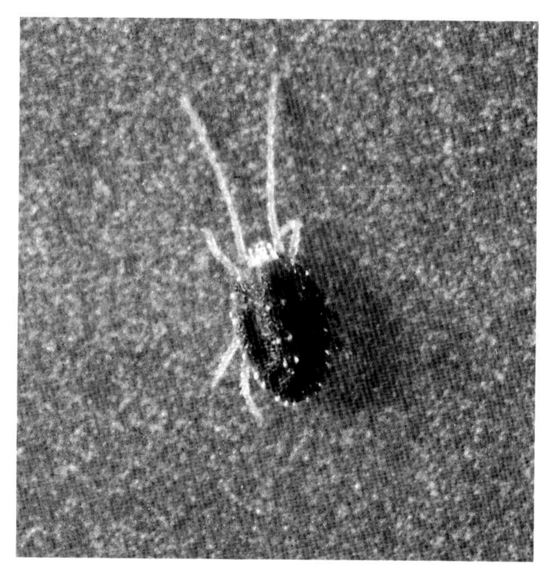

Fig.64.2 Adult female bryobia mite (× 25). Note long front pair of legs

Symptoms of attack

On fruit trees and bushes, a fine speckling develops on the foliage as a result of numerous mites sucking the sap; the leaves then lose their normal deep green colour and become pale and silvered, or bronzed (Fig. 64.1). The rosette and later leaves of fruit trees may be severely damaged. On gooseberry, damage occurs from early spring until June, often starting on the lower branches of the bush and gradually spreading upwards: the leaves become pale and eventually turn brown and wither. Heavily infested bushes may be defoliated.

Damage is similar to that caused by fruit tree and two-spotted spider mites, but silken webbing is not produced.

Description

In most species of bryobia mite the females are parthenogenetic, laying unfertilized eggs that give rise to females; males are unknown or very rare. The adult female is lead-grey, with the front end of the body and legs red or pink. The body is flattened, oval and about 0.7 mm long (Fig. 64.2). It bears minute, short, fan-shaped bristles which contrast with the long, slender bristles of the fruit tree red spider mite. The adult has four pairs of legs, of which the front pair is more than twice as long as any other pair (unlike the front pair of legs of the fruit tree red spider mite). The mouthparts are developed for piercing and sucking plant tissues.

The eggs are spherical and a slightly darker red than the winter eggs of the fruit tree red spider mite. After hatching, the mites pass through one larval and two nymphal stages before becoming adult. The larvae have only three pairs of legs and, when newly hatched, are bright red. The nymphs have four pairs of legs and resemble the adults in general appearance; the older nymphs are the same colour as the adult females.

Life history

Apple and pear bryobia

This species is an occasional pest of apple, especially culinary varieties, although serious infestations have been known on Cox's Orange Pippin and Worcester Pearmain. It also occurs on pear. Attempts to rear it on gooseberry, ivy or herbaceous plants have failed.

The winter is passed in the egg stage, the eggs being laid on the bark from late July until September. These winter eggs are slightly larger, as well as darker, than those of the fruit tree red spider mite and begin to hatch earlier, i.e. in March or early April. There are three overlapping generations during the year.

The mites feed on the leaves and may be seen moving over the upper surfaces in warm, sunny weather, but much of their time, especially in cool conditions (below about 13 °C), is spent on the bark. They moult while on the bark and the clusters of white shed skins on the underside of spurs or branches are characteristic of infestations of these mites (Fig. 64.3). Summer eggs, which resemble the winter eggs, are almost always laid on the bark, more rarely on leaves and leaf stalks. Development of this mite is relatively slow: from hatching of the winter egg to the adult stage takes 4–5 weeks, whereas the summer eggs have an incubation period of about three weeks, followed by a further three weeks for growth to maturity.

A curious feature of the life cycle is that aggregations of active stages may occur on the bark in late May and June, and again in August and September. Hundreds of mites, all in the same stage of growth, can be found clustered together, often in layers on top of one another.

Grass–pear bryobia
This species lives on grasses, including couch, and herbaceous plants such as cinquefoil, clover, hollyhock and meadowsweet; on some of these plants it breeds almost throughout the year.

Fig.64.3 White shed skins of bryobia mites on bark of apple tree (enlarged)

The winter is passed mainly as active stages, though eggs are also laid during mild weather. There are five or six generations a year, with one or two additional ones in mild winters. The length of the life cycle from egg to adult on clover and grass is about five weeks in spring, probably less in summer. In spring, usually in May, the mites may disperse to trees and shrubs, including apple, cherry, pear, blackthorn, bullace, hawthorn and rose, where two generations may occur before the mites return to herbaceous plants.

Gooseberry bryobia
Formerly known as the gooseberry red spider mite, this species was once one of the most important gooseberry pests but is now relatively uncommon. Although it can be reared on apple leaves, it has only been found in the field on gooseberry.

There is only one generation a year. The winter is passed in the remarkably prolonged egg stage. The eggs are laid under loose bark in May and June but do not hatch until the following year, starting in late February or early March and continuing until about mid-April. The earlier-hatching mites take about seven weeks to reach the adult stage.

Ivy bryobia
The bryobia mite found on ivy throughout the year will also feed on clover, where it can be found in late June and July when it tends to be more scarce on ivy. Attempts to rear it on apple have failed.

Like the grass–pear bryobia, this species breeds almost continuously, having seven or eight generations in years with mild winters. Eggs are laid mainly on the tree, wall or other support on which the ivy is climbing.

Clover bryobia
This is the species which sometimes invades buildings, including glasshouses, especially during the early part of the year. It feeds on grass and other low herbage, particularly plants belonging to the pea family (Leguminosae). Fully developed nymphs and mature

females move away from herbage, seeking the protection of cracks and crevices in tree bark or other suitable shelter in which to moult or lay eggs. The eggs laid in late spring hatch immediately and give rise to a summer population of mites. Young hatching from eggs laid by these summer mites grow slowly during the winter to reach maturity in February or March and, later, lay the spring eggs.

Frequently the mites find protection on nearby buildings and are especially attracted to south-facing walls of new structures. Often, the mites are attracted to crevices in fresh mortar, possibly because of its high humidity; as the brickwork dries out, infestations tend to decrease. Large numbers of mites may be found in crevices around windows and doors and underneath window-sills, and they may invade dwellings via windows and doors. The factors which provoke these invasions are not known. While conditions remain favourable, movement of mites between buildings and food plants will continue. In the late autumn and early winter when the temperature drops below 7 °C most activity stops.

Stone mite

A relative of the bryobia mites, the stone mite (*Petrobia latens* (Müller)), also known as the brown wheat mite, is mentioned because it feeds on grass and, when heavy infestations develop, may move on to crops. Although not common as a pest in Britain, it has been recorded on cherry and onion, and in America on apple, carrot and lettuce. It can also invade dwelling houses. Eggs are laid on stones (hence stone mite), pieces of wood, etc.

This mite is scarlet and its body has short, slender bristles instead of the minute fan-shaped bristles typical of bryobia mites. The winter egg has a characteristic shape: like a miniature pork-pie with heavily sculptured lid. Males are unknown.

Natural enemies

Many of the predators of the fruit tree red spider mite feed on several mite species and are capable of exerting an important check on populations of bryobia mites. There are numerous kinds of these predators: the more common ones are the black-kneed capsid (*Blepharidopterus angulatus* (Fallén)) and some related bugs, anthocorid bugs, a small, black ladybird and typhlodromid mites. The capsid bug *Coniortodes salicellus* (Herrich-Schäffer) is especially partial to all stages of bryobia mites. In general, however, adult bryobia mites seem to be less readily attacked than those of fruit tree red spider mite and are sometimes the more common on neglected trees. Nevertheless, on unsprayed trees where predators are present, it is rare for any of these mites to become abundant.

In commercial orchards many insect predators are killed by insecticides, especially those applied post-blossom. Most insecticides currently used in Britain against codling and tortrix moths in June and July are toxic in varying degrees to predators.

Control measures

On fruit crops

Apple and pear The apple and pear bryobia occurs infrequently in commercial orchards and probably only where an abbreviated spray programme is used. It is much easier to control than the fruit tree red spider mite. There are two reasons for this: first, there is no evidence that bryobia mites in Britain have developed resistance to organophosphorus pesticides, so these chemicals are still very effective; secondly, all winter eggs have hatched before any summer eggs are laid, so there is a short period when only mites are present and these are more easily killed than the eggs. This period occurs at the end of the blossom period of pear and just after full bloom of apple.

On pear, spraying with a recommended acaricide at petal fall will virtually eradicate this pest. On apple, good control can be obtained by spraying at either late green cluster or petal fall, though it may be necessary to spray at both times to achieve complete control or where the infestation is severe. Some of the chemicals recommended for use against bryobia mites will also control aphids and caterpillars.

A winter wash applied thoroughly when the buds are breaking will kill the winter eggs of bryobia mites but is more expensive than a spring spray.

The current practice of weed control with herbicides around the base of each tree should largely eliminate the risk of infestation by grass–pear bryobia.

Gooseberry Bryobia mite and aphid control can be combined, if an appropriate chemical is applied just before the first flowers open. With heavy infestations an earlier spray might be advisable, followed by a second 2–3 weeks later. Sprays should be applied in warm, sunny weather. Some fungicides used for control of American gooseberry mildew are acaricidal and should control bryobia mites.

A winter wash will kill the winter eggs of gooseberry bryobia.

On glasshouse crops

Bryobia mite is easily eradicated by most of the acaricides used to control two-spotted spider mite. Where biological control is being practised, localized spraying with petroleum oil should be effective.

On dwelling houses

Any mites inside a building can be removed using a vacuum cleaner. Chemical control measures are best directed at the outside of the building and adjacent grass. A recommended acaricide spray should be applied as soon as mites are seen. Emulsifiable formulations should be used to avoid the disfigurement of walls which can occur with wettable powders. Application should be made at the rate of 10–20 litres/100 m^2. The spray should be applied to the walls of the affected building up to a height of about two metres, paying particular attention to the underside of window-sills. Paths, turf or soil surrounding the building should be sprayed to a distance of two metres from the walls.

It is important that the mites should be identified correctly before treatment is applied. Some acaricides are not effective against the stone mite or another mite (*Balaustium* sp.) which occasionally invades buildings; these mites may require separate treatment.

If possible, turf or weeds should be cleared for a distance of about one metre from the walls. A gravel path appears to provide a barrier to mites but concrete does not.

65
Bulb scale mite

Fig.65.1 Narcissus leaves damaged by bulb scale mite. Note scars and 'saw-edge'

The bulb scale mite (*Steneotarsonemus laticeps* (Halbert)) is occasionally an important pest of forced narcissus bulbs and is probably widespread in many stocks. Although rarely causing extensive damage to bulbs growing in the field, it is of concern to those who grow the crop for export. Other plants attacked are hippeastrum, eucharis, *Sprekelia* and *Vallota*. The mite belongs to the family Tarsonemidae, which includes other important plant pests, e.g. the cyclamen mite (see page 338) and the strawberry mite (see page 366).

Description

Adult bulb scale mites (Fig. 65.2) are extremely small. The female is about 0.2 mm long, colourless when young, becoming pale translucent brown when older. It has four pairs of legs, the last pair ending in a single, long bristle. The male is similar in appearance to the female but smaller and with the last pair of legs forming strong, curved claspers.

The eggs (Fig. 65.2) are oval, translucent white and slightly more than half the size of the adult female. The young resemble the adults except that they are smaller and the first stage has only three pairs of legs.

Bulb mites The bulb scale mite should not be confused with bulb mites (*Rhizoglyphus* spp.), which are also common in bulbs. Bulb mites are usually found in damaged or diseased bulbs but sometimes feed on healthy tissue. They are much larger than the bulb scale mite, being visible without magnification. They are globular in shape and translucent white with two dark spots showing through the body wall.

Fig.65.2 Adult and eggs of bulb scale mite on bulb scale (× 70)

Life history

Bulb scale mites are usually found in groups near the neck of the bulb, where they feed internally in the angular spaces between the scales. They can reproduce throughout the year. The rate of development depends to a large extent on the temperature, a warm humid atmosphere being favourable. Details of the duration of the egg, larval and adult stages are not fully known but, in warm conditions during forcing (14–16 °C), the life cycle can be completed in two weeks. Mites will continue to multiply in bulbs stored at normal temperatures.

As the mite colonies increase in size during the bulb's growth, they frequently spread from the neck to the foliage. Eggs can often be seen on the leaves and flower stem before and during flowering. Occasionally they are so numerous that they are visible without magnification as a fine white powder. However, very few of the larvae hatching from eggs laid on the leaves succeed in completing their development. The principal means of spread is by adult mites that wander freely after the foliage has died down.

Symptoms of attack

Dormant narcissus bulbs

Feeding by bulb scale mite in the dormant bulb causes brown scars on one or more of the bulb scales. If an infested bulb is cut horizontally

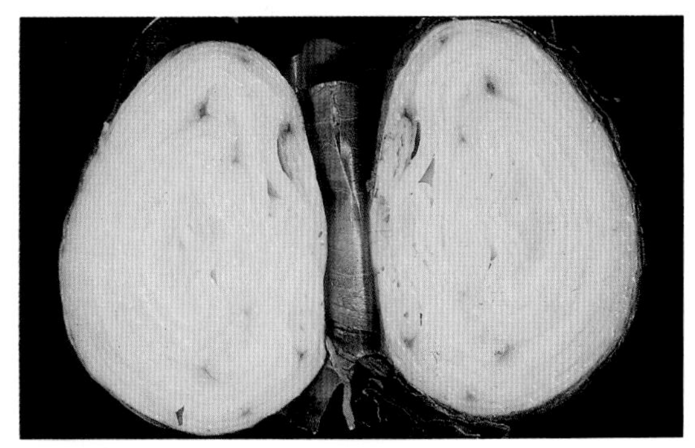

Fig.65.3 Cross-section of narcissus bulb showing brown marks caused by feeding of bulb scale mite

near the neck, these scars can be seen clearly at the angular points in the scales (Fig. 65.3). The cut should be made high in the neck region where the mites collect and, if the damaged part of a scale is examined with a hand lens, the colonies of mites will be visible.

Mites may also be present deep in the bulb. If the bulb scales are pulled apart, the brown marks can be seen to extend downwards as long, narrow scars. Mites have also been found in bulbs when brown marks were not visible. Soft, open bulbs tend to be more heavily infested than firm, solid ones, but there is no way of diagnosing the infestation from the external appearance alone.

Forced narcissus bulbs

The mite reproduces quickly in forced bulbs because of the higher temperature and can cause serious reduction in the yield and the quality of flowers. Intense feeding begins when the bulbs are brought into the forcing house.

Most of the injury is caused before or during the emergence of the leaves and flower buds from the bulb; the effect on the developing leaves and buds is often severe (Fig 65.4). The foliage of infested bulbs is abnormally bright green and does not develop its usual greyish-green bloom. As growth continues, the stem and leaf edges are scarred giving a 'saw-edge' effect (Fig. 65.1). In severely infested bulbs whole leaves become sickle-shaped (Fig. 65.4). The flower is often deformed and, occasionally, the flower bud is killed before or soon after it emerges from the bulb.

Fig.65.4 Forced narcissus showing severe damage caused by bulb scale mite

Outdoor narcissus bulbs

Infested bulbs grown in the open seldom show symptoms of mite damage even in the second year of growth. Symptoms of attack are very difficult to see and heavy mite infestations may be present without any obvious signs of damage. Damage is less severe than under glass because the mites develop more slowly. Only rarely in south-west England are severe symptoms of damage seen and then only when the weather in late winter and early spring has been exceptionally mild. In extreme cases leaf growth may be almost completely inhibited, although a small, usually imperfect, flower is produced. Infested stocks gradually lose vigour and produce smaller crops.

Hippeastrum

The first indication of mite infestation in these plants is the appearance of reddish streaks and spots on the elongating leaves. These streaks extend into the neck of the bulb and, as growth proceeds, they lengthen and form a scar. The leaves become distorted and the flower is usually poorly developed.

Control measures

Minimizing forcing losses

Bulbs for forcing should be firm and solid. They should be selected from stocks that were hot-water treated before planting and are, therefore, substantially free from bulb scale mite. Freedom from mite infestation can be checked, before boxing up, by cutting open a few bulbs and looking for the characteristic damage symptoms (Fig. 65.3). Heavily infested stocks should be rejected, but lightly infested batches will still give a useful yield of saleable flowers if treated chemically soon after housing.

A drench of a recommended acaricide, plus an efficient wetter, applied within a few days of bringing the bulbs indoors will suppress further damage by the mite. Frozen boxes should be allowed to thaw out before treatment. The drench should be directed into the centres of the plants, and the foliage should be wetted thoroughly. This treatment will not eliminate the pest but may check it, so that the foliage and flower buds will have a chance to emerge with minimal damage.

Treatment of dormant bulbs

Hot-water treatment The standard hot-water treatment of three hours at 44.4 °C eradicates bulb scale mite. Before planting, all narcissus stocks should be hot-water treated as a routine to control stem nematode (see page 451). However, hot-water treatment can seriously damage bulbs intended for forcing and only planting stock should be treated.

Following treatment, strict attention must be given to hygiene as contamination during storage is the main source of reinfestation. Provided the ground is clear of self-set bulbs, there is no risk of reinfestation in the field as it is unlikely that the mite can live in the absence of narcissus.

Fumigation. Dormant bulbs can be safely fumigated at any time during storage but this should be done only by a specialist contractor. Early treatment will minimize damage caused by mite populations building up during storage, provided the bulbs are stored properly and in a clean area. Under optimum conditions a complete kill of mites can be achieved, without damage to the bulb or flower.

Mites killed by fumigation are plump and turgid and appear alive owing to the humid conditions between the scales.

Reinfestation can occur during storage, as after other control measures.

Protection of treated bulbs After treatment the bulbs should be isolated as far as practicable from untreated stocks: even a few infested bulbs can put large quantities of clean stocks at risk. Boxes and pallets etc. used to carry untreated stocks must be dipped in a solution of a recommended disinfectant, and vehicles should be washed down before they are used for transporting treated stocks. Cross-contamination between treated and untreated stocks must be carefully avoided. Treated stocks are best planted without delay into land which has not recently carried a narcissus crop.

66
Cyclamen mite on glasshouse plants

Fig.66.1 Cyclamen with petals showing distortion, flecking and discoloration caused by cyclamen mite

The cyclamen mite (*Phytonemus pallidus pallidus* Banks) is a species of tarsonemid mite which causes considerable damage to a wide range of glasshouse plants. These include *Amaranthus*, antirrhinum, *Aphelandra*, *Aralia*, azalea, begonia, chrysanthemum, *Cissus*, *Columnea*, *Crassula*, cyclamen, delphinium, *Fatshedera*, *Ficus*, fuchsia, geranium, *Gerbera*, gloxinia, *Hedera*, *Impatiens*, *Kalanchoe*, *Nertera*, pelargonium, *Saintpaulia* and verbena. Damage is particularly severe on azalea, begonia, cyclamen (Figs 66.1 and 66.3), gloxinia, *Hedera* (Fig. 66.2) and *Saintpaulia*.

Another tarsonemid mite — the broad mite (*Polyphagotarsonemus latus* (Banks)) — can cause similar damage and is often associated with cyclamen mite. On outdoor crops, a closely related mite attacks strawberry (see page 366) and another causes severe damage to Michaelmas daisy. A further tarsonemid — the bulb scale mite — is a pest of narcissus and hippeastrum (see page 334).

Symptoms of attack

In general, attacks result in the distortion of young leaves (Figs 66.2 and 66.3). Flower buds may also become infested, causing distortion of the petals so that the flowers do not open properly; flecking and discoloration of the petals may also occur (Fig. 66.1). On some host plants such as chrysanthemum and *Saintpaulia*, the leaves become very brittle and their edges rolled or curled. Severe infestations result in stunted growth. On chrysanthemums, russeting of the stem just below the flower head may occur; damage by the chrysanthemum russet mite (*Epitrimerus alinae* Liro) can produce similar symptoms.

Description and life history

The adult female mite is very small, being 0.25 mm in length; the male is slightly smaller. Owing to their small size, these mites are best observed through a ×10 or ×20 hand lens. The adults are pale brown. Immature mites are smaller and paler than the adults. The eggs and newly emerged young are whitish. The broad mite is broader and more mobile than the cyclamen mite.

Under glasshouse conditions, eggs, young and adults of the cyclamen mite can be found on plants throughout the year; damage can therefore occur at any time. Mites and eggs are usually restricted to the terminal growth and are not found on expanded leaves. The eggs take about four days to hatch at 20 °C. The immature stages last about seven days and each generation, from egg to adult, takes about two weeks. The development of the broad mite is similar but generally more rapid.

Means of spread

The primary source of infestation in the nursery is the introduction of infested plants. Mites may also be introduced on cyclamen corms.

Fig.66.2 *Hedera* with young leaves distorted by cyclamen mite

Once present on a holding, the mites spread by crawling from one plant to another, though the rate of spread is very slow.

Natural enemies

Few natural enemies have been found attacking cyclamen mite on glasshouse plants. Typhlodromid mites are known predators of cyclamen mite on outdoor crops, but these predatory mites are unlikely to occur on glasshouse plants as routine pesticide sprays used to control aphid and spider mite infestations would destroy them.

Control measures

New plant stock brought on to the holding should be thoroughly examined for initial signs of mite damage. Plants on the bench should be

Fig.66.3 Cyclamen with young leaves distorted by cyclamen mite

spaced as wide apart as is practicable to minimize spread of the pest. Frequent inspection of plants in the glasshouse is essential. Early treatment of damage is necessary, otherwise severe financial loss may occur.

Biological control

Cyclamen crops grown under integrated pest management systems often involve the use of the predatory mites *Amblyseius* spp. to control western flower thrips (*Frankliniella occidentalis* (Pergande)). *Amblyseius* mites can also control cyclamen mite, *providing* that the initial infestation was small. Regular, weekly introductions of the predators are necessary for best results.

Chemical control

Control of cyclamen mite can be obtained by a granular application of a recommended systemic pesticide after potting, but before the plants are spaced out. *Saintpaulia* and some varieties of begonia may be susceptible to damage by such treatments. Recommended acaricide sprays may also be used to control this pest, but growers are advised to establish the tolerance of crops and new varieties by first treating a small area only.

A spray should be applied as soon as mites appear and repeated as necessary, noting the product manufacturer's recommendations for young plants and *Saintpaulia*. When spraying, particular attention should be paid to the young leaves in the crown of the plant.

Some sprays will also control broad mite as well as those spider mites which have not developed resistance to these chemicals.

67
Fruit tree red spider mite

Fig.67.1 Plum foliage damaged by fruit tree red spider mite

The fruit tree red spider mite (*Panonychus ulmi* (Koch)) is mainly a pest of apple, plum and damson, but it sometimes attacks pear. Other host plants include bush and cane fruits. It is also a pest of certain ornamental rosaceous trees, especially *Malus*, *Prunus* and *Sorbus*. Damage is caused by the feeding of the mites on the foliage.

Before 1923 the fruit tree red spider mite was unknown as a pest. Populations built up in English orchards after the introduction of routine winter spraying with tar oil washes in the 1920s. These chemicals failed to kill the mites but were very harmful to many of the insects that preyed on them, thus allowing large mite populations to develop. During the 1930s spraying methods were improved and summer sprays of insecticide became routine; throughout this period mite populations steadily increased. After the Second World War, the predators of the mite were almost completely eliminated from sprayed orchards by the use of DDT and other highly potent insecticides, so mite populations increased still further. Thus the fruit tree red spider mite became a serious orchard pest as a result of spray practices. It has proved difficult to control with chemicals because of its ability to develop resistance.

Other kinds of spider mite also infest the foliage of fruit plants. One species of bryobia mite (the apple and pear bryobia) occasionally occurs on apple and pear during the spring and summer, whereas another species (the gooseberry bryobia) attacks gooseberry (see page 329). The two-spotted spider mite is mainly a pest of protected crops, but it also attacks many outdoor plants including wall-trained fruit trees, currants, raspberry and strawberry (see page 369).

Fig.67.2 Adult female fruit tree red spider mite on underside of leaf (× 25)

Symptoms of attack

Injury is due to the feeding of large numbers of young and adult mites: they suck sap from the leaves and damage the cells of the leaf tissue. The first symptom is a minute speckling of the leaves then, as the attack progresses, the foliage loses its bright colour and becomes rather dull green (Fig. 67.1); finally the leaves assume a brownish-green or bronze colour. At this stage the leaves are brittle and groups of white skins—cast by the mites when moulting—are conspicuous on the underside near the main vein. Many of the bronzed leaves fall prematurely.

The time of appearance of leaf symptoms depends on the size of the mite population. Following very large numbers of overwintering eggs, considerable damage is visible on the spur leaves soon after petal fall and, if the infestation is allowed to go unchecked, the mites soon spread over the new growth and general bronzing may be evident by late June. Usually, leaf damage becomes obvious later in the summer, i.e. in July, August or September.

Severe attacks, especially those occurring in June and July, decrease the yield of fruit and may lead to a reduction in fruit bud formation which probably affects the following year's crop.

Some apple varieties favour the multiplication of fruit tree red spider mite more than others. For example, Discovery, Miller's Seedling and Worcester Pearmain can be seriously infested, while Cox's Orange Pippin in the same orchard has a relatively light infestation. Large applications of nitrogenous fertilizer also encourage multiplication of this pest.

Description

The adult mites are very small—about 0.4 mm long—and only just visible without magnification but easily seen through a hand lens. The female, which is conspicuous, is oval with a very convex back from which arise several long bristles (Fig. 67.2). The body is dark red, with a white spot around the base of each bristle. Immature mites and adult males vary in colour from bright red to pale yellowish-green. Newly hatched mites (larvae) have only three pairs of legs; later immature stages (nymphs) and adults have four pairs.

The presence of bristles and the convex back are features that distinguish this mite from the bryobia mites, which are much flatter and have only a fringe of minute hairs along the margin of the back. Adult bryobia mites are distinctive in having a very long first pair of legs.

Life history

The winter is passed in the egg stage. The bright red winter eggs (Fig. 67.3) are spherical and laid from August to October, mainly on the underside of the spurs and smaller branches, but to a lesser extent around the buds on young shoots and, occasionally, at the calyx (eye) or the stalk end of apple fruits. Following a heavy infestation, areas of the bark appear red owing to the thousands of eggs present.

Hatching normally begins at the pink bud stage of Cox's Orange Pippin—in late April or early May—and about half the eggs have hatched by petal fall. The remainder hatch during the following three to four weeks, i.e. up to mid-June. In a few orchards there are local strains of the mite in which peak hatch occurs two to three weeks later than normal; between these 'early' and 'late' strains there are intermediate strains in a continuous gradation. Immediately after hatching, the young mites migrate to the underside of leaves where they live and feed. Each mite moults three times before becoming adult, the different stages living an active life at first, then entering a resting phase which precedes the moult.

Fig.67.3 Winter eggs of fruit tree red spider mite on spur (× 2)

Eggs produced during the summer months are paler than the winter ones and are laid on the underside of the leaves; these summer eggs hatch in 2–3 weeks, depending on the temperature, and give rise to another generation of mites. Since the immature stages take about 14 days to complete their development, each generation—from egg to adult—takes about four weeks. Normally there are four or five generations each year. The number of generations and the short life cycle during the summer enable populations to build up rapidly in the absence of controlling factors.

Breeding comes to an end when daylight decreases to about 14 hours daily, i.e. during the first half of September. The lower temperature and shorter day induce the females to lay winter eggs on the bark; no more young mites then hatch until the following year. A poor food supply can also lead to winter eggs being laid and it is common to find winter eggs on severely bronzed trees at any time during August.

Means of spread

The mites run actively and soon spread from

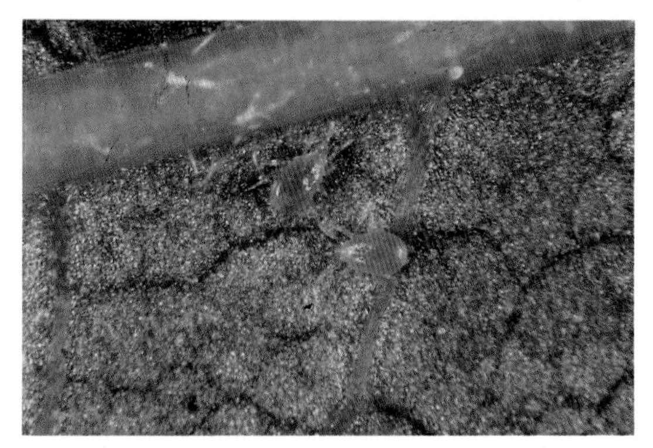

Fig.67.4 Predatory orchard mites: *Typhlodromus pyri* (× 25)

leaf to leaf on a tree. On warm, still days, especially when populations are large, some mites spin a strand of silk and, suspended from this, are carried away by air currents to other trees. The spread of the mite to most apple-growing areas in the world has probably been caused by the distribution of nursery stock carrying winter eggs.

Natural enemies

In England the predatory mite *Typhlodromus pyri* (Scheuten) (Fig. 67.4) has been shown to be the most effective natural enemy of fruit tree red spider mite on apple. Strains of *Typhlodromus pyri* resistant to certain organophosphorus, carbamate and oganochlorine insecticides are now widespread in England. Where this predatory mite is allowed to flourish by avoiding the use of certain pesticides, treatment for fruit tree red spider mite is rarely necessary.

Insect predators, especially anthocorids and certain mirid bugs such as the black-kneed capsid (*Blepharidopterus angulatus* (Fallén)) and related species, are voracious predators of spider mites, but they are attracted to feed and reproduce only when spider mites are fairly numerous. They will then reduce spider mite numbers rapidly. However, they usually over-exploit their prey so that eventually the surviving predators must leave the orchard to seek prey elsewhere or starve. Hence, although they can contribute usefully to the reduction of large spider mite populations, insect predators, unlike the predatory mite, do not provide a stable regulation of small populations.

Control measures on apple

Timing of acaricides

Acaricides should be used when necessary to prevent early-season damage by spider mites. Because winter eggs hatch over an extended period two applications are often necessary: (i) at 80% petal fall (provided that hatching is well under way) and (ii) 3–4 weeks later.

Some acaricides are toxic mainly to eggs and immature mites and are, therefore, used to best effect early in the season. In the summer, if many adult mites are present, acaricides known to kill adults should be used.

Resistance to acaricides

Failure to obtain control may be caused by poor spraying technique or incorrect timing, but if the same acaricide has been used for several years the failure could be due to development of resistance in the local mite population. Resistant mites, which have an innate greater tolerance to the acaricide, are at first very rare in the population. The acaricide kills most of the mites with a normal tolerance, but the resistant mites survive. Repeated 'selection' gradually increases the proportion of resistant mites in the population. Usually after 4–5 years (or 7–10 spray applications) they form a substantial part of the population and the failure to control them suddenly becomes apparent.

Resistance of fruit tree red spider mites to most organophosphorus compounds is now universal in sprayed orchards and is very stable. Resistance to specific acaricides has also become common in orchards where they have been used for several years. The time taken for resistance to develop depends on the acaricide.

Resistance develops specifically to each chemical group. Thus, resistance developed by selection with one or more organophosphorus pesticides confers 'cross-resistance' to most other organophosphates which have not been used, but it does not usually alter the response of the mites to other groups of acaricide with different modes of action. However, many mite populations have become 'multi-resistant' by simultaneous or successive exposure to several groups.

Control strategies

The 'selection pressure' for the development of resistance can be reduced by using acaricides less often. Two strategies for achieving this are described below.

(1) Mite management The possibilities for integrating biological control of spider mites by the predatory mite with chemical control of fungus diseases and insect pests are greater now than before. The major factors responsible for this are the commercial availability of non-acaricidal fungicides and the development in native predatory mites of resistance to some commonly used insecticides. Evidence is accumulating that (i) such resistance in the predatory mite is now widespread in sprayed orchards where these insecticides have been used for many years, and (ii) when using a spray programme that is safe for the predatory mite, the latter will often colonize the orchard and establish control of spider mites within one to two years.

To encourage the predatory mite to establish, all fungicides and insecticides in a spray programme should be selected for safety to this species. However, growers should always be satisfied that the pesticides they choose will provide good control of the pest and disease complex in their orchards.

During the establishment of the predatory mite (commonly two years) the number of spider mites will tend to increase, so that the use of a selective acaricide is usually necessary, often at petal fall during the second year. Examining the foliage with the aid of a hand lens will assist decisions on the need to spray. The predatory mite is effective when there is at least one predator for every 10 active spider mites. If the predator is less numerous than this, one application of a selective acaricide is advisable to decrease spider mites; this will have little or no effect on the predator.

Once the predatory mite is well established, acaricides are seldom needed, unless some factor (usually a pesticide) disrupts the balance; in this phase it becomes unnecessary to monitor frequently. The predator also feeds on the apple rust mite (*Aculus schlechtendali* (Nalepa)) and contributes to its control.

If resistant predatory mites do not establish naturally, they can be introduced.

(2) Use of acaricidal fungicides For control of spider mites biological control by the predatory mite, as described above, is the best policy. However, when considering disease control and fruit quality some growers prefer to use acaricidal fungicides for control of mildew. These fungicides are weak acaricides, but repeated applications at 7–10 day intervals will usually prevent small numbers of spider mites from increasing. They will also greatly reduce numbers of the predatory mite so that, when the spray programmme ends about late July, spider mites — free from predation — will tend to increase sharply until September. During this period frequent inspections should be made. If mites are increasing rapidly, a specific acaricide should be applied to depress numbers to a very low level so that few winter eggs are laid.

In the following year regular inspections should be made from petal fall and a specific acaricide applied if necessary. Experience with full programmes of acaricidal fungicides has shown that specific acaricides are needed either once per year at the end of the fungicide programme or, in some orchards, only once in every second year.

If a grower uses neither of the two strategies outlined above, e.g. he uses non-acaricidal fungicides and chooses insecticides which are highly toxic to the predatory mite, he must expect to use acaricides more frequently for spider mite and often for rust mite.

Control measures on pear

Pears are less prone than apples to outbreaks of spider mite, but the pest can cause serious damage if populations build up. The choice of acaricide hinges on control of organophosphorus-resistant pear sucker (see page 14). Some insecticides used post-blossom to control pear sucker will also suppress spider mites and are less harmful than others to predators. Post-blossom application of insecticides that are highly toxic to predators should be avoided. Fungicides commonly used to control pear scab are harmless to predators.

Control measures on plum

Plums are often considered to merit only one spray, in which case the best time is late May or early June depending on the season. Certain insecticides applied at the cot-split stage to control plum sawfly (*Hoplocampa flava* (L.)) give some control of non-resistant spider mites, but this is too early to obtain optimum effect from a single application.

As on apple, every effort should be made to conserve populations of the predatory mite.

68
Pear leaf blister mite

Fig.68.1 Pear leaves in early June showing damage caused by pear leaf blister mite

The pear leaf blister mite (*Eriophyes pyri* (Pagenstecher)) is distributed throughout Britain but seems to be most prevalent in the southern half of England. It occurs mainly on pear and occasionally on apple; a related race or species also occurs on rowan, whitebeam and the wild service tree.

This pest is very susceptible to lime sulphur and was eradicated from most orchards during the period when this fungicide was in common use. For many years thereafter infestations were uncommon, usually being noticed on single or small groups of trees, particularly in gardens. More recently, however, several extensive infestations have occurred in some commercial pear orchards.

Symptoms of attack

The damage caused by the blister mite is similar on all the above-mentioned host plants, but the following description of symptoms applies particularly to pear.

Pinkish-red blisters are sometimes present on the unfurled leaves at blossom time (Fig. 68.2). Later in spring and in early summer small greenish-yellow or yellowish-red blisters can be seen on both the upper and lower surfaces of infested leaves (Fig. 68.1); these blisters are usually more numerous close to the midrib. As they age, the blisters turn brown and finally black. Frequently the entire leaf becomes black and dies in late summer. In severe infestations the stalks of the fruitlets and the fruit itself may be attacked, resulting in malformation or premature falling of the fruit.

The related pear rust mite (*Epitrimerus piri* (Nalepa)) is free-living on leaves and fruits. When present in large numbers this species can cause browning of the underside of leaves (Fig. 68.3) and russet at the calyx (eye) end of fruits.

Fig.68.2 Unfurled leaves at blossom time showing damage caused by pear leaf blister mite

Description

The pear leaf blister mite is sausage-shaped, tapering towards the hind end, and whitish, sometimes tinged with pale brown. The adult is about 0.2 mm long. The mites are hardly visible with a hand lens, but their presence can easily be detected during the growing season by the blisters, usually 2–5 mm in diameter, on the leaves.

The pear rust mite is of similar size to the above species but is wedge-shaped. The adults are orange-brown.

Life history

The pear leaf blister mite spends the winter as an adult under the outer scales of buds. These adults have resumed activity by the time the buds start to swell in the spring and eggs are laid at the base of the inner scales. Later, the overwintered mites and their immature progeny live on the developing leaves and flower buds, where more eggs are laid. Blisters are initiated by the external feeding of adult mites, mainly on the underside of the developing leaves. The epidermal cells at the centre of the blister die and the growth of surrounding cells pulls these apart, forming a small hole. This provides a means of entry for adults of the next generation, usually during the later part of the blossom period.

From petal fall onwards the mites live in the blisters, where eggs are laid and the immature stages develop. Feeding within the blister causes blackening which is visible on both sides of the leaf, although access to blisters is through holes occurring only on the underside of the leaf. As the amount of necrotic tissue increases, some galls are vacated and the mites

Fig.68.3 Underside of pear leaf damaged by pear rust mite

enter other unoccupied blisters on the same leaf or move to the growing tips of shoots; there, feeding causes more blisters which in turn are entered as the cells die and provide access. During September and October adult mites leave the galls to seek winter shelter under the scales of fruit buds.

Several generations occur each year. The rate of development has not been studied in Britain, but in the north-west of the USA the life cycle occupies 20–35 days in spring and 10–12 days in summer.

Control measures

At present there are no acaricides with a specific recommendation for the control of pear leaf blister mite. Fruit russet caused by pear rust mite can be prevented by spraying with a recommended acaricide at white bud or petal fall. This treatment is likely to reduce the extent of blister mite damage.

69
Raspberry leaf and bud mite

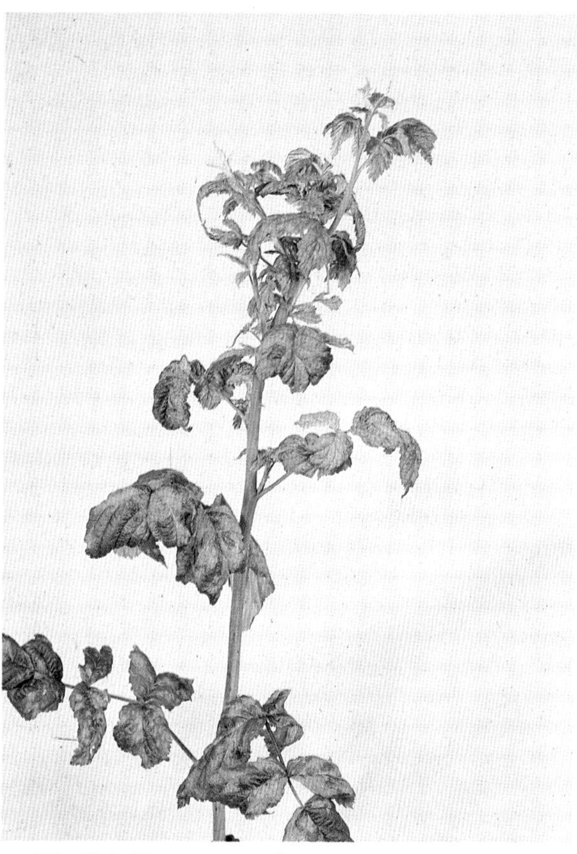

Fig.69.1 First-year raspberry cane showing yellow blotches on leaves and early-stage branching caused by feeding of raspberry leaf and bud mite

The raspberry leaf and bud mite (*Phyllocoptes gracilis* (Nalepa)) occurs throughout Britain on cultivated and wild raspberry, hybrid berries and blackberry. Where infestations are severe, vigour, yield and fruit quality are reduced. Damage is more prevalent on raspberries growing in sheltered situations.

Symptoms of attack

On raspberry and hybrid berries Irregular yellow blotches develop on the leaves of fruiting and first-year canes (Fig. 69.1) as a result of large numbers of mites feeding on the underside. In the variety Malling Jewel, mite-feeding areas become very pale green compared with the greyish bloom of the rest of the leaf surface; damaged leaves tend to be smaller and more curled than healthy ones. After a warm spring, leaf symptoms may show on the fruiting canes by June.

The blotches on the leaves can be confused with those caused by soil-borne viruses — mainly raspberry ringspot virus. However, the mite produces larger blotches than this virus (Fig. 69.2) and a close examination of the lower surface of a damaged leaf will reveal fewer and distorted leaf hairs where mites have been feeding. Another distinction between mite and virus damage is the distribution pattern in the plantation. Leaf and bud mite infestations tend to cover a fairly large area, whereas initially raspberry ringspot infection is usually confined to a few isolated plants, sometimes in small pockets within the field.

Fruit can also be damaged by mites feeding on the surface of the developing drupelets, which then ripen prematurely and unevenly (Fig. 69.3), thereby decreasing quality and making harvesting more difficult.

Plantations that have been heavily infested for several years progressively decline in vigour and yield. The growing points of first-year canes can be killed, resulting in the growth of weak side-branches which are difficult to tie in.

On blackberry Mites feed on the foliage and cause malformation of the hairs on the underside of the leaves. A severe infestation leads to the production of whitish, mildew-like blotches on the upper surface of the leaves. These blotches can be seen in late August and in September.

Description

An adult female raspberry leaf and bud mite is about 0.15 mm long. The body is worm-like with two pairs of legs near the head. In the summer, mites are translucent whitish-yellow and are difficult to distinguish amongst the leaf hairs on the lower side of raspberry leaves, even with a good hand-lens. During the winter, large colonies of pinkish-brown mites congregate beneath the bud scales in the middle region of the cane (Fig. 69.4).

Fig.69.2 Raspberry leaves showing (left) damage caused by raspberry leaf and bud mite and (right) symptoms of raspberry ringspot virus

Life history

As the buds open in the spring, mites surviving the winter migrate to the unfurling leaves where they feed and lay eggs on the underside. These eggs give rise to the first generation of mites. Many more generations are completed on the fruiting canes and the infestation eventually spreads to the new canes. Mites can be carried on air currents or insects to raspberries at least 100 metres away.

The number of generations per year varies with the season, but the population reaches a peak in early autumn, when many of the adults migrate from the leaves to overwinter in the buds (Fig. 69.4). During mild winters they feed on the green tissue of the outer bud scales, but this does not prevent the buds from opening in the spring.

Varietal susceptibility

Varieties of raspberry vary in their susceptibility to mite attack. In Scotland, Malling Jewel is very susceptible whereas Malling Promise rarely becomes heavily infested. Field observations show that Glen Clova can become infested but the effect on yield is unknown.

Natural enemies

The predatory mite *Typhlodromus pyri* Scheuten is the most common predator of the raspberry leaf and bud mite in Scotland and has been reported in raspberry plantations throughout England. Several other predatory species have been found in heavily infested mite plantations, but their effect on the mite population is unknown.

Fig.69.3 Raspberry fruits showing (lower) premature and uneven ripening caused by feeding of raspberry leaf and bud mite compared with undamaged fruits (upper)

Fig.69.4 Part of overwintering colony of raspberry leaf and bud mite in raspberry bud (× 30). First bud-scale has been removed

Control measures

Cultural control

Because many commercial raspberry plantations have small, often undetectable, populations of the mite, new planting material should be obtained only from certified nursery stocks.

Chemical control

Spraying is unnecessary unless mite damage is visible; this usually occurs in the third or fourth year after planting. One spray of a recommended acaricide should be applied to the old canes in early May to prevent migration to the young spawn. The minimum permitted interval must elapse between spraying and harvest. If necessary, this treatment should be repeated in the following year. Further treatment may not be required for several years.

A spray can be applied after fruiting, but it may not be as effective as one applied in the spring because some mites may have already entered the buds. To allow access for spray machinery in the autumn, it may be necessary to cut out the old fruiting canes and tie in the first-year canes earlier than usual.

70
Reversion disease and gall mite of black currant

Fig.70.1 Left: reverted black currant shoot with galled buds caused by black currant gall mite. Right: healthy shoot

The virus disease known as 'reversion' has been recognized for many years as a widespread and prevalent disease of black currant in Britain. Infected bushes usually grow vigorously but their cropping ability becomes seriously impaired. Plantations once affected usually become uneconomic and have to be grubbed prematurely. The reversion virus is transmitted by a common and widespread pest — the black currant gall mite or 'big bud' mite (*Cecidophyopsis ribis* (Westwood)). Bushes infected with reversion are more prone than healthy bushes to heavy infestations of the gall mite and are, therefore, a menace to other black currants growing in the vicinity.

Black currant reversion disease

Symptoms

Reversion can be recognized by either flower-bud or leaf symptoms.

Flower-bud symptoms are apparent at the late 'grape stage' just before the first flowers open. On reverted bushes the flower buds are almost hairless, which makes them appear brightly coloured, but on healthy bushes they are hairy and, therefore, have a grey, downy appearance (Fig. 70.1). It should be noted, however, that varieties differ in hairiness, but this should not cause problems in commercial crops of single varieties. Bushes should be checked for flower-bud symptoms at the beginning of flowering. This method is quick and reliable and, because there are few leaves present, it is possible to detect individual reverted branches or even single trusses. A correct diagnosis can, however, only be made when the flower buds are dry and the shoots are free from injuries caused by pests, implements, etc.

Leaf symptoms are more difficult to recognize and some experience is necessary for a reliable diagnosis. The leaf of a reverted bush is flatter and has a much less pronounced cleft between the basal lobes than a normal leaf (Fig. 70.2). A reverted leaf also has fewer main veins and fewer marginal serrations. Thus the large central lobe of a typical healthy leaf has usually five or more main veins on each side of the midrib and at least 14 serrations along each margin (Fig. 70.2). By contrast, the large central lobe of a reverted leaf has fewer than five main veins and fewer than 12 serrations (Fig. 70.2). These differences are pronounced by midsummer. They should be looked for on the vigorous shoots of the basal or extension growth and not on the atypical shoots developing alongside fruit. Diagnosis may be complicated by differences in leaf character between varieties.

Bushes in the early stages of reversion may produce leaves with a characterisic yellow pattern along the main veins, but this is not always easy to find and may be confused with yellowing due to other causes.

Fig.70.2 Left: healthy black currant leaf. Right: reverted leaf. Main veins of large central lobe of each leaf are numbered and position of cleft between basal lobes is indicated

Symptoms of reversion do not appear until the year after infection. They are then restricted to individual flower trusses and leaves on a few shoots. Flower and leaf symptoms are much more obvious two years after infection and in the third year the whole bush is likely to be reverted. The effect on cropping is variable, some infected bushes becoming virtually sterile in three years, while others crop indefinitely but at a reduced rate. There are severe and mild strains of the virus, which respectively cause sterility and reduced cropping.

Mite infestation of reverted bushes

Healthy black currant bushes appear to have considerable resistance to colonization by the gall mite. Following infection with reversion virus, this natural resistance is broken down and the reverted bushes more easily become infested with mites. It is very important, therefore, to remove and burn reverted bushes as soon as they are identified and so reduce sources of infection within the plantation. Destruction of reverted bushes is particularly important following roguing at flowering time, since the mite moves from infected to healthy bushes during the spring and early summer.

Black currant gall mite

Description and life history

The gall mites are elongate, less than 0.25 mm long and pearly-white (Fig. 70.3). They breed in the buds of black currant and possibly certain other *Ribes* species. All the main varieties of black currant are susceptible, though they vary in degree of susceptibility.

The dispersal of mites from old galls to young buds starts during the grape stage (usually in April) and may continue into June or even July. However, most of the dispersal occurs between first open flower and early fruit swell and appears to be closely associated with warm, humid periods during flowering. Dispersing mites swarm on to the outside of the galls and then crawl or leap off, some eventually reaching nearby branches. Dispersal is assisted by rain and wind and by insects such as aphids and capsid bugs to which the mites may cling. A few mites may thus be carried considerable distances.

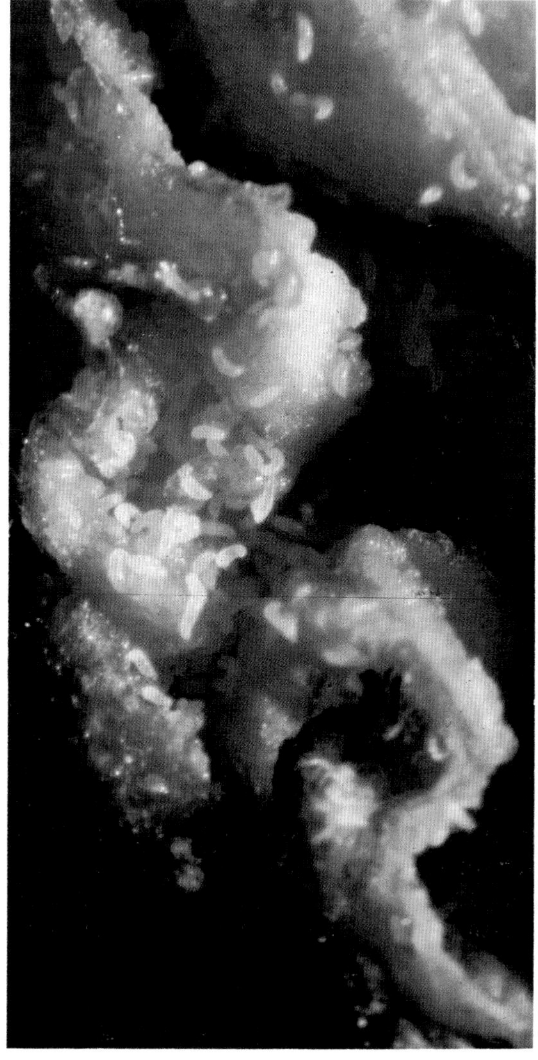

Fig.70.3 Black currant gall mites in a bud (× 35)

Except for the brief interval between leaving a gall and entering a new bud, a mite usually lives entirely within a bud. Thousands of mites and eggs may develop in a single bud. Egg laying begins in the new bud in June and reaches a peak by September. After a pause in early winter, breeding is resumed in January

Fig.70.4 Left: normal axillary bud of black currant. Right: axillary bud infested with black currant gall mite

and reaches a second peak in the spring when the mites begin to disperse from the buds.

Symptoms of mite infestation

The presence of galled buds is the best indication of infestation. The optimum time to look for these is during the dormant period when the bushes are leafless. Newly infested buds gradually change shape during the summer and by autumn they are recognizable as galls or 'big buds', being rounded, much less pointed than normal buds and usually swollen (Fig. 70.4). In early spring the galled buds swell further and begin to break, but they fail to open normally and do not produce flowers or leaves. Instead, they gradually dry up and discolour, but sometimes persist for many months as dead galls.

In addition to the production of galled buds, shoots that are infested at the apex in the spring produce severly malformed leaves by midsummer (Fig. 70.5).

Control measures

Production of health plants in the nursery

The objectives in the nursery are to produce strong bushes that are true to type, free from reversion and other diseases, and not infested with mites or insect pests. Cuttings should be taken only from healthy bushes that are without galls and that were free from reversion the previous summer.

Stool bushes used to provide cuttings should be labelled individually. The cuttings from each bush can then be kept in separate, labelled groups in the nursery and destroyed if symptoms of reversion or other virus diseases appear in the parent during the period of propagation. This practice is particularly

Fig.70.5 Malformed black currant leaf produced by shoots with black currant gall mites at growing point

valuable if bushes are to be lifted as yearlings.

Nurseries should be sited as far as possible — but never less than 100 metres—from other currants and they should be sprayed thoroughly and regularly to eradicate any mites entering the crop. Thorough inspections should be made in winter for galls and in summer for reversion symptoms.

Establishment of new fruiting plantations

Only the best material that can be obtained should be planted: two-year-old bushes and cuttings should be taken from certified stock, and one-year-old bushes should be taken from certified stock that has been grown in an isolated nursery. The entry of mites and reversion into the crop is minimized by planting as far from existing sources of infection as possible, preferably upwind considering the winds prevailing in April and May, and by routine sprays and inspections. It is important to ensure that the mite population is at a very low level when the plantation comes into cropping.

Chemical control in nurseries and plantations

Chemical control of the gall mite is difficult because the mites are generally hidden within galled buds. It is, therefore, necessary to protect the new growth during the spring when mites are leaving the galls. Complete protection from mite attack is almost impossible to achieve because the dispersal period of the mites extends over several weeks when the bushes are growing rapidly. Also, in older fruiting plantations it is very difficult to obtain an adequate spray cover, owing to the large size of the bushes and the density of the foliage.

Despite these problems many commercial plantations are virtually free from mite infestation. This has been achieved by planting certified stock, regular spraying and intensive roguing. Virus infection still occurs on a small scale and indicates that any relaxation of control would lead to a decline in the health of fruiting plantations.

Acaricides applied for control of gall mite may also give some control of black currant leaf midge (*Dasineura tetensi* (Rübsaamen)), aphids and capsids. Acaricides may be used alone or in combination with certain fungicides used for disease control, but they should not be applied during the main part of the flowering period because of the risk to bees. To avoid any hazard from residues on the fruit at harvest, the number and timing of acaricide applications to fruiting bushes are strictly limited. Crops grown under contract may be subject to additional restrictions imposed by processors.

Some fungicides are also toxic to the mite, so that repeated applications of them greatly reduce mite populations. In lightly infested plantations where such fungicides are used for a full disease control programme (usually four applications at fortnightly intervals from the grape stage), only one application of a specific acaricide may be necessary for mite control. This should be applied three or four weeks after first open flower. Provided the minimum permitted interval before harvest can be maintained, an application four weeks after first open flower is preferable because both fungicide and acaricide can then be applied together and, at this time, flowering is almost over so there is less risk to pollinating insects.

Sprays must be applied thoroughly, and preferably at high volume, to ensure good wetting of foliage and young shoots, including those arising from the base of the plant.

Control in gardens and allotments

Black currants are widely grown in gardens and allotments, and both reversion disease and galled buds are frequently present. Many gardeners are apparently unaware of the presence of either the disease or the mite, or of their effect on cropping. Diseased bushes should always be grubbed before healthy replacements are planted, otherwise there is a very serious risk that the young bushes will become infected with both virus and gall mites in the first season after planting. This procedure means that no crop is available in the year after grubbing, but the disadvantage is offset by the extended cropping life of the replacement bushes. It is advisable always to buy bushes that have been certified by the Ministry (see below) and to plant them as far as possible from any other black currants.

There is a limited choice of chemicals available to the gardener for the control of black currant gall mite. One of these should be applied as a spray on three occasions at fortnightly intervals, beginning at late grape or first open flower. In addition, the infestation can be further reduced by picking off any galled buds remaining after the bushes have been pruned in winter.

Inspection and certification of black currant bushes

A voluntary scheme for the inspection and certification of black currant bushes is organized by the Ministry of Agriculture, Fisheries and Food to ensure satisfactory standards for planting material. Applications are accepted only for bushes which are at least two years old and which have not been planted as fruiting bushes for more than three years. Bushes which have been stooled are also eligible, but maiden bushes in the first year of development are not.

Bushes are inspected while growing and certificates are issued only for those stocks which, subject to a small tolerance, are found to be true to type and which show no symptoms of reversion at the time of inspection. Certificates are not issued for stocks which are evidently unhealthy, lacking in vigour, substantially affected by gall mite or grown within 100 metres of uncertified stocks of black currants.

At the close of the inspection season a Register of Certified Stocks is issued for the use of buyers. Growers are advised in their own interest to plant only certified stock.

71
Spider mites on protected crops

Fig.71.1 Cucumber leaf damaged by spider mites

Two species of spider mite, namely the two-spotted spider mite (*Tetranychus urticae* Koch) and the carmine spider mite (*Tetranychus cinnabarinus* (Boisduval)), are serious pests of plants grown under glass or polythene. Plants damaged include cucumber, dwarf French bean, pepper, strawberry, tomato, carnation, chrysanthemum and rose; also arum lily, hydrangea, orchids and many pot plants as well as stone fruits and vines.

The mites may enter the glasshouse or polythene structure from weeds or cultivated plants outside. They may also be introduced on purchased plants, uncleaned staging or in hollow canes used as plant supports. Once established, the mites are difficult to eradicate, particularly the hibernating females of two-spotted spider mite which may be concealed in holes and crevices in the structure framework.

Symptoms of attack

The first sign of attack is usually a very fine and more or less regular speckling on the upper surface of the foliage, often low down on the plant. This speckling is caused by the mites feeding on the underside of the leaves; only on carnation do the mites feed and breed to any extent on the upper surface. On plants with leathery leaves, an infestation may not be noticeable until it has extended to the upper parts of the plant.

As the mites multiply, the foliage loses its healthy green colour (Fig. 71.1) and new growth becomes hardened and stunted. In bad attacks the mites spin a thick web over the plants (Fig. 71.2). Heavy infestations may severely weaken the plants and reduce their yield.

Description

The general appearance of spider mites in the egg, immature and adult stages is shown in Fig. 71.3. The adult female is about 0.5 mm long.

Fig.71.2 Spider mite webbing on chrysanthemums

The two species occurring under protection in Britain differ in colour and habits. A brief description of each is given so that growers may recognize which species is present and plan control measures accordingly. In spring and summer both species may occur together.

Two-spotted spider mite
This is the species usually found on cucumber, tomato, dwarf French bean, pepper, year-round chrysanthemum, pot plants and glass-house roses. It also infests various crop plants outdoors (see page 369).

The summer females are usually green with a large dark blotch on either side of the body. The males and active immature stages are yellowish to pale green. Starved mites may be reddish. The eggs are small, spherical and translucent white.

In September, with the arrival of shorter days (less than 14 hours of daylight) and reduced plant vigour, reproduction gradually ceases in most individuals and only the adult female mites survive. These turn brick-red and wander off the plants to seek hibernation sites, such as ridge capping, ventilator fittings, drip irrigation nozzles, crevices and holes in the house structure, and in litter.

When the houses are heated for the new

Fig.71.3 Eggs and active stages of two-spotted spider mite on underside of leaf (× 30)

crop in late winter or early spring, the surviving females leave their winter shelter and those that find fresh host plants resume breeding. This type of infestation is particularly serious in propagating houses, where the high density of plants increases the chance that the newly active mites will find suitable hosts. Where the crops are, like cucumber, grown at high temperatures, the mites breed during January and February despite the short day-length. If the early infestation occurs on crops grown at lower temperatures, the progeny, when mature, re-enter hibernation until the day-length exceeds 14 hours.

The strains of mites associated with year-round chrysanthemum often become adapted to continuous reproduction throughout the year, irrespective of day-length. Mites may also be found throughout the year on some glasshouse roses.

Carmine spider mite

This species is usually found on carnation, arum lily and other ornamental plants under protection.

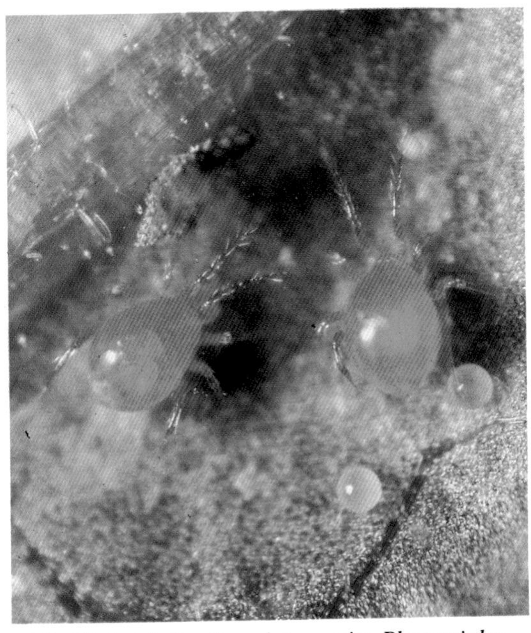

Fig.71.4 Adults of predatory mite *Phytoseiulus persimilis* and eggs of spider mite on underside of leaf (× 30)

The females are carmine red with a large dark blotch on either side of the body, while the males and active immature stages are yellowish. The eggs are small and spherical with a reddish tint.

At favourable temperatures, breeding occurs throughout the year and the mites leave the plants only when starved.

Hypertoxic strains

Virulent strains of both the above spider mite species have been detected in recent years and have caused severe damage to protected tomatoes. Damage symptoms are variable, but only a few mites are needed to produce severe chlorotic blotching of foliage (Fig. 71.5), culminating in flaccid, desiccated leaves. Speckling of the leaves associated with typical spider mite damage occurs when mixtures of strains are present.

Fig.71.5 Tomato leaf showing chlorotic blotching caused by spider mites of virulent strain

Life history

Each spider mite female may lay 100 eggs or more in a period of up to three weeks. On the underside of the leaves the mites secrete a loose web of silken threads beneath which the eggs and later stages are protected. After hatching, the mites pass through three immature stages, each divided into an active and a resting phase of equal duration, before becomming adult.

Both species pass through a continuous succession of generations from March to September. The following data for the average length of life cycle of the carmine spider mite from egg to adult show how much the rate of development is controlled by temperature: 62 days at 10 °C, 13 days at 21 °C, 8 days at 26.5 °C and 6 days at 35 °C. The two-spotted spider mite develops more rapidly, e.g. 55 days at 10 °C and 12 days at 21 °C. Likewise, temperature also determines the need for, and timing of, repeated treatments to control the mites.

Prevention of hibernation

Nearly all infestations of the two-spotted spider mite in the early part of the year arise from mites which hibernated in the house structure the previous autumn. It is therefore desirable to remove the old crop from the house while still green, although the economics of cucumber growing may not always allow this. As the condition of the plants deteriorates, the proportion of mites entering hibernation increases; once in their winter quarters it is not possible to eradicate the mites. It is important, therefore, to continue effective control measures right up to the time the crop is removed from the house. (On roses, effective acaricides should be applied immediately after the season's picking is complete and before winter pruning.)

Artificial extension of the autumnal daylight to 16 hours, or interruption of the night period by an interval of two hours' artificial light, markedly decreases the number of mites entering hibernation, even on heavily infested plants.

Biological control

A predatory mite (*Phytoseiulus persimilis* Athias-Henriot), originally imported accidentally from South America, usually gives efficient control of spider mites if introduced as soon as the feeding marks of the pest are seen.

The predator (Fig. 71.4) is a fast moving, orange-brown mite somewhat larger than the spider mites. Because of its colour the predatory mite is sometimes confused with the winter form of the two-spotted spider mite. At 21 °C each female predator lays 50–60 pinkish, ovoid eggs at the rate of three or four per day; they hatch in two to three days. There are three immature stages and the complete life cycle takes about seven days — about half the time of the spider mite life cycle.

The predatory mite is very efficient at searching out its prey and each female will devour up to five adult spider mites, or 30 eggs and young, per day. It does not feed on plant material and its survival is entirely dependent on the maintenance of a small population of spider mites. However, experience has shown that some predators will survive for at least three weeks after the pest has been eliminated.

Unlike its host, the predator does not normally hibernate during the winter months and must be reintroduced each year as required.

Protected vegetable crops

Predators can effectively control spider mites on cucumber, tomato and sweet pepper, if managed correctly. The technique originally developed was the 'classical' or 'pest-in-first' method, whereby spider mites were introduced to the plants to provide a food source for the predators until natural infestations developed. However, most growers are reluctant to introduce spider mites into their crops and prefer to introduce predators at the first signs of spider mite damage. This technique is successful as long as the initial number of spider mites is minimal. If necessary, a compatible specific acaricide (i.e. an acaricide that is not harmful to the predator) can be used to reduce spider mite infestations before predator introduction.

The recommended number of predators to be introduced varies with the crop and the degree of spider mite infestation. In general, rates of four predators per plant on cucumber and two predators per plant on tomato and sweet pepper are effective if introduced to small populations of spider mite. Introductions should be made on every fifth plant and the plants should be touching to allow predator migration. Further introductions or spot treatments with a compatible acaricide may be necessary for patches of more severe infestation, or if the predators fail to establish well. Careful monitoring of the balance between predators and spider mites is essential for successful control. Even if this balance is satisfactory early in the season, sometimes the tops of the plants become suddenly and severely infested in high temperatures in midsummer. This can be corrected by an application of a compatible specific acaricide to the tops of the plants.

Strawberries

On strawberries grown in 'walk-in' polythene tunnels, predators should be introduced in late March or early April, shortly before flowering, at a rate of one per plant. A further introduction should be made after replanting (with cold-stored runners) during the summer months, but not after autumn replantings when spider mite is usually in diapause (resting stage). Mite populations should be monitored regularly to ensure that the numbers of prey and predator are in balance; if necessary, a compatible acaricide should be used to avoid severe damage.

Chrysanthemums

An integrated pest control programme has been developed which includes the introduction of one predator to every 10 chrysanthemum plants (or one to every three plants of susceptible varieties) four weeks after planting. This is usually sufficient to control spider mites for the life of the crop, but the programme should be carefully monitored.

Other ornamental crops

An introduction of one or more predators to each plant can be made at the first sign of spider mite attack. If preferred, on pot plants, regular weekly introductions of one predator per 20 plants can be made and the rate adjusted if necessary at the time of spider mite infestation.

Hypertoxic strains

The predatory mite is less successful in controlling virulent strains of spider mites. This is because there is a delay of at least a fortnight between introduction of the predator and an effective decrease in spider mite numbers, during which time virulent strains cause severe damage. Also, the predator is most effective at relatively high pest densities, which do not occur with populations of virulent strains. More predators should therefore be used if the presence of a virulent strain is suspected.

An alternative predator currently being tested on tomato is the predatory midge *Therodiplosis persicae* Kieffer. This may be more effective against the hypertoxic strain of spider mite, although a specific acaricide may still be necessary to reduce damage symptoms.

Integration with pesticides

The predatory mite is very susceptible to many pesticides commonly used on protected crops, so that care must be taken in the choice of chemicals used to control other pests and diseases. Most broad-spectrum pesticides are harmful to *Phytoseiulus* eggs and adults; only some fungicides, specific acaricides, aphicides and short-persistence insecticides can be used safely in integrated control programmes. If in doubt, advice should be sought on integrating pesticides with biological control agents.

Chemical control

Several chemicals are recommended for the control of spider mites on protected crops. The choice will depend on the crop, the growing medium, the interval before harvest, whether resistant mites are present and whether biological control agents are being used for spider mites or any other pests.

Resistance

The continuous use of certain acaricides over a long period has selected strains of mites which are so resistant to poisoning that they can no longer be effectively controlled by them. Strains resistant to some of the specific acaricides and to carbamate and organophosphorus compounds exist on many nurseries. Mites resistant to one acaricide may also be resistant to other acaricides belonging to the same chemical group. Growers should consider carefully whether repeated and frequent chemical treatment is necessary as a routine. Both cucumber and tomato can tolerate considerable mite damage without yield being affected; more restricted acaricide usage related to mite damage would help reduce resistance and give a more economic control. Where repeated treatments are necessary, growers should consider using two or three unrelated acaricides alternately, rather than relying on one. However, product label recommendations regarding the time intervals before applying other pesticides after organotin acaricides should be strictly followed. Alternatively, biological control with *Phytoseiulus* predators should be considered (see above).

Specific acaricides

Specific acaricides are often more effective for control of spider mites than are broad-spectrum pesticides. They vary in their mode and speed of action; some affect female fertility or egg laying, others kill eggs and/or young or adult mites.

Other acaricides

Organophosphorus sprays were widely used in the past for the control of spider mites on protected crops, but they are rarely effective nowadays because of widespread resistance. However, sprays of some of the newer insecticides belonging to a different chemical group have acaricidal activity and should be effective.

Some granular systemic pesticides with broad-spectrum activity can give useful early control of spider mites on some crops, e.g. tomatoes and pot plants. However, spider mites on some nurseries are resistant.

Hypertoxic strains

When dealing with virulent strains of mites, speedy action is essential because of the severe damage that can be induced rapidly by only a few mites. Chemical control with an effective acaricide is preferable because it acts much faster than biological control.

Application of sprays

The development of resistance to acaricides has increased the need for efficient application. Poor control with sprays is often due to inadequate coverage of the foliage. Infested weeds round the house should be sprayed at the same time as the plants inside.

Wet sprays should not be applied to plants that are dry at the roots or wilting, nor during hot sunshine; spraying should be completed under conditions which ensure that the plants do not remain wet overnight. Sprays should be applied as soon as possible after trimming the plants.

Seedlings and young plants are much more liable to injury than established plants and risk of injury to all plants is greater when growth is forced. Whenever possible, spider mite control should be started before the flower buds begin to colour. Flowers for sale should be cut before spraying, otherwise they may be damaged or their appearance spoilt by deposits.

72
Strawberry mite

Fig.72.1 Strawberry plants damaged by strawberry mite

The strawberry mite (*Phytonemus pallidus fragariae* (Zimmerman)) is potentially a very serious pest of strawberry. Formerly, attacks were severe only during hot summers, but in recent years damage has been more widespread irrespective of seasonal temperatures. Everbearing varieties appear particularly prone to attack.

Symptoms of attack

Injury is caused when large numbers of immature and adult mites feed on the contents of the surface cells of very young strawberry leaves. As the young leaves expand, this feeding makes the surface of the leaves rough and wrinkled. In some varieties, notably Cambridge Favourite, infested leaves develop a characteristic down-curling of the leaf edges (Fig. 72.2). In others, e.g. Redgauntlet, the leaf-edge curl is less marked or even absent. These symptoms may be confused with those of leaf nematode (see page 414), but with leaf nematode there is never any leaf-edge curl. Also, nematode damage is most evident in the spring, whereas mite damage does not usually appear until July, after the first or main crop has been picked.

The number of mites may increase throughout August and September. Attacked plants become stunted and the young leaves often turn brown and die (Fig. 72.1); patches of stunted plants may occur. The appearance of patches of stunted, infested plants, particularly in the autumn, is a good indication of an established infestation. Flower bud formation can be reduced, thus affecting the yield of the following year's crop of single cropping varieties, or the late summer or autumn harvest of everbearing varieties. On infested everbearing varieties the later fruits may turn brown and dry up.

Fig.72.2 Underside of strawberry leaf showing characteristic leaf-edge curl caused by strawberry mite

Description and life history

The adult mites are hardly visible without magnification, the female being about 0.25 mm in length and the male 0.2 mm. They are best examined using a ×10 or ×20 hand lens. They are pale brown and may be found within the unopened young leaflets. Immature mites are somewhat smaller and whitish. The eggs are oval and whitish.

Small numbers of adult female mites overwinter in the crowns of strawberry plants between the bases of the petioles. The eggs are laid from early spring onwards and take about 10 days to hatch at 15 °C. The larval and resting nymphal stages last about eight days, so that each generation from egg to adult takes about 18 days. A succession of generations occurs throughout the summer, but populations do not normally build up appreciably until late July. Though males occur, they rarely constitute more than 5% of the population and do not normally overwinter; parthenogenesis, i.e. reproduction by females without fertilization, is the general rule.

Local dispersal occurs when mites crawl from plant to plant where the foliage touches. They are also carried on the runner plants developing at the tips of stolons which elongate and grow away from the crowns of infested mother plants. In addition, mites may be spread by other means such as insects, wind, trays, punnets and other equipment, or on people's clothing.

Natural enemies

Typhlodromid mites are the commonest known predators of the strawberry mite. Few of these predators occur in strawberry fields which have received routine sprays of broad-spectrum pesticides. The predators are also affected by the trend towards a shorter crop replacement cycle because a shorter cycle allows them less time to become established naturally.

Control measures

Cultural control

Many new infestations are the result of planting infested runners. When planting new fields, only runners certified under the Ministry of Agriculture,, Fisheries and Food's certification scheme should be used (preferably one of the higher grade stocks). New stocks should always be planted as far away from infested fields as possible. One-year-old plants are not usually severely attacked unless the runners were substantially infested. The older the plant the more severe the damage is likely to be; therefore maiden cropping will help to reduce infestations.

Chemical control

Spraying should not be necessary until after harvest of the main crop as damage rarely occurs before picking. Where old foliage is burnt or cut off immediately after picking, spraying is best delayed until the plants are beginning to produce new growth.

The choice of acaricide depends on the type of crop and the severity of the infestation. Some acaricides, for example, may be used only on non-fruiting crops or after harvesting the fruit; others are only effective for controlling light infestations.

High-volume spraying is essential whichever chemical is used. The spray should be directed at the crowns of the plants.

Runner beds grown near infested fruiting beds should be examined from mid-July and, if necessary, sprayed before the taking of runners but no later than the end of August.

Established crops may be treated with a granular acaricide after the completion of picking for the year but while growing conditions still allow translocation to take place.

73
Two-spotted spider mite on outdoor crops

Fig.73.1 Nursery hops showing leaf speckling caused by two-spotted spider mite

Two-spotted spider mite (*Tetranychus urticae* Koch), also known as red spider mite, is primarily a pest of protected crops (see page 361), but it can damage several crops outdoors, particularly in hot, dry summers. It has a wide host range and may spread from crop to crop by way of weeds or hedgerow plants. Strawberries, hops, beans (especially dwarf French and runner beans), black currants, cane fruits, wall-trained fruit trees (particularly apricot, nectarine and peach) and many ornamentals are frequently attacked.

Symptoms of attack

The mites feed on the underside of the leaves, sucking the sap. The first symptoms of attack are usually small areas of whitish spots that are visible on the upper leaf surface and indicate the positions of the colonies of mites living on the underside. As the infestation increases, the leaves become more speckled and later some bronzing appears (Fig. 73.1).

This species of mite spins a fine silken web, at first covering a small area of leaf where the mites are living but later sometimes exending

Fig.73.2 Webbing of two-spotted spider mite on strawberry leaf

to cover the foliage completely (Fig. 73.2) and also the stems.

Damage may appear at any time from April onwards, depending on the number of over-wintering mites and the weather. Attacks become more severe in hot, dry weather from June to early September.

On unprotected strawberries, severe infestations do not usually develop until after picking and are more likely to occur in the second or third years of a fruiting bed. Bronzing and withering of the foliage by mites cause severe stunting of the plants. Plants under cloches and polythene tunnels often become heavily infested during flowering or fruiting.

On black currants, the bronzing of the foliage and premature leaf-fall caused by severe attacks reduce the vigour of the bushes.

On hops, severe infestations can cause defoliation and browning of the cones, resulting in loss of yield and quality.

Description and life history

Two-spotted spider mites are extremely small. The length of the adult female (Fig. 73.3) is about 0.5 mm and the males are even smaller. After mating in the autumn the females become brick-red and begin searching for winter quarters. They hibernate in large numbers just below the surface of the soil, in cracks and crannies in the poles in hop gardens (Fig. 73.4), in the canes used for runner beans, and in the crowns and dead curled leaves of strawberries and other plants. The practice of using straw as a mulch for crops such as strawberries also provides a suitable hibernation site.

As soon as the plants begin to grow in spring, the mites resume activity and start feeding on the young leaves, Minute, round, translucent eggs are laid on the underside of the leaves and hatch after about two weeks, the incubation period varying with the temperature. The young mites moult after feeding for a short time and pass through two more immature stages before becoming adult.

The adults of the summer generations are at first yellowish or greenish with dark green markings, but later they may become darker. During the season there may be up to seven generations, which overlap one another. In hot, dry weather the increase in numbers is exceedingly rapid.

Natural control

Under natural conditions, predatory insects and mites destroy large numbers of spider mites and may become abundant when heavy infestations occur. Anthocorid bugs and

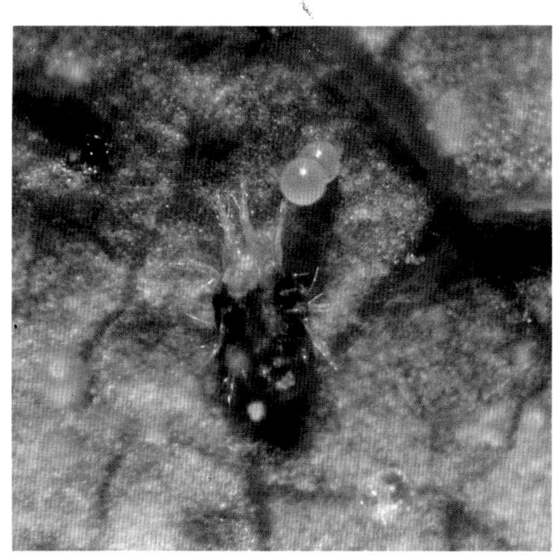

Fig.73.3 Adult female two-spotted spider mite and eggs on hop leaf (× 45)

typhlodromid mites are the most important predators, while some thrips species, a staphylinid beetle (*Oligota flavicornis* (Boisduval & Lacordaire)) and a small black ladybird (*Stethorus punctillum* Weise) also play a part in destroying the mites. Some outbreaks of spider mite are the result of killing these predators by the use of broad-spectrum insecticides.

The death of overwintering mites may be another important factor in the reduction of spider mite populations. This mortality may be due to unsuitable hibernation sites or to predation, particularly by anthocorid bugs.

Biological control

The predatory mite *Phytoseiulus persimilis* Athias-Henriot is widely used for the control of two-spotted spider mite on protected crops (see page 363). This predator has recently been used successfully on various crops outdoors. Introductions of the predator into strawberry and raspberry crops have been particularly successful and the method is being developed for runner beans, hops and hardy nursery stock.

Good spider mite control has been obtained on soft fruit with introductions of 25 000 predators per hectare, but the rate of introduction should be varied according to the crop, season and number of spider mites.

The key to successful control with predators is as follows.

(1) Introduction early in the season *but* after the risk of penetrating frost is over.

(2) Careful choice of pesticides for control of other pests and for diseases. (Many insecticides, acaricides and fungicides are toxic to the predator.)

(3) Thorough monitoring of the numbers of spider mites and predators after introduction. If necessary, more predators should be introduced or, if spider mite numbers become too great, curative acaricide treatments should be applied.

On perennial crops where predators have been used successfully for one season, there are usually very few spider mites in the following year.

Fig.73.4 Hibernating two-spotted spider mites on chip of wood from chestnut hop-pole

Chemical control

Resistance to acaricides

Failures of chemicals to control mites may be due to poor application or incorrect timing, or to the presence of strains of mites that are resistant to the chemicals. The appearance of resistance is associated with repeated use of the same, or similar, acaricides, usually over a period of years, although resistant mites may be introduced on strawberry runners or other planting material. Mites resistant to one chemical are often cross-resistant to closely related compounds, though there is little evidence of cross-resistance to chemicals in other groups. Multi-resistant strains of mites, which are resistant to two or more chemical groups, may occur.

Choice of chemicals

The choice of an effective acaricide is usually limited by the presence of resistance. The majority of spider mite populations are resistant to most organophosphorus compounds and resistance to some specific acaricides has also been recorded.

Care should be taken to avoid using chemicals that are liable to be phytotoxic to the particular variety of plant being treated. Some varieties of black currant, for example, can be damaged by certain acaricides.

Any hazard from residues on harvested produce should be avoided by using a chemical that requires a suitable minimum interval between application and harvest; this interval may vary according to the crop. If the harvested produce is destined for processing, the processor's requirements should also be checked as they often stipulate which chemicals may be used.

Application of sprays

Routine protective sprays are required only on crops where damaging infestations occur frequently. The most common examples are strawberries, raspberries, black currants, wall-trained fruit trees, hops, runner beans and certain hardy ornamentals such as buddleia, ceanothus, *Hedera* and magnolia. Otherwise, sprays should be applied only when the pest or damage is seen.

Field-grown strawberries should only be treated as a routine if damaging infestations occur frequently. Otherwise, treatment should be applied at the first signs of damage. Protected strawberries should be treated as a routine, particularly those grown in 'walk-in' polythene tunnels.

On single-cropping varieties, one or two sprays should be applied before harvest: the first shortly before flowering and the second between flowering and harvest. Crops in flower should not be sprayed. An aphicide may need to be included in the pre-flowering spray. Further sprays may be required during the summer if an infestation develops.

On double-cropping varieties grown under protection, a further spray may be required shortly before, or just after, the second flowering period.

On continuous-fruiting varieties, it is necessary to use a chemical that has short persistence and is safe to flowers and pollinating insects.

On black currants, certain fungicides used against powdery mildew should reduce spider mite populations. If necessary, a specific acaricide should be applied just after flowering or immediately after picking the crop.

Other bush and cane fruits grown in the open seldom show symptoms of infestation until July or August. In average summers, infestations are usually too small to cause economic damage but, in a warm summer, treatment may be necessary. Raspberries are particularly susceptible to damage and autumn-fruiting varieties are the worst affected. On these varieties a pre-blossom treatment should be applied. Where necessary, summer-fruiting varieties should be treated post-blossom.

On wall-trained fruit trees, a spray in the first half of May, followed by another about two weeks later, should keep spider mite under control throughout the summer. However, it is necessary to examine crops frequently so that additional sprays can be applied if required.

On hops with a history of damaging infestations, two acaricide applications should be made early in the season to control the mites emerging from hibernation and the mites of the first summer generation. The first spray should be applied immediately after the first training of the bines or, if the lower leaves are burnt off early (in late May), immediately after this latter operation. The second spray should be applied two weeks later. Care should be taken to ensure that the ground is weed-free at the time of spraying, otherwise mites infesting the weeds may later move on to hops. In hop gardens which are not sprayed early in the season, infestations may develop in August.

On beans, control of spider mite can be difficult as few chemicals are currently available for use on this crop. Some growers with facilities for steam sterilization steam their runner bean canes before re-using them. If sterilization is not practised, a careful watch should be kept in the year following a mite attack so that, in the event of a recurrence, a spray can be applied at an early stage.

On hardy ornamentals, spider mite can be troublesome, particularly under protection. Susceptible plants should be grouped together if possible in one bed, or in one area of the nursery, so that they can be inspected for mites and sprayed together. Plants that have recently been grown or propagated under glass should be hardened off before treatment in order to reduce the risk of phytotoxicity to soft or tender foliage. Routine treatment of susceptible plants may be desirable if spider mite is a frequent problem on the nursery. The first spray outdoors should be applied in late May with further applications as necessary. Earlier treatment is required under protection.

Violets are very susceptible to spider mite attack and may be severely damaged. In bad cases it is best to remove and destroy all affected foliage before spraying. New foliage will appear quickly if the plants are growing in satisfactory conditions. However, spraying soon after planting, and again during dry periods in the summer, should give good control and such defoliation should not then be necessary. An alternative method is to dip the plants in a solution of a recommended acaricide before planting.

Application of granules

Application of acaricide granules to ornamentals, strawberry runner beds, or cropping strawberries after harvest will control spider mite infestations.

OTHER ARTHROPODS

Myriapoda and Crustacea

74
Millepedes and centipedes

Fig.74.1 Beet seedlings showing typical damage caused by millepedes feeding at or just below soil surface

Millepedes and centipedes are allied to insects but differ from them in many characters, the most obvious being the presence of legs on almost all of the many body segments.

As some millepedes are harmful to a wide range of plants while centipedes are generally beneficial, it is important to distinguish between them. Both live in moist surroundings and tend to avoid light. They are rather alike in general form, but millepedes (Figs 74.2 and 74.3) have two pairs of legs to almost every body segment while centipedes (Figs 74.5 and 74.6) have only one pair. Millepedes usually move slowly and, when touched, some roll up; centipedes are more active.

Millepedes and centipedes are sometimes confused with wireworms, which are the larvae of click beetles (see page 216). Wireworms, however, possess only three pairs of legs which are situated just behind the head. Millepedes and centipedes are more often confused with symphylids (see page 382) to which they are related. Symphylids are delicate, white creatures that are less than 9 mm long with 12 pairs of legs when fully grown; they are very active with long, rapidly palpating antennae. Springtails of the genus *Onychiurus* can be confused with immature stages of some millepede species, or with symphylids, but springtails are insects and never have more than three pairs of legs.

Fig.74.2 Flat millepede: *Brachydesmus superus* (× 12)

MILLEPEDES

There are 48 species of millepede known to occur in Britain, but many are rarely found as they are mainly confined to localized habitats. Two main groups, i.e. flat millepedes and snake millepedes, contain the most commonly occurring injurious species; they are often found together in the same habitat.

Description and life history

Flat millepedes

This group is characterized by the possession of paired lateral projections on the ring-like body segments (Fig. 74.2); these projections contribute towards a flat dorsal surface — hence the name 'flat millepede' (also 'flat-backed millepede'). These millepedes are particularly common under stones and surface litter, but in unfavourable conditions they may retreat deeper into cracks within the top soil.

Brachydesmus superus Latzel (Fig. 74.2) is the most common injurious species in fields, gardens and glasshouses. The adults are 8–10 mm long and generally a light brown colour. The female lays about 50 eggs, mainly during the spring and summer, within a specially constructed dome-shaped nest built on a solid base. The first-stage larva has seven segments and only three pairs of legs. After six moults, or seven in the case of some males, each within earthen chambers, the adult stage

Fig.74.3 Snake millepede: *Blaniulus guttulatus* (the spotted millepede) (× 7)

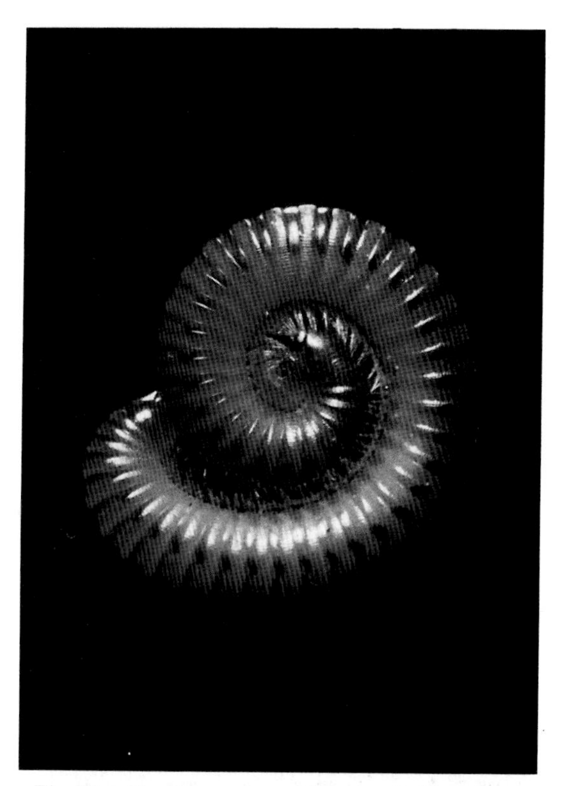

Fig.74.4 Resting snake millepede: *Cylindroiulus punctatus* (× 10)

is reached. The adult has 19 body segments with 29 pairs of legs in the male and 28 in the female. The life span is about 12 months.

Millepedes of the genus *Polydesmus* are similar in appearance and colour to *B. superus* but are generally larger, being up to 25 mm long, and have 20 body segments when adult. The female may lay up to 300 eggs. The life history occupies 2½–3 years.

The glasshouse millepede (*Oxidus gracilis* (Koch)) is a related tropical species which sometimes occurs in heated glasshouses in the British Isles.

Snake millepedes

This group is so called because the appearance and movements of the millepedes resemble those of a snake and enable them to move easily in the soil. They have a smooth rounded body of 30–60 segments (Fig. 74.3). When at rest they assume a spiral shape (Fig. 74.4) and are often found deep in the soil. The commonest species that are injurious to crops are recognized by their small slender bodies, no wider than 1 mm, which bear a row of orange to red spots along each side, one on each segment.

The spotted millepede (*Blaniulus guttulatus* (Fabricius)) (Fig. 74.3) is the most common snake millepede in arable fields, nurseries and glasshouses. It is up to 18 mm long and is easily recognized in the adult stage by its bright orange-red spots. *Boreoiulus tenuis* (Bigler) is less common and the smallest of the group, often no more than 10 mm long.

Archiboreoiulus pallidus (Brade-Birks) is similar to the spotted millepede both in size and overall appearance except that its pale orange spots are less distinctive. *Cylindroiulus punctatus* (Leach) (Fig. 74.4), which is not a pest, is particularly numerous in woodland litter, especially beneath rotting bark and in decaying wood, where it moults and lays its eggs.

The life history of snake millepedes is essentially similar to that of flat millepedes. However, the egg clusters are not protected by such elaborate chambers in the soil (some species may even lay eggs singly) and there can be up to 15 stages in the life cycle. The age at which sexual maturity is reached varies from species to species; some die after the first reproductive period, whereas others continue to moult after reproducing. An example of the latter is *C. punctatus*, which reproduces during the eighth stage, after three years, and has two or more further annual broods during later stages.

Plants attacked

The food of millepedes is almost entirely vegetable. The roots of most kinds of plants, as well as bulbs and tubers, are eaten. Sugar beet (Fig. 74.1), peas and beans (usually the germinating seeds and the seedlings), carrots, potatoes and strawberries are among the crops especially subject to millepede injury. Cucumbers are attacked by the glasshouse millepede, which can cut down mature plants.

Fig.74.5 Centipede: *Geophilus* sp. (× 3)

Millepedes frequently feed on plant tissues injured by mechanical means or other pests, but they also attack the soft parts of undamaged healthy plants. When numerous, they may be troublesome pests, but more usually of garden and allotment than of field or glasshouse crops. However, millepedes are regarded as rather serious pests of sugar beet seedlings in England, Belgium and France, where they often occur with symphylids and/or springtails (see page 378).

There is nothing very distinctive about the type of injury caused. As damaged tissues soon begin to decay and may then attract other soil animals, it is sometimes difficult to establish the identity of the original culprit.

Control measures

Millepedes are difficult to control and an infestation is seldom discovered until much damage has already been done. They are abundant in most soils containing organic matter and can always be found in leaf mould, garden rubbish heaps and other decaying vegetable matter. The conditions under which they attack living plants are not known. Various measures for checking damage and decreasing the numbers of millepedes have, however, been found useful and some suggestions are made below. Where damage by millepedes is secondary, attention should be given to controlling the primary cause of damage.

Outdoors

In ground in which millepedes are known to be troublesome, a recommended soil insecticide worked into the soil before sowing or planting should give some protection from attack. Care should be taken to avoid using insecticides

Fig.74.6 Centipede: *Lithobius* sp. (× 2½)

likely to taint crops such as carrots or potatoes grown within 18 months of application.

Granular pesticides applied in the sugar beet seed furrow as a general prophylactic treatment against soil pests and aphids will decrease the amount of millepede damage.

Under glass

Millepedes can be killed by soil sterilization, but control can never be complete since the young stages tend to live deep in the soil or go there to moult within their protective chambers. Where an infestation is found in a growing crop, special treatment may be desirable, e.g. in small areas of valuable crops that warrant the extra expense.

A soil insecticide applied for control of symphylids and springtails may also control millepedes.

CENTIPEDES

As centipedes are not injurious to plants, only a brief description of them is necessary. They are carnivorous, feeding mainly on insects, small slugs and worms. Two groups are well represented in Britain; these may be distinguished as follows.

The **Geophilomorpha,** e.g. *Geophilus* species (Fig. 74.5), have long, narrow bodies with at least 37 pairs of short legs. The species vary from red or brown to very pale yellow. They move in a serpentine manner and are well adapted to movement in humus and soil. They are, therefore, usually found in leaf mould and in the surface layers of the soil, where they are sometimes mistaken for wireworms (see page 217). *Necrophloeophagus longicornis* (Leach), a common species closely related to *Geophilus* species, has a bright yellow body with a darker head. It is found in a wide range of habitats and is common in gardens.

The **Lithobiomorpha,** e.g. *Lithobius* species (Fig. 74.6), are shorter and more compact than the Geophilomorpha and have 15 pairs of long, powerful legs. The head and body are brown. The body is less flexible than that of other centipedes but members of this group can run faster than *Geophilus* species. They are unable to penetrate the soil readily and are usually found under stones or logs. The common garden centipede (*Lithobius forficatus* (L.)) is the most familiar and widely distributed centipede in Britain. It is frequently found in gardens and sometimes enters buildings. It is most active at night.

75
Symphylids

Fig.75.1 Sugar beet seedlings damaged by symphylids. The variation between rows is characteristic and is associated with differences in the number of symphylids in the surface layers of soil

Symphylids are small, white, active creatures which can be very numerous around growing roots in glasshouses and in some field soils. They are related to insects, millepedes and centipedes, but most closely resemble centipedes. They feed on living and dead plant material and micro-organisms. The main species found in glasshouses is the glasshouse symphylid (*Scutigerella immaculata* (Newport)) (Fig. 75.3); this is not confined to glasshouses, but other species, especially open ground symphylids (*Symphylella* spp.), are often more abundant in outdoor crops.

Symphylids occur in most parts of Britain and in the Channel Islands. They are most numerous in silty soils, particularly those with a high organic matter content and abundant moisture. In Britain they were originally considered as pests of glasshouse crops only. However, they also attack outdoor crops on market gardens and smallholdings and, since the mid-1960s, they have become a major pest of sugar beet seedlings in some fields, particularly in the Fens and the Yorkshire Wolds, where plant establishment may be seriously affected (Fig. 75.1). Changes in sugar beet crop husbandry-have probably accentuated the damage.

The crops most commonly attacked in Britain are soil-grown tomato, lettuce and chrysanthemum in glasshouses and sugar beet in the field. Attacks also occur on a wide range of other crops including asparagus, bean, brassicas, celery, cucumber, parsley, pea, pepper, potato and strawberry. Anemone, rose, sweet pea and pot plants are also attacked. In the Channel Islands symphylids are a particular pest of tomato and potato.

Nature of damage

Not all species of symphylid cause crop damage and, even if many symphylids are seen in the soil, they are unlikely to be causing damage *unless* there are definite symptoms on the roots. Such symptoms are best seen if a sample of roots is washed in water (see Examination of plants, page 385).

Several crops are susceptible to symphylid attack. Young plants can be killed but the survivors usually outgrow the damage.

General symptoms

A common type of damage is the removal of root hairs from young growing roots; this leads to the disappearance of many small roots and consequent stunting of the plants. In the glasshouse, severely damaged plants wilt readily. Damage is sometimes mistaken for that caused by other problems such as excess salts, waterlogging and acidity.

On many crops damage also shows as tiny black marks on the roots where a hemispherical piece of tissue has been scooped out (Fig. 75.2). These small lesions may aid attack by fungi and other organisms causing root rots. Where this occurs, the original cause is liable to be overlooked.

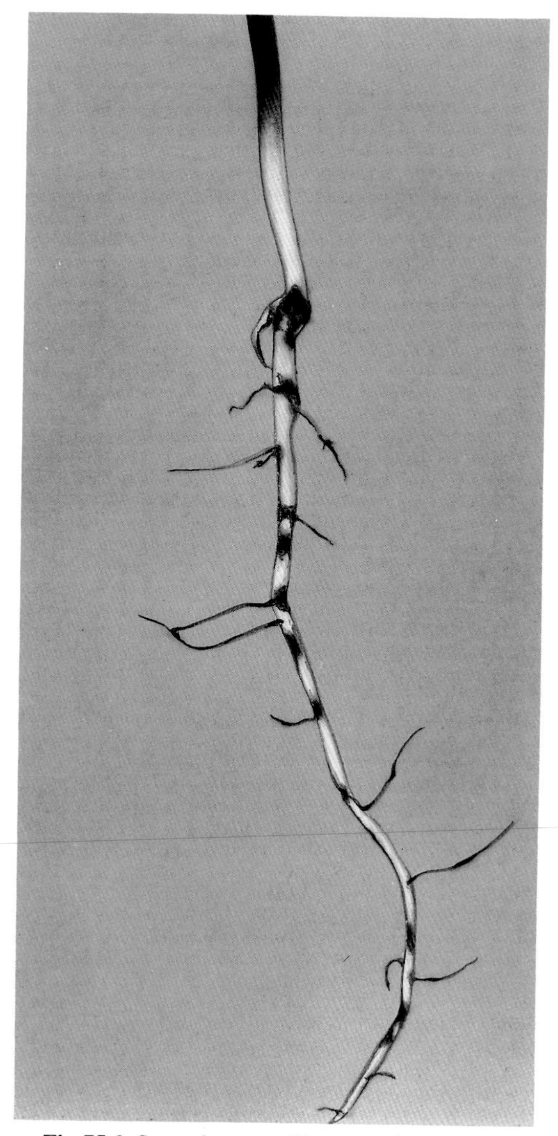

Fig.75.2 Sugar beet seedling root damaged by symphylids

Specific symptoms

In addition to the general symptoms, there are some that are specific to the crop.

Tomato leaves take on a bluish tinge and the plants become very stunted.

Lettuce plants do not form a heart; injured plants often die as a result of secondary root rots and botrytis.

Chrysanthemum growth is stunted, often in patches in bed-grown crops. Roots become thickened and gnarled, sometimes with a reddish tinge.

Anemones acquire small holes in the leaves and brown, oval lesions along the base of the stem.

Sugar beet Very young seedlings can be killed as a result of the symphylids feeding on the hypocotyl or because of damage caused by a secondary pest or disease. In older seedlings the damage to the fibrous root system stunts their growth.

Description

Adult symphylids (Fig. 75.3) are delicate, white creatures less than 9 mm long. They are extremely active, possess long mobile antennae and 12 pairs of legs when fully grown.

Immature symphylids, which are small and have 6–12 pairs of legs according to age, are often confused with immature millepedes and certain species of white springtail that are common in glasshouse soils. Young millepedes (Fig. 75.4) have three pairs of legs at first and grow more, whereas springtails (Fig. 75.5) never have more than three pairs. Springtails also differ from symphylids in being shorter and moving only slowly. Young millepedes have shorter antennae than do symphylids and lack pointed cerci (projections) at the hind end.

Life history

Symphylids lay their eggs in batches of 2–20 in the upper layers of soil in glasshouses but in the deeper layers outdoors. The eggs, which are tended by the adult, hatch after 1–3 weeks and small symphylids with six pairs of legs emerge. The first moult occurs within two days and the others at intervals of 2–6 weeks. At each moult a further pair of legs is added until the adult number of 12 is attained. After a further two or three moults the symphylids are sexually mature and begin to lay eggs, but they continue to moult at approximately monthly intervals throughout their lives.

Eggs and small larvae can be found in the soil all the year round, but breeding is at its peak in the spring and early summer. The complete life cycle from egg to adult never takes less than three months and usually considerably longer. Specimens have been kept alive in laboratories for as long as four years.

Fig.75.3 Glasshouse symphylid (× 18)

Symphylids migrate from the surface soil when conditions there become unsuitable; they may penetrate nearly two metres below the surface if there are adequate fissures in the soil and subsoil. This migration, though partly

Fig.75.4 Young millepede (second stage of *Brachydesmus superus*) (× 35)

seasonal, is mainly due to adverse conditions in the surface soil, such as drying out and high temperature, and to the absence of growing plants. When the surface soil is warm and moist and a suitable crop is present, symphylids migrate to the upper layers. Normally their movement through the soil is confined to existing cracks and crevices as they are unable to penetrate packed soil unaided. In glasshouses, symphylids are most active in the surface soil in late winter and early spring, when they attack tomato and lettuce crops. Outdoors they usually come to the surface soil during late spring and again in the early autumn.

Infestations in glasshouses

Infestations may arise from the introduction of manure or infested plants to a glasshouse. An infestation is seldom spread evenly over the entire area of a glasshouse but is in characteristic patches, which may persist from year to year and will gradually enlarge.

Fig.75.5 Springtail (*Onychiurus* sp.) (×55)

Certain rotations of crops in glasshouses appear to encourage the build-up of large numbers of symphylids, particularly a tomato and lettuce rotation. Nevertheless, even under a favourable cropping rotation, it may be several years before an infestation becomes serious. But once a large population of symphylids has built up, it is extremely difficult to eradicate, owing to migration to the lower soil levels. A reservoir of the pest is thus maintained at the lower levels from which the surface can be reinfested after control measures have been taken.

Examination of plants

The extent of a symphylid attack can be determined by lifting poorly growing plants and quickly lowering them, together with the soil surrounding the roots, into a bucket of water. The soil should then be gently kneaded or stirred under the surface of the water, so that the symphylids come out of the soil and float to the surface of the water where they can be counted. Springtails will also float to the surface; they may be causing similar damage, but this is significant only if very large numbers are present.

Glasshouse soil that is suspected of being infested before being sown or planted up can be examined by the same method. When there is no crop present the symphylids are below the soil surface, so a few spadefuls of soil should be taken to a depth of about half a metre and each dropped into a bucket of water. The numbers of symphylids coming to the surface after the soil has be broken up will give a useful indication of the degree of infestation.

Many growers whose glasshouse crops have suffered attacks in the past plant out a few extra tomato plants between the rows. One or two of these plants can be dug up and dropped into a bucket of water at daily intervals to determine whether a serious attack is building up on the roots. If 10 or more symphylids are found among the roots of a single plant, it will be well worth applying control measures. Damage has been observed, however, when as few as three symphylids per plant have been found by the bucket method.

For sugar beet the critical level is about one symphylid per three seedlings, but numbers cannot be determined before sowing because the symphylids are still deep in the soil. No control is possible once the seedlings are damaged, so control measures must be applied to fields with a previous history of symphylid damage.

Control measures

In glasshouses

Steam sterilization as normally practised in a glasshouse appears to have little effect on this pest. Soil fumigants or sterilants applied when most of the symphylids are in the upper layers of the soil will help to prevent serious outbreaks.

For tomatoes, pot plants and flower crops, a recommended soil insecticide applied as a drench within three days of planting out will protect the young plants until they are well established. Cucumbers, and also tomatoes, can be protected by applying the soil insecticide to the planting hole two or three days before setting out the plants.

On field crops

For field infestations on some crops, e.g. sugar beet, a recommended soil insecticide may be used when damage is expected. The insecticide should be worked into the soil during seedbed preparation. Care should be taken to avoid using chemicals likely to taint susceptible crops such as carrots and potatoes.

Some granular pesticides applied as seed-furrow treatments for protection against damage by various pests will give good control of symphylids.

76
Woodlice

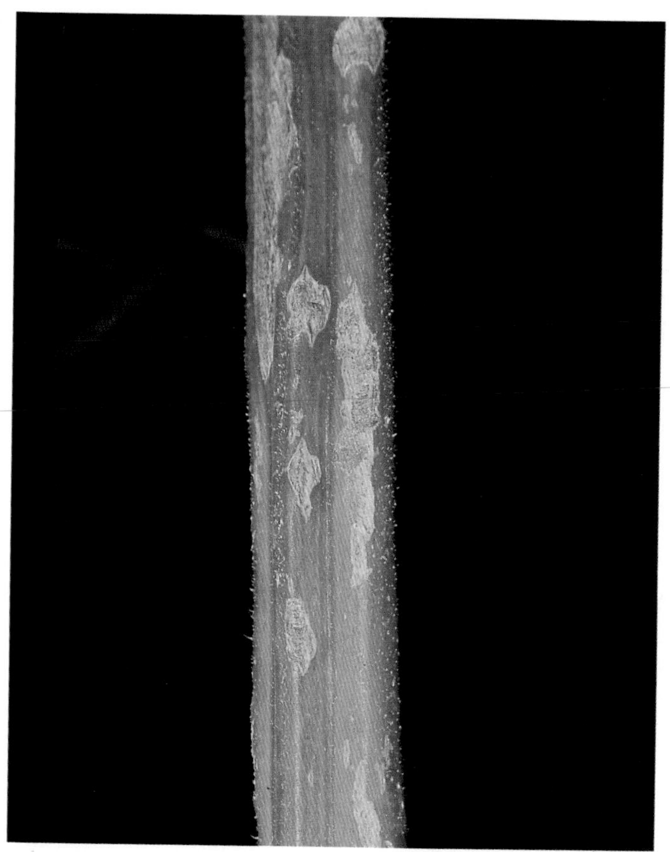

Fig.76.1 Base of cucumber stem showing woodlouse damage

Although related to insects and millepedes, woodlice are crustaceans and have a closer affinity with crabs and lobsters. They are one of the few groups of the Crustacea to have left the sea and successfully colonized the land. However, they are unable to survive in dry conditions and most species are restricted to moist habitats. Woodlice can be distinguished from insects and mites by the possession of seven pairs of legs. Their oval shape, as well as the much smaller number of legs, distinguishes them from millepedes (see page 378) and centipedes (see page 381).

Woodlice feed mainly on dead and dying plant material, but they sometimes eat green food and have a strong liking for germinating seedlings. Most serious injury occurs in glasshouses, where they attack a wide range of plants, especially cucumber grown in straw-based beds.

There are 42 species of woodlouse known to occur in Britain, some of which are found only in heated glasshouses. Thirty species are regarded as native, but only about eight of these are likely to be noticed generally.

Fig.76.2 Woodlouse (*Porcellio* species)

Fig.76.3 Cucumber leaf showing woodlouse damage

Habits

Woodlice avoid light and are active at night. By day they shelter under stones, plant debris or damp wood, where they tend to occur in large aggregations. They readily climb rough surfaces, such as walls and tree trunks, in search of food. In frosty weather they tend to move down into the soil and are generally inactive during the winter. Woodlice are great travellers and enter buildings under doors or through air vents, particularly when the first sharp frosts of autumn occur.

Woodlice may be abundant in the looser kinds of compost heaps, where the heat produced by bacterial decay protects them from winter frosts. Some species are confined to this habitat outdoors in Britain. Woodlice are scarce in arable fields except round the edges.

Description

The woodlice known as 'slaters', such as *Oniscus* and *Porcellio* (Fig. 76.2), are flattened and tend to cling to surfaces, whereas the 'pillbugs', such as *Armadillidium*, are more humped and roll up into a ball when disturbed. Most of the common species are slate-grey, but a bright rose-pink one is quite common. The size of the adults ranges from 2.5 to 18 mm. The species usually found attacking cucumber is the blunt snout pillbug (*Armadillidium nasatum* Budde-Lund).

Life history

The fertilized eggs are laid in a brood pouch under the female's body; there they hatch and develop. By the time the young are released, after about three weeks' incubation, they resemble miniature adults. Outdoors in Britain woodlice breed only in the summer; they produce between six and 200 young in a season according to species and climate. Most species breed during the year after birth. The maximum life span is between one and four years, but the great majority of young are dead within two months of being released and thus fail to reach maturity. The reasons for this high juvenile mortality are unknown, but susceptibility to predators is probably a factor. Woodlice moult, and hence grow, all their lives; occasionally long-lived individuals of very large size can be found.

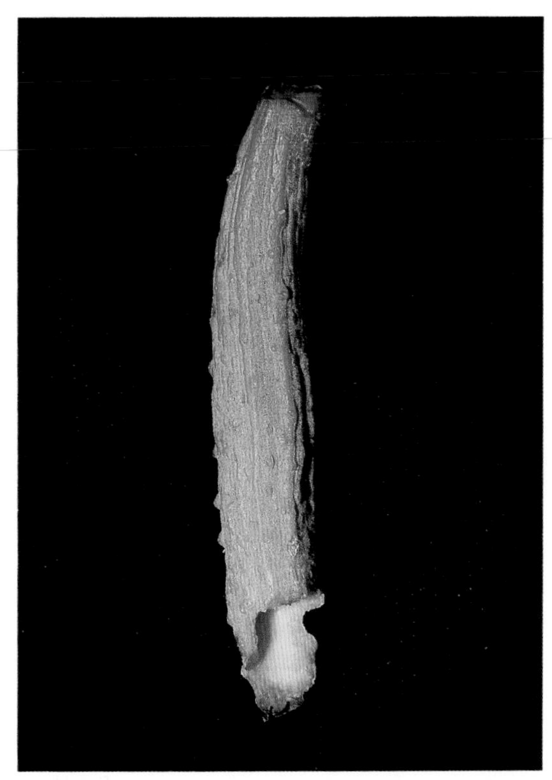

Fig.76.4 Young cucumber fruit showing woodlouse damage to tip

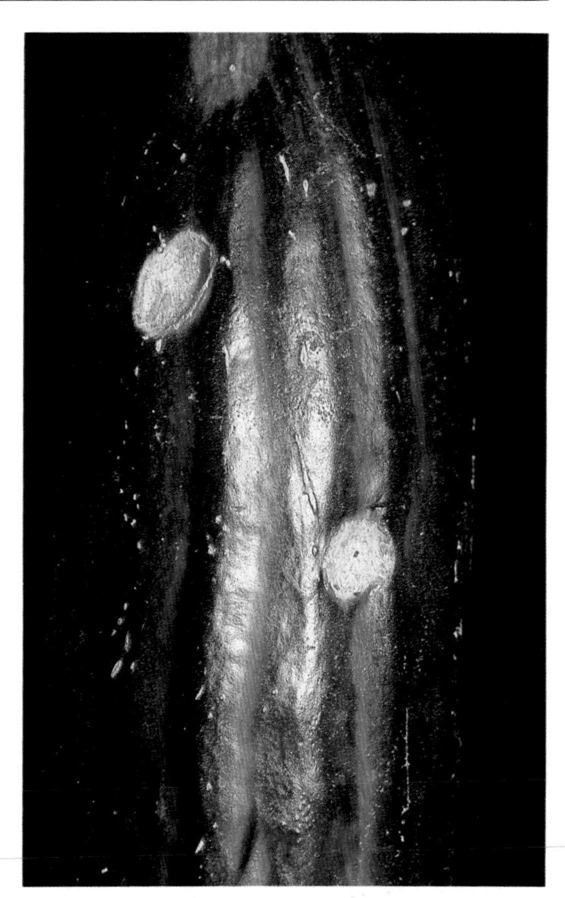

Fig.76.5 Close-up of mature cucumber fruit showing woodlouse damage

Economic importance

On balance, woodlice do more good than harm, since their activities accelerate the breakdown of dead plant material and the return of nutrients to the soil, leading to an increase in productivity. In compost heaps woodlice are entirely beneficial as they speed up plant decay. Their activities in pastures are also beneficial.

Woodlice can be a problem in gardens and glasshouses, particularly in those which are less well tended. They sometimes cause a significant amount of damage to seedlings and to young bedding and cucumber plants. Mature healthy plants are not usually attacked, with the notable exception of cucurbits. Traditional, straw-based cucumber beds, because of their open organic composition, provide an ideal habitat for woodlice. Once infested, such beds are likely to remain so for the life of the crop. However, woodlice often live in cucumber beds without adversely affecting the crop and control measures may not be necessary. Crops growing in soil-less media are less vulnerable to woodlouse attack. With the increase in the use of such media for cucumber production, woodlice have become less of a problem on this crop.

Nature of damage

On cucumber

Woodlice eat irregular holes in the leaves (Fig. 76.3). The lower leaves resting on the beds may become perforated by a network of holes similar to that seen in attacks by French 'fly' (*Tyrophagus longior* (Gervais)). Stems are often gnawed (Fig. 76.1), and may be completely girdled, at or just above soil level; they can be weakened in severe attacks. This damage renders the plants more susceptible to fungal pathogens. Sometimes woodlice climb higher into the crop, where flowers and young fruit (Fig. 76.4) may be damaged. In serious infestations the growing points may be completely removed. The damage resembles that caused by slugs and snails (see page 467), but there is no slime trail. Surface-feeding millepedes also cause similar damage (see page 380); they often occur with woodlice. Feeding marks by woodlice on mature fruit (Fig. 76.5) can cause a significant reduction in quality.

On other plants

The leaves and growing points of young ornamentals such as carnations can be damaged and the aerial roots of orchids chewed.

Where mushrooms are grown in glasshouses, holes are sometimes chewed in the caps (see page 287).

Control measures

Hygiene

Large populations can build up where day-time shelter sites are plentiful. Removal of the shelter is the key to control. In glasshouses, where damage is more likely, cleanliness is essential. Damp refuges such as old seed boxes, pots and plant debris should be removed, cracks in damp brickwork repaired and decaying structural timber replaced. Control in gardens and allotments is also a matter of tidiness, in particular keeping compost areas well away from seedbeds; this will also give protection from slugs and snails.

Chemical control in glasshouses

Because woodlice are mobile, reinfestation can easily occur, so chemical control measures are of limited value unless combined with strict hygiene. In cucumber beds, chemical control measures can do no more than keep down the numbers of woodlice living near the surface.

The use of volatile pesticides should be avoided if biological control agents are being used to control other pests. The vapours of such chemicals may prevent the successful establishment of parasites and predators.

Smokes Some smoke formulations are recommended for the control of woodlice, although they are likely to kill only those which are active on the surface.

Baits Broadcast applications of certain pelleted baits used for slug control will also give incidental control of woodlice.

Control in dwelling houses

Woodlice invade houses, especially in the autumn and winter, but do not breed or cause damage indoors. To prevent further invasion, outside sources of infestation such as plant debris and rotting wood should be removed and a dust formulation of a recommended pesticide applied round the entry points into the house, e.g. door, ventilation bricks and waste-water openings. This done, there should be no need for pesticide applications inside the house.

NEMATODES

Nematoda

77
Beet cyst nematode

Fig.77.1 'Beet-sick' patch in a field

Beet cyst nematode (*Heterodera schachtii* Schmidt) can cause 'sickness' in sugar beet, mangels and red beet where host crops of the nematode have been grown too frequently in the rotation. Beet cyst nematode occurs, and is a potential threat to intensive sugar beet cropping, in all temperate countries. Where the nematode is established it is important to lessen the risk of crop damage by adopting suitable crop rotations.

Beet cyst nematode was first found attacking beet in Britain in 1934 in a field near Chatteris, Isle of Ely. Frequent crops of mangels had been grown in this field as well as seed crops of cress, mustard and turnip, all of which would have been capable of increasing the nematode population. Many fields in the Isle of Ely are now known to be infested. Infestations are also frequently found on light land in west Norfolk and west Suffolk and less frequently in brassica-growing areas in the West Midlands. Infestations are even less frequent in other beet-growing areas, but all areas with peaty or light sandy soils should be regarded as potentially at risk, especially where there is a high water table.

Effects on crops

When few nematodes are present in the soil the damage caused is insignificant as only a few rootlets are killed. It is thus possible to grow good crops of beet and mangels in fields with a small population of nematodes and to remain unaware that the land is infested. As the nematode population increases with repeated cropping, so the crop yield decreases. Serious losses of yield can occur when the nematodes are so numerous that they kill the tap root early in the season.

Symptoms of attack

Usually the first sign of 'beet sickness' observed by the farmer is the appearance of a small patch or patches in the crop (Fig. 77.1) where the plants are stunted and the foliage looks unhealthy. The outer leaves wilt in the sunshine, turn yellow prematurely and die; the heart leaves may remain green but are undersized. Later in the season the foliage may partly recover, but the roots remain stunted.

Severely attacked plants are easily pulled from the soil, the tap root being small and short, or even absent, and there is an excessive development of lateral roots sometimes known as 'hunger roots' (Fig. 77.2). Most of these lateral roots are dead and the small lemon-shaped female nematodes can be seen protruding from them (Figs 77.2 and 77.3).

Patchiness and reduced vigour in beet and mangels are not always caused by beet cyst nematode. On light, sandy soils free-living nematodes associated with Docking disorder (see page 409) are a more frequent cause of stunted growth from early in the season onwards. Virus yellows commonly causes yellowing and late-season stunting. Plants on dry soils frequently wilt in patches during a drought without any pest or disease being present.

Description and life history

The beet cyst nematode lies dormant in the soil throughout the winter as larvae within the eggs, which are enclosed in brown cysts. These cysts are lemon-shaped and up to 1 mm in length; they are the dead remains of female nematodes whose skin forms a protective covering for the eggs. A large cyst may contain as many as 650 eggs. In the spring some of the eggs hatch and the young nematodes escape into the soil. About half of the eggs hatch each year, but some may remain alive in the cysts for many years.

The young nematodes penetrate the rootlets of susceptible plants and may kill them. The plant responds by forming additional roots, many of which may also be killed.

As the female nematodes mature, they swell to assume the characteristic lemon shape and burst out of the rootlets so that only the head and neck remain embedded. After fertilization by the males (which by contrast are active and eel-like), the females are soon filled with developing eggs. The females then die and their body walls become hard and leathery, changing from creamy-white to brown as they become cysts. Under favourable conditions the beet cyst nematode can complete two generations during a growing season in England: the first flush of females appears on the roots in June/July and the second in September. The highly protective cysts are responsible for the persistence of this nematode in the soil, for the ease with which it is spread and for the difficulty experienced in combating it once damaging infestations have built up.

Plants attacked

Crops attacked include sugar beet, fodder beet, mangel, red beet, spinach, turnip, swede, cabbages of all kinds, cauliflower, broccoli, Brussels sprout, kale, oilseed rape and all other kinds of rape, cress and mustard. Certain weeds, such as charlock, chickweed and shepherd's purse, are also attacked.

Both culinary and fodder radishes are inefficient hosts of beet cyst nematode and are not considered important in building up infestations.

Brassicas are seldom appreciably damaged by beet cyst nematode, but most of them are efficient hosts of the nematode and can cause an increase in the number of cysts in the soil. This increase can be as great as, or even greater than, that caused by sugar beet. Oilseed rape and other brassica seed crops are particularly efficient hosts and can lead to a quick build-up of beet cyst nematode; they should not be used as break crops for cereals in rotations that include sugar beet.

Means of spread

Nematodes can travel only very short distances by their own efforts; they are spread mainly by the movement of soil. Thus beet cyst nematode is spread from field to field by tractors, trailers, farm implements, workers' boots, and also by animals. Cysts can be carried in soil adhering to seed potatoes or to plants for transplanting. Strong winds that blow soil dust to neighbouring fields may be responsible for local spread of cysts. Not all of these means of moving soil can be prevented so that some nematode spread is inevitable.

Beet cyst nematode introduced with soil is unlikely to be easily detectable until numbers are increased by the growing of beet or other host crops. Damaging populations of this nematode can always be traced to too frequent cropping with host plants; what is too frequent on a light soil may, however, be safe on a heavy soil.

Control measures

There are currently no recommendations for chemical control of beet cyst nematode. However, research continues into the use of nematicide treatments as well as other methods of control such as the use of resistant varieties and biological control techniques.

Crop rotations

At present the only economic method of control is crop rotation.

Until 1983 a clause in the contract between

Fig.77.2 Severely stunted sugar beet root showing absence of tap root, excessive development of lateral rootlets, and white females of beet cyst nematode (about life size)

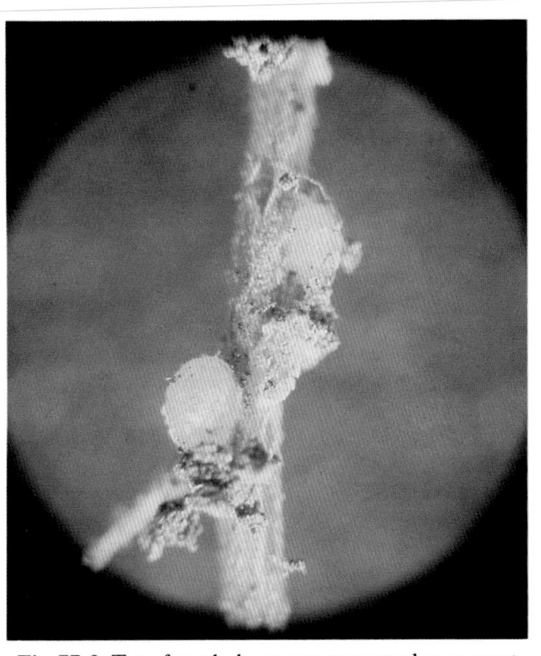

Fig.77.3 Two female beet cyst nematodes on root (× 20)

British Sugar and the grower stipulated that sugar beet should not be sown on land that had grown a host crop of beet cyst nematode during either of the two preceding years. This clause was removed from the 1983 contract, but in 1988 a modified clause was introduced to limit cropping with *sugar beet* to one year in three. The nematode is no less of a threat than in the past and an interval of two years between host crops should still be regarded as a minimum. On peaty soils, and on all fields known to be infested, an interval of three or, preferably, four years between host crops is recommended.

The Beet Cyst Nematode Order 1977 replaced the Beet Eelworm Orders 1960 and 1962, which restricted the frequency of growing any host crops of the nematode in a 'scheduled area' of the fens of eastern England and in all other fields known to be infested. With the introduction of the new Order, the earlier Orders were revoked and all notices and licences issued under them were cancelled. It is, therefore, important for growers to ensure that their rotations are planned to minimize the risk of crop loss from this pest and to seek advice whenever necessary. The 1977 Order provided reserve powers to restrict the planting and growing of crops of the beet and cabbage families on land declared by notice to be infested with beet cyst nematode, if infestations in any particular area become sufficiently serious to make statutory control necessary. These reserve powers were transferred to The Plant Health (Great Britain) Order 1987. An annual survey is made by the Ministry of Agriculture, Fisheries and Food in the fen areas of eastern England where beet cyst nematode is common, in order to detect any important changes in the incidence of this pest.

Farmers who normally adopt a suitable rotation in which host crops are widely spaced have little to fear from beet cyst nematode.

Soil from sugar beet fields

Where beet cyst nematode is already widespread, such as in the fen soils of eastern England, it is extremely unlikely that apparently clean fields could have escaped contamination with the nematode. In these districts there is little point in trying to prevent the spread of the nematode in small quantities of soil. However, rhizomania, a serious soil-borne virus disease of sugar beet, was detected in England for the first time in 1987 and is the subject of regulatory control. For this reason, precautions should be taken to prevent the spread of soil from beet-growing farms. Soils should not be moved from one farm to another and all visiting vehicles and machinery should be cleaned free from soil. In addition, untreated waste, particularly from imported vegetables, should not be spread on the land (see The Disposal of Waste (Control of Beet Rhizomania Disease) Order 1988).

78
Cereal cyst nematode

Fig.78.1 Field of oats infested with cereal cyst nematode: pale patches in the crop indicate areas of poor growth

Cereal cyst nematode (*Heterodera avenae* Wollenweber) is a widely distributed pest of cereal crops, occurring in Britain, Ireland, various other northern European countries, Israel, India, Australia and N. America. In Britain it is more damaging to oats than to other cereals and is prevalent mainly on light-textured soils. For many years crop losses occurred chiefly on sandy soils in the West Midland counties and adjoining areas of Wales. After the Second World War damage became more widespread, especially on the thin chalk soils of Wiltshire and Hampshire, following wartime ploughing of large areas of downland. With the subsequent decline in the growing of oats, this pest became less important. However, it still reduces yields of barley and wheat, especially on light and inherently lower yielding soils; it is seldom important on heavy-textured soils.

Intensification of barley growing from the 1960s onwards led to fears that cereal cyst nematode might become a serious problem on this crop. However, after initial increases, infestations declined in many areas, chiefly because of fungal parasites of the nematode

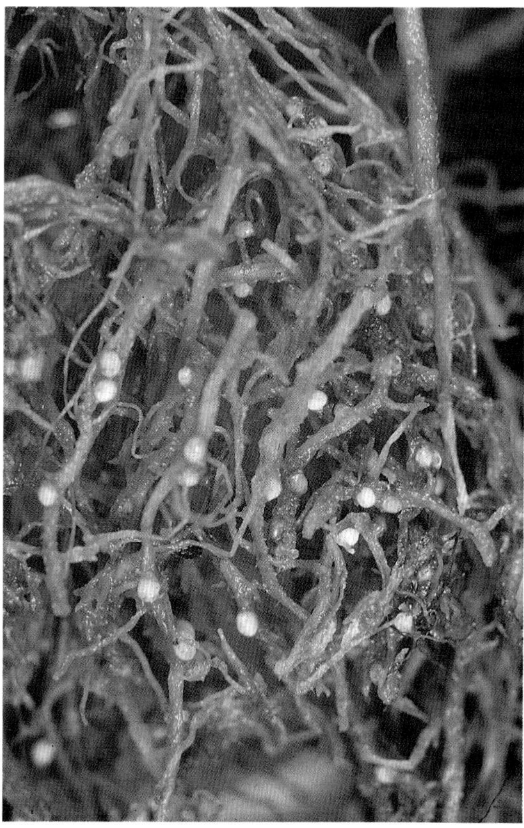

Fig.78.2 Root system of wheat plant infested with cereal cyst nematode, showing typical branching and white nematode cysts

which are favoured by intensive cereal growing, especially in moist summers.

Oats may also be attacked by another species of nematode — the stem nematode, which causes 'tulip root' (see page 441). Both species are sometimes found on the same crop.

Description and life history

Cereal cyst nematode occurs in the soil as dark brown, lemon-shaped cysts about 1 mm long. Each cyst consists of the dead and extremely tough skin of a female nematode containing large numbers of young larvae each coiled within an egg. Thus protected, young nematodes can live in the soil for several years, although some hatch and escape into the soil each spring. If these hatched larvae do not find a suitable food plant, they probably die within a few months.

When a susceptible crop is grown in infested soil, a large number of the hatched larvae invade and feed within the rootlets, which may be stunted and killed. On autumn-sown cereals this larval invasion may begin before the onset of winter. The plant responds by forming fresh rootlets which are in turn infested, so that a characteristic bushy or matted, shallow root system results (Fig. 78.4). The larvae feed and grow in the roots, developing into slender males, which become free in the soil, and lemon-shaped translucent females, which rupture the root surface but remain attached to it by the head end. Mating occurs outside the roots and, shortly afterwards, the males die.

Soon after fertilization, the female also dies and her skin changes from white to dark brown as the tough, leathery cyst is formed. While the cysts are on the roots, however, the brown colour is obscured by a white incrustation (the subcrystalline layer) and, on heavily infested plants, the cysts can easily be seen under a lens as tiny white bodies scattered along the rootlets (Fig. 78.2). Females or cysts may be seen during June and July or, on winter oats, sometimes as early as April. Later, when the stubble is ploughed in and rots, the cysts become free in the soil. There is only one generation a year, even on autumn-sown cereals.

Symptoms of attack

A crop of oats infested with cereal cyst nematode usually shows patches of stunted, pale yellowish-green plants sometimes tinged with red or purple. This patchy distribution and unhealthy colour are the most characteristic features of the early stages of the attack (Figs 78.1 and 78.3). The effect is mainly one of severe nitrogen starvation, resulting from interference with normal root action. Yellowing of plants may also be caused by mineral deficiencies in the soil, especially lack of nitrogen, and by certain fungus and virus diseases. However, with oats particularly, the severity of the damage in the bad patches, and the dramatic contrast between the sickly plants in these patches and plants in the good parts of the field, usually distinguish cereal cyst nematode from other troubles. The affected patches soon become choked with weeds, often showing up the trouble with striking effect if, for example, the dominant weed is charlock or poppy. Weed grasses, such as couch and wild oats, may gain serious hold in such patches and, being hosts, may have some effect in maintaining the nematodes. It is particularly important to control these weed grasses when a nematode-resistant cereal variety is being grown, otherwise the benefit of the resistance will be reduced. The effects on wheat and barley are similar to those on oats but normally much less severe.

The typical bushy root system of oats attacked by cereal cyst nematode is another useful recognition feature (Fig. 78.4). A bunch of badly infested plants, lifted with a trowel, generally shows a spongy mass of root which holds an abnormal quantity of soil. The root systems of attacked wheat and barley are similarly affected (Fig. 78.2). (Soil acidity may have a similar effect on barley.)

The presence of the nematodes in the roots may be confirmed in the early stages by a simple staining technique in a laboratory or, later, by the cysts which become visible on the root systems (Fig. 78.2).

Fig.78.3 Close-up of patch of oat plants damaged by cereal cyst nematode

Effects on crops

Weather conditions and local soil variations have a marked effect on the damage caused by cereal cyst nematode. Crop losses, especially in barley, are often very difficult to relate to the degree of soil infestation, probably because the effects of the nematode are inseparable from those of other soil and seasonal factors such as drainage, drought and diseases. Damage is generally more severe when a dry summer follows a wet spring.

In Britain, oats suffer more frequently and severely than other cereals but are grown much less than formerly. Wheat and, to a lesser extent, barley may also be badly affected by heavy infestations on some soils, but rye is

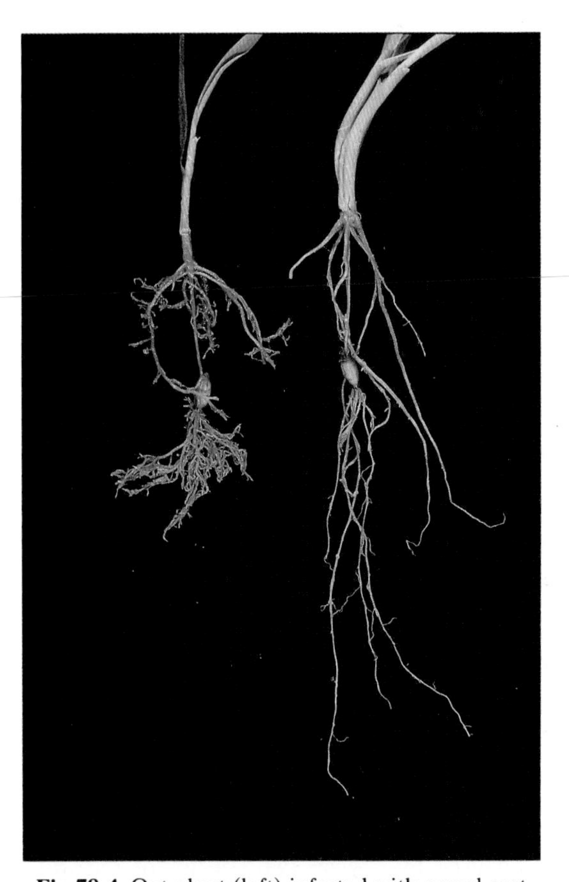

Fig.78.4 Oat plant (left) infested with cereal cyst nematode, showing typically stunted and branched root system compared with uninfested oat plant (right)

scarcely affected. Maize is a poor host, i.e. although the roots are readily invaded by the larvae, very few, if any, new females develop. However, the crop appears to be hypersensitive to larval damage and is severely stunted even by very small populations.

As would be expected, yields of oats are often greatly reduced by attacks of this nematode. The panicles are frequently much smaller than normal and may consist of undeveloped flowers and undersized grain. In fields where oats would be quite badly damaged, barley and winter wheat often show little sign of attack but yields may be decreased. Winter oats are less damaged than spring-sown oats. Slightly damaged crops may, nevertheless, result in big increases in the nematode population; conversely, when a plant is severely damaged, the pest often has less chance of multiplying as its food supply is reduced.

Cereal cyst nematode can also live on most of the commonly grown grasses, such as perennial and Italian ryegrass and meadow fescue, but populations usually decline steadily when grasses are grown. In experiments both timothy and cocksfoot have shown resistance to this nematode.

The relationship between host crops and infestations of cereal cyst nematode may be summarized as follows.

Descending order of *susceptibility to damage:* spring oats and maize, winter oats, spring wheat, winter wheat, barley, rye and grasses.

Descending order of *ability to increase infestation:* winter oats, spring oats, barley, spring wheat, winter wheat, rye, maize and grasses.

Natural enemies

Even under successive crops of susceptible cereals, populations of cereal cyst nematode often reach a peak, when obvious crop loss occurs, and then decline rapidly to low levels. This reduction in population is now known to be caused mainly by two species of parasitic fungi: one destroys the young developing females and the other kills the larvae within the eggs.

Prevention and control

To prevent losses from cereal cyst nematode, the first rule is to avoid excessive cropping with oats; the second is to maintain good soil structure and fertility, and especially to keep up the humus level. As with many other pests and diseases, good husbandry and correct crop rotation can do much to prevent losses.

It should be remembered that fields cropped continuously or intensively with wheat or barley without apparent trouble from cereal cyst nematode may contain populations sufficiently large to cause damage to oats, especially spring oats on light land. Although winter oats are more tolerant of cereal cyst nematode, they can allow an infestation to rise to a serious level. Therefore, where oats are grown on light land as a break to control cereal fungus diseases after several barley or wheat crops, they should be followed by a non-host crop before cereals are grown again. In areas where cereal cyst nematode is prevalent, farmers contemplating growing oats after a long sequence of barley crops should have a soil sample taken so that the risk can be assessed. Mixed cereal crops which contain a large proportion of oats should be classed with oat crops when considering cereal cyst nematode.

Weed control On land subject to cereal cyst nematode it is worth taking extra care to control wild oats by cultural or chemical means.

Resistant varieties Certain oat varieties are resistant to cereal cyst nematode and greatly inhibit the production of females within the roots. However, they are invaded by the larvae to a similar degree to a susceptible variety and can suffer serious damage and loss of yield from heavy infestations. They are, therefore, best used on lightly infested fields where they will yield well and also reduce the infestation to a negligible level. If in doubt about their use, the infestation level should be checked by soil sampling.

Similar resistance to the two most common

cereal cyst nematode races or pathotypes has been incorporated in spring barley. A resistant variety will outyield a comparable susceptible variety on heavily infested land and in addition will effectively reduce the nematode population in the soil. Usually one crop of resistant barley is sufficient to reduce the nematode infestation to a level that is safe for all cereals, except possibly nematode-susceptible oats.

Where mixed cereals are grown, a mixture of resistant spring oats and barley should be considered for use on infested land.

In heavily infested fields, even resistant barley, although reducing the infestation, may suffer a loss of yield; to avoid this it may be advisable to rest the field from cereals for one or two seasons. Even one season under a non-host crop can reduce the infestation and a resistant barley variety thereafter will reduce it further. Alternatively, two consecutive crops of resistant barley may be grown (accepting some yield loss in the first crop) where this suits the farming programme better.

Infestations of cereal cyst nematode usually decline fairly quickly under non-host crops or fallow (as much as 60% in one year), in contrast to infestations of potato cyst nematodes (see page 426).

Since most grasses seem able to maintain only small populations of cereal cyst nematode in the soil, the number of nematodes usually decreases when fields are put down to short leys. A minimum two-year, preferably three-year, ley is advised on heavily infested land.

When cereal cyst nematode damage is expected after sowing oats or barley, an additional 12.5–25 kg of nitrogen per hectare applied as a split dressing is valuable in encouraging growth. Consolidation by rolling also aids crop recovery. Usually, however, damage is noticed too late for remedial action to succeed.

If a spring oat crop fails because of cyst nematode attack, it should be ploughed in early before the cysts mature and the field used for a catch crop such as kale in the same season. By thus preventing a new nematode generation from completing its development, the risk to a subsequent cereal crop will be much reduced.

In lightly infested fields, barley or winter wheat will usually be successful, but care must be taken to prevent build-up of the nematode population before introducing oats into the cropping.

79
Chrysanthemum nematode

Fig.79.1 Chrysanthemums with foliage showing typical effects of attack by chrysanthemum nematode

Chrysanthemum nematode (*Aphelenchoides ritzemabosi* (Schwartz) Steiner & Buhrer) is a potentially serious pest of florists' chrysanthemum. It occurs much less frequently than in the past but may be found where chrysanthemums are still grown as a seasonal crop. It is easily spread on plants or cuttings but rarely occurs on nurseries where cuttings are obtained from specialist producers. All varieties of florists' chrysanthemum (species and varieties of *Dendranthema*) and related plants grown for cut flowers, e.g. shasta daisy (*Leucanthemum maximum*), may be attacked.

The same species of nematode can infest the leaves and buds of many other plants, e.g. aster, buddleia, calceolaria, delphinium, *Doronicum*, peony, phlox, pyrethrum, *Saintpaulia*, weigela and zinnia. It is also a common pest of strawberry (see page 413) and will attack buds of black currant although it rarely causes serious damage to this crop. It can live successfully on various common weeds such as buttercup, chickweed, cleavers (goosegrass), groundsel, sowthistle and speedwell.

Description and life history

Chrysanthemum nematodes are colourless, slender and microscopic, the adults being about 1 mm long. They live and feed either among the young leaves round the growing point of the plant or within the tissues of older leaves, which they enter through the leaf pores (stomata). They breed quickly during the plant's growing season, the life cycle from egg to egg taking 10–14 days.

As many as 15 000 nematodes may be present in one chrysanthemum leaf. They are active, moving over the surface of the plant, only when a film of water is present. Infestation spreads rapidly when the plants are thoroughly wet. Spread from plant to plant is by leaf contact or rain splashes.

Adult nematodes coil up in dried, dead leaves and can survive in this state for 18 months. However, in moist conditions they are unlikely to survive for more than about three months.

At the end of the season when the plants die down, nematodes remain on the stool and in the surrounding soil. The stool is by far the most important source of infestation unless infested plots are allowed to become weedy. Provided the land is ploughed and kept free from weeds, any chrysanthemum nematodes remaining in the soil are unlikely to survive the winter. Cuttings from infested stools are likely to be infested and are the usual means of spread from one nursery to another.

Symptoms of attack

The first sign of attack is a yellowish-green blotching of the leaves (Fig. 79.2) or, on thick-leaved varieties, a bronzing or purpling. The blotches have a translucent or water-soaked appearance when the leaves are moist. Initially, the discoloured areas are generally sharply limited by the main veins, but they darken through brown to dark grey or almost black and gradually spread to adjacent areas on the leaf until the whole leaf is dead and shrivelled. Most nematodes are found in the slightly discoloured areas; very few remain in leaves that have become dark brown or black.

The damage appears first on the leaves at the base of the stem and extends upwards until the whole plant is affected (Fig. 79.1). Nematodes may invade the flowers and cause deformed, undersized blooms.

The lower leaves of traditionally grown chrysanthemums often die for reasons other than nematode attack; such dieback is normal at the end of the season but may occur prematurely when growing conditions are unsatisfactory. Bright yellow, angular leaf blotches that could be confused with nematode damage are often associated with this kind of dieback, but there is generally a sharp transition from healthy to dead leaves with a few of the yellow blotchy ones between. The black leaf blotches caused by *Septoria* spp. may also be confused with nematode symptoms, but these fungus spots are not bounded by the leaf veins and are often rounder in shape.

Resistant varieties Chrysanthemum varieties differ in their reaction to the nematode. The resistance found in a few varieties is due to their greater sensitivity to the presence of the

Fig.79.2 Chrysanthemum leaf showing early stage of blotching caused by chrysanthemum nematode

nematode: the affected parts of the leaves turn brown quickly, the nematodes are cut off from their food supply and they fail to multiply. Resistant varieties show typical, well-defined leaf blotches only in the early stages of attack; as the nematodes do not generally multiply in these areas, infestation does not spread up the plant. Affected leaf areas later become black with a yellow border and the discoloration is not always limited by the veins. As a rule no nematodes can be found in the leaves at this late stage.

It is not practicable to list the resistant varieties of chrysanthemum because varieties in commercial use are constantly changing and new ones being introduced.

Effects of bud infestation

Where the apical bud is infested, the young leaves that are formed round it are irregular in shape and often have brown scars on their surface. Similar scars may be present on the stem below the bud and on the leaf stalks. The leaves may also be thicker than usual, are often strap-shaped and have irregular raised areas or ridges on the upper surface. Sometimes the bud is killed; this is often followed by an outgrowth of side shoots whose buds may be similarly infested. This type of damage is favoured by the humid conditions of protected cultivation and is particularly common in the stool beds. Distortion of young leaves caused by nematodes feeding in buds can be confused with thrips damage and with the damage associated with contamination of the water supply by hormone-type weedkillers. These possibilities should be considered when chrysanthemum nematode damage is suspected.

Microscopical examination is necessary to confirm the presence of this nematode; in cases of doubt, samples of affected leaves should be examined by an expert.

Control measures

Cultural control

Precautions should be taken to ensure that newly introduced cuttings or plants are healthy and come from a reputable source. Thorough roguing of infested or suspect plants should also be practised.

The 'standing ground' for plants in containers before they are moved into the houses should be kept weed-free and changed each year. If this is difficult, the containers may be placed on ashes laid on corrugated iron or plastic sheeting on the ground; these sheets should be well cleaned each year and fresh ashes used.

When signs of infestation are seen, it is advisable to refrain from overhead watering, if this is practicable, and to keep the foliage as dry as possible. Containers should be well spaced so that the leaves of adjacent plants do not touch one another.

Where the grower has a small but valuable stock, it is possible to keep plants free from infestation by smearing the base of the stems with narrow bands of petroleum jelly or tree-banding grease, the bands being renewed as stem girth increases. The bands must be applied early in the season.

Cutting off green shoots from stools before boxing, as close as possible to the stem without damaging axillary buds, removes most of the nematodes and increases the number of clean cuttings produced. This procedure, together with thorough washing of the stool, can give quite good control. The stools should be housed under glass or in plastic structures where the soil has been sterilized previously (see below).

Hygiene

Hygiene is most important: all plant remains should be removed and destroyed after cutting back, and dead leaves should not be allowed to blow about the holding. Soil, pots, boxes, etc. must be known to be free from infestation, or sterilized before use. Planting-out ground for outdoor varieties should be kept free from weeds. A winter fallow of 2–3 months helps to prevent reinfestation from the soil.

Where chrysanthemums are grown outside or in temporary structures, the soil should be sterilized at least once every three years with steam or a chemical which also controls weeds.

Chemical control

Control of the nematode is usually achieved by applying a recommended nematicide a few weeks after planting out. However, provided a suitable interval is left between treatment and handling, a nematicide can be applied either to stool beds at the onset of a flush of new cuttings or to boxes or beds of well rooted cuttings.

Hot-water treatment

Hot-water treatment is not practicable for year-round chrysanthemums because of the small size of the stool. It is, therefore, seldom if ever used nowadays on commercial holdings, few of which still grow chrysanthemums traditionally as a seasonal crop. Nevertheless it remains an effective and safe method of controlling chrysanthemum nematode.

The dormant stools should be immersed in water at 46 °C for 5 minutes, or 43.5 °C for 20–30 minutes. The shorter treatment is usually more convenient and involves less risk of damage to the more sensitive varieties.

It is preferable to use purpose-built tanks with efficient circulation and thermostatic control, but good results can be obtained using simpler equipment. Growers who wish to set up or renew equipment for hot-water treatment should consult their local advisers.

Before treatment, stools should be cut down so that the stems are 7.5–23 cm long. If stem cuttings are to be taken, the best results are obtained when the stems are left fairly long. Before placing them in the tank (or other container), the stools should be washed thoroughly to remove all soil; as the water used for this purpose may contain nematodes, it should be poured down a drain or disposed of in such a way that it will not again come into contact with soil or plants. Any remaining green shoots should be stripped off and then the stools placed in the tank for the prescribed time. On removal, the stools should be plunged at once into cold water, then planted out, boxed or potted in sterilized soil or soil known to be free from nematodes.

Hot-water treatment may delay the production of cuttings for a few days, but the retarding effect can be largely eliminated by careful attention to the following points: box up promptly after treatment, grow on at about 10 °C (cool-house conditions) and water sparingly.

Treated stools should not be bedded or boxed in very acid, e.g. peaty, material, or in a compost of low pC or high CF, otherwise they will deteriorate and may die a month or two after treatment. The nutritional level of the bedding soil or material is of no consequence unless the stools are to be kept for several months — even pure, sharp sand will do.

80
Docking disorder of sugar beet

Fig.80.1 Sugar beet crop affected by Docking disorder

Docking disorder of sugar beet is characterized by irregularly stunted plants, frequently with fangy roots, on light, often structureless, sandy soils. Seedling emergence is normal and uniform but, as early as the two rough-leaf stage, differences in growth rate appear in areas of irregular shape and extent. Stunted plants ('chicks') commonly occur haphazardly — singly, in groups or in lengths of row — among the large plants ('hens') (Fig. 80.1). The large plants appear extremely vigorous in comparison, becoming ten or twenty times bigger than the small ones, which may show symptoms of nitrogen, magnesium or other deficiencies. The stunted plants may appear to grow normally from July onwards, but root yield is less than from unaffected plants. If a large proportion of plants are affected and recovery does not occur or is delayed until September, root yield may be only about 8 tonnes/ha. Many of the stunted plants cannot be harvested and the yield is derived mainly from the scattered large ones.

Causes

Docking disorder is caused by stubby-root nematodes (*Trichodorus* and *Paratrichodorus* spp.) or needle nematodes (*Longidorus* spp.), which feed on the developing roots and impair their function or kill them. Several other interacting factors can influence the distribution and degree of damage and the ability of the plants to recover. The disorder almost always occurs in sandy soil because this is the most suitable habitat for the ectoparasitic nematodes and possibly also because this type of soil compacts readily (so that in a wet spring sugar beet grows slowly in it) or because nutrients leach too readily from it.

The disorder was named after the parish of Docking in west Norfolk where it was first observed as an unexplained trouble of sugar beet. Stubby-root nematodes are numerous at Docking, as they are in other areas of poor sugar beet growth in East Anglia, Lincolnshire and east Yorkshire. However, needle nematodes are more important in some fields in Norfolk, Suffolk, Nottinghamshire and Shropshire. Very often stubby-root and needle nematodes occur in the same field.

Soil acidity, soil compaction, waterlogging or excessive fertilizer in the seedbed can intensify the gross symptoms of nematode damage. These factors and also lightning, insects or fungi may cause root stunting and fanginess but not typical Docking disorder.

Other crops affected

Carrots, parsnips and kale are damaged by needle nematodes and the first two by stubby-root nematodes also. Other crops, especially cereals, may grow poorly in patches where the nematodes are numerous, but the damage is not so obvious as in the tap-rooted crops.

Root symptoms

Sugar beet seedlings attacked by stubby-root nematodes typically have stubby-ended lateral roots that turn grey-brown and later black as they die and decay. The tap root is often stringy and scurfy and may be killed a few centimetres below the surface. Other roots take over its function and grow horizontally or diagonally but rarely straight down (Fig. 80.2); these thicken and so produce a fangy storage root in sugar beet (Fig. 80.4), parsnips and carrots. However, identical fangy storage roots in sugar beet have been produced by compacting these light soils with tractor wheelings in the final stages of seedbed preparation.

Stunted seedlings from patches where there are many needle nematodes usually have normally colourerd tap roots, but many short lateral roots with darkened and swollen tips, often grouped in tufts (Fig. 80.3). Occasionally a lateral root that has grown normally may thicken and extend abnormally in a horizontal direction just below the surface. From May onwards, provided that the soil is moist, needle nematodes can just be seen without magnification on the roots and in the adjoining soil. They look like fine hairs and often coil when disturbed. Stubby-root nematodes are too small to be seen without magnification.

Root symptoms can give some indication of which nematode is responsible for the damage, but examination of soil conditions in the field and of soil samples is necessary for confirmation. In the root zone of stunted plants there may be as many as 9000 stubby-root nematodes or 2000 needle nematodes per litre of soil; usually there are fewer, but still many more than in the soil around vigorous plants nearby. Because nematode populations decline as the soil dries during the summer, samples for diagnostic purposes should be of moist soil taken as soon as possible after symptoms have been noticed.

Factors affecting incidence

The most important factor is the soil type, which influences the number and type of nematodes in the seedbed at the time of crop germination. The number of nematodes is also

Fig.80.2 Beet seedling with root damaged by stubby-root nematodes

Fig.80.3 Beet seedling with root damaged by needle nematodes

influenced by previous cropping whereas soil moisture affects their activity. There is often a strong interaction between nematode damage and the other factors, such as soil compaction, that affect root development.

Soil

Docking disorder is confined to sandy soils. These are usually of low organic matter content, but Docking disorder can also occur on some sandy peats in which needle nematodes are numerous. Sandy soils compact readily during seedbed preparation and this has been shown to affect root growth adversely, especially in wet springs. The subsoil may be chalky (e.g. west Norfolk and west Suffolk), or sand over clay (e.g. north Humberside). Sandy soil provides the most suitable environment for the nematodes and variations in its texture influence nematode activity. The patchiness so characteristic of Docking disorder is generally linked with differences in soil texture; increased clay content usually leads to fewer nematodes and better plant growth.

Plant nutrients

Damaged root systems are unable to take up adequate nutrients and the plants show deficiencies, especially of nitrogen and magnesium. Harmful effects of nematodes are probably accentuated by the shortage of mineral reserves in sandy soils and the ease with

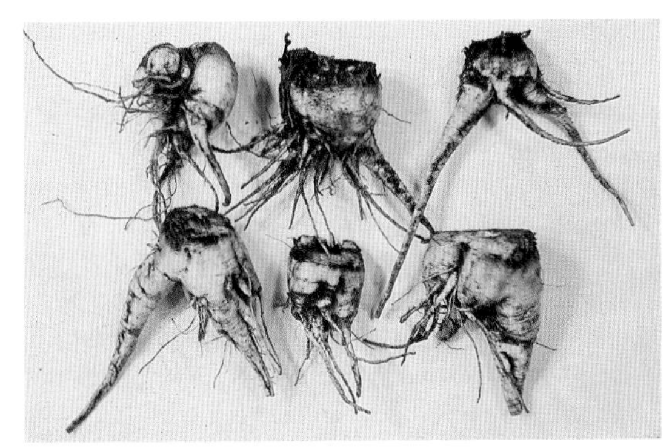

Fig.80.4 Beet with fangy roots following attack by stubby-root nematodes

which nitrogen is leached. The apparent infertility of the 'poor' sands on which Docking disorder occurs may be largely a combination of an incomplete and inefficient root system and a nutrient-deficient soil. Crops severely stunted in May and June sometimes recover in July and yield satisfactorily if there is adequate organic and inorganic fertilizer. Where stunting is associated with stubby-root nematodes, recovery in the growth of tops can be misleading because storage roots remain fangy and usually yield poorly.

Previous crops

So far as is known the nematodes can feed on the roots of any crop or weed. However, they multiply more on some crops than on others, for example large populations of needle nematodes often develop under grass leys.

Season

The area of sugar beet affected varies greatly from year to year, reflecting mainly the differences in rainfall. High rainfall in May enables the nematodes to move readily through the soil and to congregate on and damage the plant roots. Rain also decreases soil nitrogen content by leaching, lowers soil temperature and may even cause temporary waterlogging and root death if the soil has been cultivated when too wet.

Herbicides

There is circumstantial evidence that too much herbicide applied to the beet crop may accentuate Docking disorder, presumably because it retards seedling growth. However, Docking disorder can occur when herbicides are not used.

Virus and other diseases

Two virus diseases of beet, 'yellow blotch' (caused by tobacco rattle virus and transmitted by stubby-root nematodes) and 'ringspot' (caused by tomato black ring virus and transmitted by needle nematodes), are found occasionally in fields where Docking disorder occurs. They indicate the presence of the nematodes but, in most fields, too few plants are infected to influence yield and there are many stunted plants that are uninfected.

Weakly parasitic fungi, especially *Rhizoctonia* spp., are often found on roots injured by stubby-root nematodes; fangy roots can be produced by infecting plants experimentally with *R. solani*.

Control measures

Cultural control

In fields where there is a history of Docking disorder, yield losses may be minimized as follows.

Improving soil condition Organic matter is valuable, either applied or from crop residue. Marling gives long-term improvement and is not too costly if suitable marl is readily available nearby; it helps to avoid blowing, which seriously damages or even removes the plants, and probably also decreases nematode activity.

Applying adequate nutrients Soils subject to Docking disorder are often deficient in nutrients. Fertilizer programmes are best based on the results of soil analysis. If possible, and provided P and K levels are adequate, all fertilizer except N should be applied before ploughing. Sodium is beneficial but may be leached if winter rainfall is heavy. Magnesium is likely to be essential. Generous applications of organic manure release nutrients slowly and improve soil conditions.

Nitrogen must be applied in the spring and after ploughing; because of the possibility of loss by leaching and the risk of damage to germinating seeds it is best to split the application. The first N application (of about 40 kg/ha) should be made immediately after sowing; the remainder (about 85 kg/ha in the absence of organic manure or previous legume) is best applied on the soil surface as soon as all seedlings have emerged but no later than the four-leaf stage. If a full dressing of N was applied before sowing (which until recently was the usual practice), a further top dressing of 25–40 kg/ha may be necessary in very wet springs but no later than the end of May.

Providing good growing conditions Minimal seedbed cultivation decreases costs and avoids harming the almost structureless soil. In the spring, rolling and Dutch harrowing should be adequate to level and firm the soil sufficiently for satisfactory drilling. Alternatively, a zero-cultivation technique can be used by sowing into ploughed and furrow-pressed soil. This method offers some protection against wind erosion, which may also be decreased by inter-row crop windbreaks (provided they are removed early enough), factory waste lime applied to the soil surface, or proprietary soil conditioners. Sowing before 15 March is not advantageous and may even be harmful.

Chemical control

It is not practicable to advise treatment on the basis of the number of nematodes in soil samples taken before sowing sugar beet; the extent of damage depends greatly on the amount of rainfall in May. The need for treatment is best decided by the grower according to the history of Docking disorder in the field.

Granular nematicides, applied in the furrow at sowing, are recommended to control damage and increase yields. These materials also give partial control of damage by other soil-inhabiting pests and early aphid infestation.

81
Nematodes on strawberry

Fig.81.1 Strawberry plant infested with stem nematode, showing typical leaf crinkling

In Britain strawberry is damaged by three species of nematode that live on the plant above ground and several species that feed on the roots.

Two species of leaf nematode (*Aphelenchoides fragariae* (Ritzema Bos) Christie and *Aphelenchoides ritzemabosi* (Schwartz) Steiner & Buhrer) infest only the buds and leaves. The stem nematode (*Ditylenchus dipsaci* (Kühn) Filipjev) may invade all parts of the plant except the roots. Root-lesion nematodes (particularly *Pratylenchus penetrans* (Cobb) Filipjev & Schuurmans Stekhoven) may enter the roots from the soil. Other nematodes, e.g. a dagger nematode species (*Xiphinema diversicaudatum* (Micoletzky) Thorne) and needle nematodes (*Longidorus* spp.), live only in the soil and feed on the roots from the outside. The plant may be damaged if these root-feeding nematodes are very numerous, but the more injurious role of the dagger and needle nematodes is in transmitting certain viruses that cause disease.

Symptoms of attack

Damage by either leaf nematode or stem nematode is most noticeable on newly formed leaves soon after fresh growth starts and is generally most obvious during early spring, particularly after a wet autumn and winter. It is much more difficult to find symptoms after more rapid growth in late spring and early summer, but they may reappear for a short time when leaf growth starts again in the autumn. Strawberry plants on which leaf nematodes are present may lack symptoms — sometimes for up to two seasons. Symptoms of attack by the leaf nematode *A. fragariae* usually appear when an exceptionally cold spring is followed by a rapid increase in temperature. Unless these conditions occur, infested plants can remain symptomless.

Expert examination is usually necessary to confirm the presence of nematodes on strawberry plants because symptoms are frequently complicated by the presence of virus or other disorders.

Leaf nematodes

These nematodes feed in the buds and between unfolded leaves, causing the leaflets to become puckered and distorted (Fig. 81.3). Many of the leaflets show rough, greyish or silvery 'feeding areas' near the base of the main veins (Fig. 81.2). Affected leaves often have an almost hairless stalk which is strongly tapered towards the base of the leaflets. The teeth on the margin of the leaflets are frequently distorted and reduced; leaflets without any serrations are quite common. Leaflets are sometimes reduced, both in number and size, and may be virtually absent so that only very short tapered leaf stalks remain.

Severe attacks are rarely seen, but when they occur the main crown is usually killed and weak secondary crowns are formed. These secondary crowns often bear small spindly leaves with thin, almost hairless stalks and rather dark-coloured leaflets. The teeth along the edges of these leaflets are often larger, though fewer, than normal. Similar effects can, however, occur when the crown is injured by other causes.

Flowering and fruiting are seriously affected by leaf nematode infestation, and the flower trusses are often killed in the early stages of their development. When the main crown is killed, flowering on the secondary crowns is later than usual.

The leaf distortions caused by these nematodes are sometimes confused with the effects of the strawberry mite (see page 366); the mite does not, however, produce such marked 'feeding areas' as do nematodes. Nematode symptoms can also be confused with virus symptoms or be difficult to distinguish when both are present.

The condition known as 'cauliflower disease' occurs in a few strawberry varieties when *A. ritzemabosi* is associated with certain strains of a bacterium (*Corynebacterium fascians*). This bacterium causes 'leafy gall' disease of various

Fig.81.2 Strawberry leaf from plant infested with leaf nematode (*A. ritzemabosi*), showing greyish 'feeding areas' near the main veins of two leaflets

plants. Although 'cauliflower disease' is primarily bacterial, the typical 'cauliflower' effect is not produced unless the nematode is also present. The disease is characterized by severe stunting and swelling of leaf and flower stalks, and eventually by suppression of the leaf blades.

Stem nematode

Stem nematode is a more serious pest than the leaf nematodes because of the wide range of other field crops that may be attacked and because it persists much longer in the soil. It is particularly serious when strawberries are replanted annually in the same ground.

The leaflets of infested strawberry plants (Figs 81.1 and 81.4) are deeply crinkled, with their margins turned down towards the under surface, and are usually more brittle then normal. In severe attacks the leaf outline becomes rounded or irregular and has fewer marginal teeth than usual. In some varieties leaves are a darker and duller green than those of healthy plants and there may be a tendency to reddening of leaf stalks. On the underside of the leaflets, the lower parts of the main veins near the junction with the stalk are generally pale, enlarged and puffy. Leaf and flower stalks may be thickened, stunted and spongy in texture, and often show a brown core inside when cut lengthwise. Very few nematodes in a leaf stalk can cause marked leaf distortion. Fruit is much reduced in size and when ripe may show pale patches and be soft and easily squashed. Runners from infested plants often show symptoms at an early stage and the stolons may be shortened and thickened.

Affected plants are stunted and make poor growth. Runners planted in heavily infested soil may be killed in their first season or, if the main crown is killed, the plant may develop several weak secondary crowns.

Root-feeding nematodes

Root-lesion nematodes sometimes occur in strawberry roots and the surrounding soil. *Pratylenchus penetrans* is particularly damaging to strawberry and its effect may be enhanced by interaction with root-feeding fungi. Other species of *Pratylenchus* may invade strawberry roots but are not known to cause damage.

When present in the soil in large numbers, the dagger nematode and needle nematodes may damage plant roots by their feeding. Damage is characterized by gall-like swellings at the tips of primary and lateral roots and thickening of other parts of the root system. Severe stunting of strawberry has occurred in association

Fig.81.3 Strawberry plant infested with leaf nematode (*A. ritzemabosi*), showing leaf distortion and silvery 'feeding areas'

Fig.81.4 Strawberry leaves from plant infested with stem nematode, showing crumpled leaflets and thickened leaf stalks

with *Longidorus elongatus* (de Man) Thorne & Swanger (Fig. 81.5): root systems of affected plants were greatly reduced with short, swollen laterals.

The dagger nematode and needle nematodes are generally more important as vectors of certain soil-borne virus diseases. The dagger nematode transmits arabis mosaic virus and strawberry latent ringspot virus. Arabis mosaic virus causes chlorotic mottling on the leaves of many strawberry varieties, some of which are severely stunted by the virus, while others are only slightly affected. Strawberry latent ringspot virus produces similar symptoms in the strawberry variety Cambridge Vigour.

Needle nematodes transmit raspberry ringspot virus and tomato black ring virus. In strawberry these viruses cause leaf mottling and depress growth and yield. *Longidorus elongatus* transmits the Scottish strain of both viruses. *Longidorus macrosoma* Hooper and *Longidorus attenuatus* Hooper transmit the English strain of raspberry ringspot virus and tomato black ring virus respectively.

Other external root-feeding nematodes that occur in soil may sometimes contribute to poor growth of strawberry plants.

Fig.81.5 Strawberry plant (left) stunted by feeding of needle nematode (*Longidorus elongatus*) compared with undamaged plant (right)

Other plants attacked

Leaf nematodes

Both species of leaf nematode attack other cultivated and wild plants. One of the species (*A. ritzemabosi*) is the chrysanthemum nematode (see page 404), a serious pest of chrysanthemum and other ornamental plants; it also attacks black currant and can live successfully in many common weeds such as chickweed, cleavers, groundsel, sowthistle and speedwell. The other species (*A. fragariae*) is a pest of mint, ferns and various flowers, including begonias, lilies and violets.

Stem nematode

Stem nematode occurs on a very wide range of plants. It exists as several 'races' or 'strains' which are distinguished by their host plants. At least three of these races can infest strawberry: they are the narcissus, oat and red clover races. Thus any, but not necessarily all, of these crops may be attacked if grown on land where stem nematode has occurred on strawberry. Where the oat race is concerned, other plants that may be attacked include broad and field beans, mangel, oats, onion, parsnip, rhubarb, rye and vetch (see pages 440 and 458). When strawberry plants are infested by the red clover race (see page 443), the swellings of leaf and flower stalks tend to be localized and almost gall-like. Stem nematode can infest certain common weeds and large numbers have been found in chickweed growing among infested strawberries.

Raspberry, black currant and gooseberry are not attacked by stem nematode.

Root-feeding nematodes

Root-lesion nematodes have wide host ranges and the extent to which they damage their hosts varies enormously. *Pratylenchus penetrans* is known to damage narcissus and is suspected of damaging raspberry canes.

Dagger and needle nematodes feed on a range of plants and are less host-specific than races of stem nematode. Nevertheless, they reproduce better on some crops than on others, though this is not necessarily related to the nematodes' ability to feed on and transmit virus to the crops. The dagger nematode and *Longidorus macrosoma* thrive best under perennial crops: the former reproduces well on ryegrass and can damage it directly; the latter reproduces well on raspberry, black currant and rose, and causes damage to rose. *L. elongatus* also reproduces well on ryegrass and can damage it directly, but it can also damage arable crops. *L. attenuatus* reproduces on many arable crops and is particularly damaging to tap-rooted crops (see Docking disorder of sugar beet, page 409).

Description

Most plant nematodes are slender, transparent, microscopic creatures. The adults of leaf nematodes and the stem nematode are about 1 mm long, those of the former being a little shorter and conspicuously thinner than those of the latter. Adults of dagger and needle nematodes vary from 4 to 12 mm in length; those of other root-feeding species may be less than 1 mm.

Life history and spread

Adult leaf nematodes and stem nematodes breed continuously, laying many eggs in or on plant tissues. The life cycle for both types of nematode is completed in only 2–3 weeks. Eggs of root-lesion nematodes are laid within roots and in the soil. The life cycle takes up to three months. Dagger nematodes and needle nematodes may take two or three years to reach maturity; they lay their eggs in the soil during the summer months only.

Leaf nematodes on strawberry usually live as external parasites among the folded leaves of the crown and runner buds and around the growing points, where they feed on the delicate tissues and cause damage which becomes apparent as the young leaves expand. These nematodes are also found occasionally in the leaf tissue.

Leaf nematodes move about the crown in films of surface moisture but probably rarely travel any appreciable distance from the plant through the soil. Small colonies living in the buds of runners are carried out from the parent plant, so that propagation from infested plants is one of the chief means of spread.

Stem nematodes live mainly within the plant tissues and may invade all parts of the plant except the roots. They migrate from infested plants and can be spread by water draining on the surface and through the soil. They are able to persist in the soil for several years, especially in heavy soils. Strawberry buds that are to form runners may become infested at an early stage before they grow out from the parent crown, or rooted runners may be infested from the soil at a later stage.

Root-lesion nematodes are parasites of the root cortex, though in the later stages of attack *Pratylenchus penetrans* may penetrate and damage the vascular tissue of some roots.

P. penetrans is usually found in sandy soils. It can be spread with soil, by movement of soil water and in rooted nursery stock.

Dagger and needle nematodes live in the soil, penetrating as far as one metre below the surface. They have wide host ranges but will also feed on the roots of plants that do not induce reproduction and, therefore, do not support large nematode populations. These nematodes can withstand many months of starvation, resuming breeding when food becomes available. Thus, absence of a suitable host does not bring about a rapid decline in the nematode population.

X. diversicaudatum is widespread in Britain on a range of soil types from sandy loams to heavy clays and organic soils. *L. elongatus* is also widely distributed, occurring on loamy sands, sandy loams and organic soils. *L. macrosoma* is mainly confined to southern and central England on loam and silty loam soils.

L. attenuatus is mainly confined to East Anglia and occurs on light sandy soils.

Control measures

Cultural control

Runners from nematode-infested plants should never be used for propagation. It is essential to plant runners of reputable origin such as those certified by the Ministry of Agriculture, Fisheries and Food. It is advisable to rogue weak and stunted plants during the spring, especially those showing the leaf distortions already described.

Because strawberry plants may be attacked by nematodes from various crops, care should be taken to avoid planting strawberries on land where other plants are known to have been infested. Conversely, it is unwise to plant susceptible crops where nematodes have occurred on strawberry. Expert advice should be sought to determine, if possible, the source of the infestation and the race of nematode concerned. Broad bean can be infested with many stem nematodes without showing obvious symptoms of attack: widespread and serious damage has occurred in strawberry crops grown after beans. Stem nematode races that affect strawberry can be carried in the seed of broad bean, field bean, onion, red clover and teasel. These crops should not be grown in the same field as strawberries.

Clean and hygienic methods of cultivation help to prevent the spread of nematodes; weeds should be controlled because they may harbour infestation. Leaf nematodes are unlikely to survive more than 3–4 months in weed-free fallow soil, but stem nematodes are much more persistent and will not be eliminated by fallowing.

Hot-water treatment

Nematodes infesting strawberries have been controlled successfully without damage to the plants by treating runners in a hot-water tank at 46 °C for 10 minutes. Strawberry mite and any insect pests present are also controlled by this treatment. It is important to plunge the runners into cold water immediately after removal from the tank. To prevent adverse effects on growth, planting conditions should be good and treated runners *must* be well firmed in.

Chemical control

Granules A granular pesticide is recommended for control of leaf nematodes. It will also reduce the symptoms of stem nematode when applied to 'maidens' within 7–10 days of planting the runners, but it is not effective against this species when applied later in the season or to older plants. This treatment is likely to control other pests including some root-infesting nematodes.

Fumigants A fumigant nematicide may be used to treat small areas of soil from which plants infested with stem nematode have been lifted. Fumigants are less effective in controlling this pest in fields where the infestation is widespread and severe. Use of a soil fumigant checks the spread of soil-borne virus diseases of strawberry since the nematode vectors are controlled in the upper layers of soil. Control is, however, difficult below 30 cm. Soil fumigants will also control other nematode pests in the soil. When using a fumigant, soil temperature and correct soil preparation are particularly important.

82
Pea cyst nematode

Fig.82.1 Pea crop damaged by pea cyst nematode

Pea cyst nematode (*Heterodera goettingiana* Liebscher) is a serious pest of peas and probably the most difficult to control. It was reported in Germany in 1890 and was first found in Britain in 1912. Since that time it has appeared in widely scattered localities, especially in gardens and allotments.

For many years this pest has been locally troublesome in areas where peas are grown for sale in the pod on the fresh vegetable market, notably the Vale of Evesham. The area of market peas is now much smaller than formerly and, as a result, serious damage is less common in such areas. Pea cyst nematode is a potential risk in the intensive pea-growing areas of eastern England and populations have built up on some specialist vining farms. An increase in the area of dried peas could aggravate the situation.

Description and life history

Pea cyst nematode has a similar life history to that of the related cyst nematodes which attack beet, potatoes and cereals (see pages 395, 422 and 399). The cysts are the resting stage of the nematode. They are brown, lemon-shaped bodies about half the size of a pin's head, each having a tough leathery skin and containing as many as 200–300 eggs. Each egg contains a larva which is ready to hatch when conditions are suitable. The unhatched larvae within the protective cyst can remain alive in the soil for several years in the absence of susceptible crops. Thus infestations are very persistent and difficult to starve out.

Each year some of the larvae hatch even in the absence of host plants but, when peas or other host plants are grown in infested soil, larvae emerge from the cysts in larger numbers and invade the roots. The nematodes grow and develop within the roots at the expense of the plant and eventually lemon-shaped females are formed which burst through the surface of the roots. Many of these females ('white cysts') can be seen on the surface of the roots of attacked plants carefully lifted from about mid-June onwards (Fig. 82.2), although some of them remain almost completely embedded. The females become full of eggs and then die, their skins turning brown and leathery. The new generation of cysts thus formed corresponds roughly with senescence of the plants. After the haulms have withered, mature cysts are to be found in the soil.

Females and cysts should not be confused with the nitrogen nodules which occur normally on the roots of peas, beans and related plants (Fig. 82.2). Nodules are usually much larger and more irregular in shape than cysts and are often pinkish.

Plants attacked

The only commercial crops known to be attacked by this nematode are peas (both culinary and field varieties), broad beans, field beans and vetches or tares; runner beans and dwarf French beans are not attacked, neither are sweet peas.

This nematode can multiply on certain wild vetches, e.g. hairy tare (*Vicia hirsuta*), but little is known about the practical importance of weeds as hosts.

Effects on crops

Peas are the worst affected crop. Infested plants make poor growth and the haulms turn yellow prematurely (Fig. 82.1). The yield is reduced to a varying degree, depending on the severity of the attack. Heavily infested plants are spindly and severely stunted with small leaves; such plants often die before any pods that have formed have a chance to fill.

In many infested fields the effects of the pest are first seen as patches of bright yellow plants which contrast with the green of the normal plants (Fig. 82.1). As the season advances, these patches often increase in size and new ones appear. This is not due to movement of nematodes but to the nematodes in these areas taking longer to affect the crop. If peas continue to be grown on the same land, the patches increase in size until the whole field is badly infested and subsequent crops fail completely. It may then be 10 years or more before a good crop of peas can be grown in that field; failures have been known 12 years after the last host crop.

Although broad and field beans do not

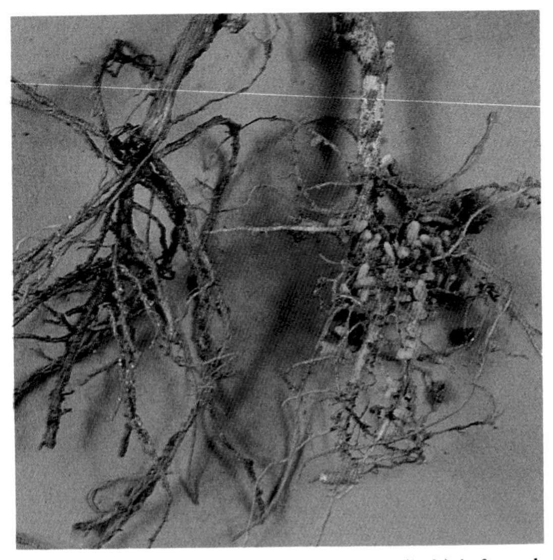

Fig.82.2 Root systems of pea plants: (left) infested with pea cyst nematode and (right) uninfested. Note white female nematodes and absence of nitrogen nodules on infested roots

generally cause such a great increase as peas in the nematode population of the soil, they are similarly affected. However, beans can tolerate much higher levels of attack before they show damage; yellowing is also less marked. Some long-established populations highly dangerous to peas have been maintained by beans in certain areas.

Poor patches in any host crop may also result from certain fungus diseases. Suspected nematode attacks should, therefore, be confirmed by examining the roots for females or cysts and obtaining expert advice if in doubt.

The root systems of plants attacked by pea cyst nematode are poor and the nitrogen nodules are suppressed and often absent (Fig. 82.2). There is no excessive formation of small side roots like that caused by other cyst nematodes. Nematode attack is often accompanied and aggravated by foot rots caused by certain fungi.

Prevention and control

Much has still to be learned about the behaviour of the pea cyst nematode, the rate at which it builds up under different host crops and in different soils, and the rate at which infestations may decline under different systems of cropping. Nevertheless, some general advice can be given, most of which amounts to good husbandry, common sense and vigilance.

Care should be taken to avoid overcropping with peas, field beans, broad beans, vetches, or any mixed crops containing them. None of these susceptible crops should be grown on the same land more often that one year in four. Where sufficient land is available, it is much safer to extend the rotation to seven or eight years.

Once a field has become heavily infested, normal rotations are of little or no use for control and a much longer rest from susceptible crops will generally be necessary before the land is fit to grow peas again. Long leys might sometimes be useful where field peas are affected, though land growing market crops is usually too valuable to grass down.

Pea growers should be constantly on the watch for yellow patches in their pea crops.

Chemical control on peas

Yield losses on infested soil can be minimized by treatment with a recommended nematicide. The nematicide should be broadcast as an overall treatment during final seedbed preparation and thoroughly incorporated afterwards. Some reduction in the incidence of pea early browning virus (transmitted by the free-living nematodes *Trichodorus* and *Paratrichodorus* spp.) will also be achieved.

83
Potato cyst nematodes

Fig.83.1 Potato crop infested with potato cyst nematode and showing patchy growth on low-yielding land

There are two distinct species of potato cyst nematode (*Globodera rostochiensis* (Wollenweber) Behrens and *Globodera pallida* (Stone) Behrens) and both are major pests of the potato crop in temperate regions of the world. One or both species have been well established for many years in the main ware-potato growing districts of England and infestations are common in gardens and allotments throughout the country. As the cysts of these pests are easily spread, it is unlikely that any cultivated area in the arable districts of Britain is entirely free from them, although in the absence of potato crops the cysts would remain undetected and gradually lose their viability.

Frequent cropping of slightly infested land with maincrop potatoes can lead to a rapid rise in the level of infestation and to serious yield losses. However, the use of resistant potato varieties, where appropriate, combined with suitable nematicides and crop rotation should largely prevent this situation from arising in future. With the introduction of these integrated control measures, the long rests from growing potatoes hitherto required to reduce high nematode densities to a safe level should seldom be necessary.

Description and life history

The cyst is the only stage of development which is easily visible without magnification. Initially, it is attached to a potato root but later occurs free in the soil. It is spherical, about 0.5 mm in diameter and, when mature, has a tough, dark reddish-brown skin which is the dead remains of a female nematode. The cyst contains numerous young nematodes or larvae, each of which is coiled within an egg shell (Fig. 83.3). One moult occurs in the egg so that, on hatching, the larvae are in their second stage.

A newly formed cyst usually contains 200–600 larvae but, as a small proportion of them hatch and escape into the soil each year, the number gradually decreases. If the larvae fail to find the roots of a host plant they die, but even after 10 years or more there may be some left unhatched in the cysts.

When potatoes are planted in infested soil, substances produced by the roots diffuse into the soil and cause a large proportion of the nematode larvae to hatch. These invade the potato roots and feed in them. If the attack is severe, the roots may be seriously damaged and sometimes killed. The plant responds by forming numerous additional rootlets which may also be attacked.

After the larvae have entered the roots the females swell and eventually become almost spherical; as they grow they split the surrounding tissues and burst through to the outside, only the head end remaining embedded in the root. Soon afterwards the females are fertilized by the slender, eel-like males that have emerged from the roots.

When infested plants are lifted, the females or cysts can be seen as small bead-like objects attached to the roots (Fig. 83.2). They are glistening white at first but finally become dark reddish-brown; the colour change coincides with the death of the females. When the crop is lifted, many mature cysts remain in the soil and are a source of infestation for future potato crops.

Species

The two species of potato cyst nematode are distinguished by (i) the colour of the immature

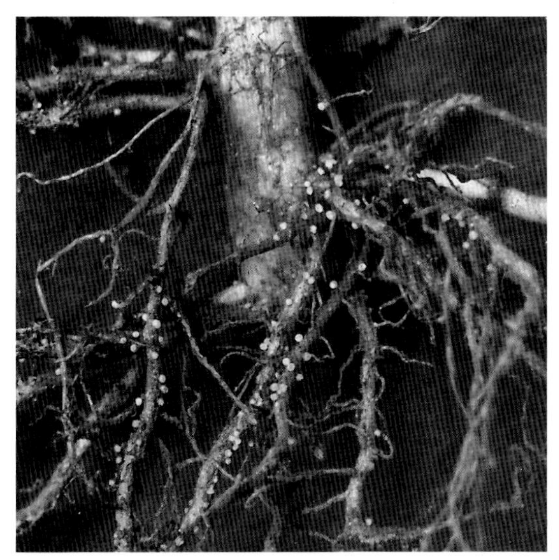

Fig.83.2 Potato root bearing mature female potato cyst nematodes that will later form cysts

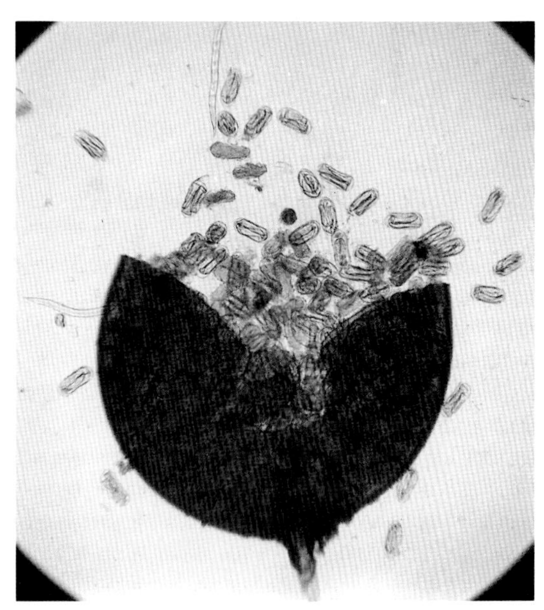

Fig.83.3 Ruptured cyst of potato cyst nematode showing eggs and two free larvae (× 75)

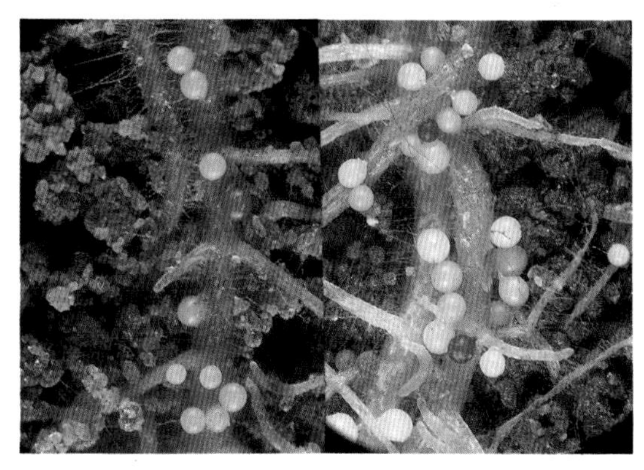

Fig.83.4 Females and brown cysts of potato cyst nematodes on potato roots (× 7). Left: yellow potato cyst nematode. Right: white potato cyst nematode

cysts, (ii) structural details visible only under high magnification and (iii) their inability to interbreed.

Yellow potato cyst nematode (*Globodera rostochiensis*). Females of this species pass through a prolonged golden-yellow phase (Fig. 83.4, left) before turning brown to form cysts. It is the golden nematode in American literature.

White potato cyst nematode (*Globodera pallida*). Females of this species do not have a yellow phase but remain white or cream-coloured (Fig. 83.4, right) for much longer before finally turning brown.

Pathotypes

Each species of potato cyst nematode has different strains or 'pathotypes'. In the UK only one pathotype of the yellow species is known (Ro 1, formerly known as pathotype A); this is unable to multiply on any of the potato varieties with resistance to Ro 1.

In continental Europe however, other pathotypes of this species occur which are capable of breaking this resistance; they would be particularly dangerous if introduced into Britain because they are indistinguishable from type Ro 1 in all respects except for their ability to infest resistant potatoes.

At least three pathotypes of the white species occur in the UK (Pa 1, Pa 2 and Pa 3). These are distinguished by their behaviour towards certain wild potato species and hybrids used in breeding for resistance. Pathotype Pa 1 appears to be rare, while Pa 2 and 3 are common and occur mainly as mixtures. They all multiply freely on Ro 1 resistant potatoes and are easily distinguished by colour from pathotype Ro 1 of the yellow species, provided that they are seen at the right stage of development. Commercial potato varieties with some resistance to the white potato cyst nematode (Pa resistance) are becoming available. However, unlike the Ro 1 resisters in which the resistance is total, varieties bred for resistance to the Pa pathotypes are only partially resistant and allow some multiplication of this species.

Other plants attacked

Potato cyst nematodes attack tomato as well as potato (see page 428). Egg-plant is also a host but is seldom grown in Britain. Cysts have been found on the hedgerow plant bittersweet (*Solanum dulcamara*) and the introduced weed *Solanum miniatum*. The related black nightshade (*Solanum nigrum*) is only slightly susceptible. It is in fact rare to find cysts on the roots of weed hosts in the field in Britain.

Effect on potato crops

Potato cyst nematodes cause loss of potato yield which may or may not be accompanied by obvious symptoms in the haulm. Crop failures occasionally occur at low nematode densities. Conversely, on some good soils a large yield may be obtained when the nematode density is high; the yield loss may then pass unnoticed.

On most soils the yield loss caused by potato cyst nematodes in fields with low or moderate infestations averages about 6 tonnes/ha for every 10 'full' cysts per 100 g of soil. (A 'full' cyst contains 50 or more eggs.) With higher infestations the losses average about 3 tonnes/ha per 10 'full' cysts per 100 g, although the *total* loss will usually be greater than in fields with lower levels of infestation.

The presence of cysts on the roots of stunted potato plants in patches does not necessarily mean that the nematodes are the chief cause of the trouble; rhizoctonia and other fungus diseases are often contributory causes of poor growth.

Changes in nematode population

In the absence of control measures the population usually increases when potatoes are grown and declines when the field is fallowed or other crops are grown. A small nematode population in the soil may be increased up to 25 times by one crop of maincrop potatoes. Early potatoes do not cause such a rapid increase and first earlies may even decrease nematode numbers. This is because the crop is often lifted before there has been time to produce a new generation of cysts. Generally, the later the crop is lifted after mid-June the more cysts will be present in the soil afterwards. When a crop fails, for whatever reason, the nematode population may actually decline.

Statutory regulations

The Potato Cyst Eelworm (Great Britain) Order 1973 came into effect on 1 August 1973 to comply with an EC Directive on the control of these pests. The provisions of this Order were later incorporated in The Plant Health (Great Britain) Order 1987. The Order is intended particularly to provide safeguards against the spread of potato cyst nematodes with seed potatoes. It requires that seed potatoes intended for marketing are produced only on land on which no potato cyst nematode has been found by official soil examination. On land declared by notice to be infested, the Order prohibits the growing of seed potatoes, and of ware potatoes except under licence, and the planting or removal of any plants (including bulbs) for transplanting. Licences may be issued to grow potatoes, other than seed potatoes, on land under notice provided that (i) they are a variety resistant to the species or pathotypes of nematode present, or (ii) an approved nematicide treatment has been applied, or (iii) early potatoes are grown and lifted by 30 June.

The restrictions on infested land will be lifted when potato cyst nematode is no longer found by a further statutory soil test. Applications for re-testing will be accepted after an interval which varies according to the crop to be grown.

The arrangements for advisory soil testing at the request of the farmer (see page 426) are not affected by the statutory controls. Advisory soil samples are taken when the farmer or grower seeks such a test and are used solely for advising him on his cyst nematode problems. Farmers are urged to continue to make the fullest possible use of the advisory services in the interests of effective control of these serious pests.

Prevention and control

Spread of infestation

Cysts of potato cyst nematode occur in enormous numbers in infested soil. There may be several million cysts per hectare before they are likely to be detected in soil samples. By the time an infestation has reached a level sufficient to cause serious yield loss, many millions of cysts are present in each hectare. Cysts are therefore easily spread whenever soil is moved from place to place, e.g. on farm machinery, on contractors' vehicles, on workers' boots, on roots and other farm produce. A particularly common means of spread is the clamping of potatoes from infested land on other fields; this often leads to a heavily infested area in an otherwise clean or slightly infested field. Cysts are also spread by wind and water; the spectacular 'blows' which occur in fenland areas are particularly important in this respect.

As it is practically impossible to stop the greater part of this spread, there is little point in taking elaborate precautions. However, infested soil carried on transplants or on seed potatoes may lead to quick establishment of well-distributed infestations; therefore transplants should not be raised in infested fields unless they are to be planted on infested land. On land scheduled as infested under The Plant Health (Great Britain) Order 1987 (see page 425), the growing of seed potatoes and plants for transplanting is forbidden.

Crop rotation for ware growing

Once potato cyst nematode has gained entry to a farm that grows ware potatoes regularly, it will eventually spread to all fields on the farm. The aim must then be to establish an effective system of control so as to minimize yield losses.

Normally, large nematode populations build up as a result of overcropping with potatoes. When infested fields are rested from potatoes for long periods the nematode population decreases gradually. After some years nematode numbers are usually small enough to allow potatoes to be grown again, but unexplained failures at low nematode densities are occasionally recorded after many years' rest from potatoes; these failures may be associated with other root pathogens.

Early potatoes can be grown more frequently than maincrop varieties without building up damaging populations of cyst nematode. In some districts where potatoes (including those grown as covered crops) are lifted by mid-June at the latest, it may even be possible to practise annual cropping.

Groundkeepers

In some soils, large numbers of groundkeepers (self-set potatoes) remain, particularly in a wet autumn when conditions are bad for lifting. In general they probably have little effect on nematode numbers as their root systems are usually small compared with those of a commercial crop. However, groundkeepers are undesirable for other reasons, so every effort should be made to reduce them to a minimum by efficient harvesting. Some machines are equipped with a crushing device to destroy the small tubers that pass through the web but, even so, large numbers are often left in the soil. Many of these may be killed by frost, provided they are kept near the surface of the soil by avoiding ploughing or deep cultivation.

Soil sampling

Soil sampling can be used to show that potato cyst nematode is present in a field but can give no guarantee that it is absent. Under The Plant Health (Great Britain) Order 1987, soil sampling is used to detect the presence of potato cyst nematode species for statutory purposes, especially in connection with seed-potato growing and exports. In the advisory context it is used to enable sound advice to be given on suitable rotations for potatoes and to help farmers choose fields on which nematicides and resistant potato varieties (see below) may be used successfully.

Resistant varieties

The breeding of potato varieties resistant to potato cyst nematodes has made considerable progress. Several varieties on the market, e.g. Maris Piper and Cara, are resistant to the pathotype Ro1 of the yellow potato cyst nematode; others resistant or partially resistant to the white species are becoming available.

Varieties resistant to Ro1 are of most use in the southern half of England, roughly south of a line from Stafford to Holbeach, including the important potato-growing areas in the eastern counties. Here the majority of fields originally contained mainly pathotype Ro1, but the incidence of pathotypes of the white potato cyst nematode is increasing as a result of the use of Ro1-resistant varieties. In other parts of the country including Yorkshire, Lancashire and most of Lincolnshire, infestations are mostly of the white potato cyst nematode, which can breed on these varieties. Although the varieties bred for resistance to pathotypes of the white species are only partially resistant, they do limit the multiplication of the nematode, especially when used with a nematicide.

No potato variety is immune from attack. Nematode larvae are attracted to and invade the roots of all varieties but, in a resistant variety, males are formed as usual, whereas females do not usually complete their development. Both resistant and susceptible varieties suffer yield losses. Nevertheless, resistant varieties often yield fairly well even in the presence of moderate infestations of potato cyst nematodes. Used wisely, resistant varieties are a means of reducing nematode populations.

Chemical control

Nematicides are available in granular form for use against the two species of potato cyst nematode and their pathotypes.

The nematicides should be applied to the soil and well mixed into the top soil as close to the time of planting potatoes as possible. The nematode larvae hatching from the cysts are prevented from invading the potato roots at least until late in the growing season. Losses due to the nematodes are substantially decreased; breeding and build-up of the pest are often very much reduced. These chemicals are useful mainly in fields moderately infested with potato cyst nematode, where an economic yield response can be expected.

Soil treatment with either a soil sterilant or a fumigant also increases yields on infested soils but should be applied in the previous autumn before potatoes are planted. This treatment does not prevent the residual nematode population from increasing when the potato crop is grown.

Integrated control

The use of an appropriate resistant variety together with a granular nematicide is recommended on those moderately infested fields where the nematode species is known. The resistant variety will minimize the build-up of the nematode, while the nematicide will reduce yield loss and decrease the multiplication of the other nematode species if mixtures are present.

Previously, where large nematode populations were present, seven or more years' rest from potatoes was needed to reduce the infestation to a level at which cropping with potatoes was economic. By integrating resistant varieties and nematicides, rotation assumes less importance in nematode control and rotations of one potato crop in four, or even one in three, have become practicable.

84
Potato cyst nematodes on tomato

Fig.84.1 Tomato plant (right) infested with potato cyst nematode, showing stunting of plant and yellowing of lower leaves. Healthy plant on left

Potato cyst nematodes, well known as pests of potato (see page 422), are also serious pests of tomato. For many years they have been established in the main tomato-growing areas of England and Wales. Intensive growing of tomatoes in soil, either under glass or outdoors, can easily lead to nematode populations building up to a damaging level unless regular control measures are taken.

Description and life history

The nematodes occur in the soil as spherical reddish-brown cysts, about 0.5 mm in diameter. The cyst is the dead remains of a female whose tough skin encloses 200–600 young nematodes, each of which is coiled within an egg shell.

When tomatoes are planted in infested land, the roots produce substances which diffuse into the soil and stimulate the nematode larvae within the cysts to hatch. These larvae emerge from the cysts, move through the soil and invade the tomato roots. Attacked roots become slightly swollen and stunted. Additional rootlets are produced to replace those damaged and these in turn may be invaded.

The larvae feed and develop into males and females within the roots. As the females grow and swell they burst through the root surface, leaving only the head and neck embedded, and are fertilized by the slender, eel-like males that have already left the roots.

When infested plants are lifted, the females and cysts can be seen as tiny, spherical objects attached to the roots. The female nematodes are glistening white at first but finally become dark reddish-brown cysts, the colour change coinciding with the death of the female.

Work on resistance in potatoes revealed the existence of what were at first thought to be different strains or pathotypes but are now known to be two distinct species of potato cyst nematode, each of which has different pathotypes. One species — the yellow potato cyst nematode (*Globodera rostochiensis* (Wollenweber) Behrens) — is distinguished by the females having a prolonged golden-yellow phase before they turn brown to become cysts. The other species (*Globodera pallida* (Stone) Behrens) does not pass through this yellow stage and the females remain white much longer before turning reddish-brown.

When the crop is harvested and the plants lifted, many cysts are left in the soil and provide a source of infestation for the next tomato crop.

Other plants attacked

Apart from potatoes and tomatoes, these nematodes do not attack any other commercial crop except the egg-plant (aubergine). They have been found on two weed species: bittersweet (*Solanum dulcamara*) and *Solanum miniatum*, the latter being closely related to black nightshade (*Solanum nigrum*), which is only slightly susceptible. However, potato cyst nematodes are seldom found on the roots of weed hosts and these are unlikely to be an important source of infestation.

Symptoms of attack

The first indication above ground of potato cyst nematode attack is the appearance of areas of uneven plant growth. As the attack progresses, affected tomato plants appear relatively thin and spindly with small, light fruit trusses. The lower leaves become pale and finally yellow (Fig. 84.1) and the plants show a marked tendency to wilt on sunny days. In severe attacks the plants may fail to produce fruit and may not grow above 45 cm tall. If an infested plant is lifted in the early stages of an attack, the presence of the nematodes is indicated by small bead-like swellings along the roots. These swellings are much smaller than the galls caused by root-knot nematodes (see page 436). Later, the characteristic females and cysts can be seen on the roots.

Plant damage can be much greater when roots are invaded first by nematodes and later by root rots. Several other factors can cause patchiness in the tomato crop: thus the presence of nematodes on the roots does not necessarily mean that they are the sole cause of the trouble.

Disease-resistant rootstocks

Tomato rootstocks resistant to certain soil-borne diseases are available and the grafting of scion material of different varieties on to them provides an alternative method of tomato growing. Some of the rootstocks incorporate resistance to root-knot nematodes, but none is yet resistant to potato cyst nematodes. However, as root damage is often due to a combination of nematode and fungus diseases, the more vigorous plants resulting from disease resistance can often withstand levels of nematode infestation that would otherwise be damaging. Nevertheless, under these conditions potato cyst nematodes are able to multiply so a heavier soil infestation is left after cropping.

Prevention and control

Soil used for pricking out seedlings can be a source of infestation; therefore, either sterilized soil or a soil-less compost should be used for this purpose. Once nematodes have gained entry to a tomato house, expensive and time-consuming sterilizing operations are necessary to control them or, alternatively, cultural methods can be used which do not involve growing the plants in the glasshouse soil.

Preparation of soil for sterilization

After the glasshouse has been cleared of plants and the normal hygiene measures taken, the soil should be deeply cultivated to obtain a reasonably fine tilth that will allow easy penetration of steam or sterilant gases. Organic manure should not be applied before chemical sterilization as it tends to retain gases for a long period. These may be released too slowly to function as nematicides but may, nevertheless, be poisonous to plants; this could make it necessary to delay planting if it is shown, for example by the 'cress test' (see page 431), that fumes are still present in the soil.

Steam sterilization

Thorough steaming of the soil can be an efficient and lasting method of sterilization provided a sufficiently high temperature (at least 80 °C) is achieved and penetration of the soil is adequate to reach infective nematode cysts.

Chemical sterilization

The high cost of steaming makes the cheaper chemical methods seem more attractive. However, they only achieve partial sterilization and, because of the inferior control of root rots, tomato roots growing in chemically treated soil usually deteriorate more quickly after mid-season than those growing in steamed soil.

The ideal chemical soil sterilant should kill all pests, diseases and weed seeds, leave the soil quickly, be harmless to crops, safe to handle, suitable for use on a wide range of soil types and inexpensive. None of the chemicals in current use fulfils all these requirements and most have at least one undesirable quality. It is, therefore, most important to select the chemical best suited to a particular site, pest and disease situation, or cropping programme.

No soil sterilant at present available can achieve 100% kill of potato cyst nematodes; surviving larvae invade tomato roots and usually produce enough new cysts at the end of the season to damage the following year's tomato crop. Therefore, chemical soil treatments must normally be applied annually where tomatoes are grown in the glasshouse soil.

Methods of application Various methods are available as follows.

1. Injection of the undiluted liquid can give adequate sterilization to a depth of 30 cm. Success depends on accurate calibration of the equipment and correct spacing of the injection points. The chemical must have a high vapour pressure to penetrate the soil both vertically and horizontally. Surface sealing of the soil by rolling or with water is important to prevent undue loss of gas at the surface.
2. Drenching can be used if the chemical has a high contact toxicity or breaks down to chemicals with a high vapour pressure. On application the diluted chemical will only soak 5–8 cm of soil but can penetrate to about 20 cm as a gaseous breakdown product. However, the application of large quantities of dilute chemical is difficult except on light, well-drained soils.
3. Direct application of a gaseous nematicide under impervious plastic sheeting can achieve fumigation to a depth of 15–30 cm.

4. Broadcast application of granules or prills over the surface of the soil should be followed by thorough mixing by digging or rotavation.

Post-treatment cultivation After treatment with soil fumigants or with sterilants that release toxic gases, it is important to work the soil thoroughly and ventilate the glasshouse to accelerate the dispersal of gaseous residues. Sometimes it may be desirable to raise the temperature of treated houses to help in removing the last traces of gas.

All soil fumigants can damage growing crops and small amounts of some fumigants can volatilize from glasshouse soil long after treatment, scorching the plants or producing abnormal growth. The 'cress test' will establish whether appreciable amounts of fumigant are retained in the soil. For this test several samples of the treated soil should be taken from a depth of 30 cm, put into jars, moistened and cress seed sprinkled on top. For comparison, two or more similar jars of untreated soil should be sown. The jars should then be sealed and kept in a warm place; germination should occur in about 48 hours. If germination on the treated soil is retarded compared with that on untreated soil, fumes are still being released and the test should be repeated at seven-day intervals until the cress germinates evenly and quickly; it should then be safe to plant tomatoes.

Cultural control

Pot and ring culture Tomato plants can be grown in pots or containers of sterilized compost. With ring culture the containers are open at the base and are stood on beds of moist gravel or cinders overlying polythene sheeting to prevent the tomato roots from penetrating infested soil below. The plants are fed with nutrients via the containers ('rings') and additional water is taken up by the roots that grow through into the gravel or cinder bed.

Soil-less compost Peat composts are widely used in bolsters or 'grow-bags': these are long plastic bags filled with compost, the plastic preventing the tomato roots from penetrating the glasshouse soil, as in ring culture. Such bags can also be used in plastic-lined troughs.

Nutrient-film technique and rockwool The nutrient-film technique (NFT) consists in growing plants in troughs through which a complete nutrient solution circulates, bathing the roots that are thus isolated from the glasshouse soil.

Rockwool is a synthetic material that is made into cubes in which seedling tomatoes can be produced. When the seedlings are sufficiently well grown they are placed on strips of the same material. The whole strip is wrapped in plastic sheeting to provide an enclosed environment for the roots; water and nutrients are supplied to the roots, which are isolated from the glasshouse soil.

85
Potato tuber nematode

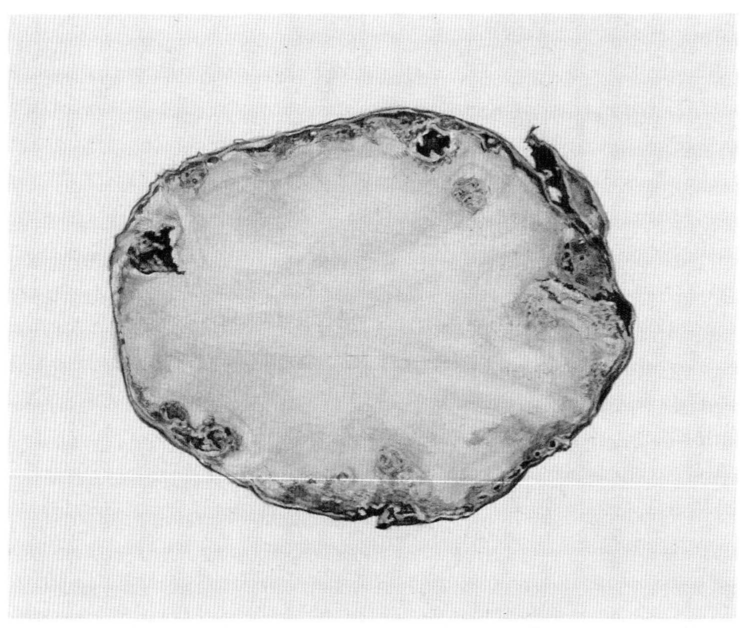

Fig.85.1 Potato tuber cut open to show internal damage caused by potato tuber nematode

Several species of nematode are known to attack potato crops in Britain. The most important are the potato cyst nematodes (see page 422), which are serious and widespread pests attacking the roots. Potato tuber nematode (*Ditylenchus destructor* Thorne) is much less serious and attacks only the tubers; it is less common now than in the past.

Description and life history

The nematodes are tiny, transparent, eel-like creatures which cannot normally be seen without magnification. Potato tuber nematode closely resembles stem nematode (see pages 440, 447, 454 and 458) in size, shape and structure, and can be distinguished from it only by specialist examination under a microscope. It differs, however, from stem nematode in being less able to survive dry conditions.

Breeding is continuous when conditions are favourable and there is an ample food supply; all stages (eggs, young and adults) are present at the same time in an infested tuber.

Plants attacked

Crops

Few farm crops other than potatoes are likely to be seriously affected by potato tuber nematode. Mangels, sugar beet and carrots have been shown to be capable of serving as hosts, but attacks on these crops in the field have not yet been found in Britain. In tests in Ireland, infested carrots showed severe dry rotting of the crowns.

Potato tuber nematode is sometimes troublesome to bulb growers as a pest of Dutch, English and Spanish varieties of bulbous iris (see page 434) and has occasionally been found damaging the corms or tubers of dahlia, gladiolus and *Tigridia*.

Weeds

Corn mint (*Mentha arvensis*) is the chief weed host of potato tuber nematode in Britain. The nematode invades and multiplies in the creeping rhizomes or runners of this plant and, in some districts at least, the presence of this plant appears to be the main factor determining whether the nematode can persist and become a serious pest. Potato tuber nematode has also been found in the runners of field (or perennial) sowthistle (*Sonchus arvensis*), which may thus play a similar though less important part in the persistence of infestations. Other common weeds, such as coltsfoot, daisy, broad-leaf dock and greater plantain, can also serve as hosts.

Corn mint and field sowthistle are among the herbicide-resistant weeds that are more prevalent under cereal cropping; this may explain the observation that the pest tends to persist more under cereal growing than under potatoes.

Fig.85.2 Potato tuber showing external symptoms of attack by potato tuber nematode

Effects on potatoes

Unlike potato cyst nematodes, which can cause serious stunting and reduction in yield, potato tuber nematode generally causes no obvious symptoms in the growing plant. However, if heavily infested tubers are planted, the young plants may be unable to make sufficient growth to establish themseves and will die.

A potato tuber heavily infested with this pest shows, typically, slightly sunken areas on the surface, with cracked and wrinkled skin that is detached in places from the underlying flesh (see Fig. 85.2). The decaying tissue, seen through the larger cracks, usually has a dry and mealy appearance and varies from greyish to dark brown or black. Much of the discoloration is due to invasions by fungi or bacteria that follow the nematode attack. In the early stages, however, none of these more obvious effects may be seen. The nematodes enter potato tubers either through the natural pores (lenticels) or the eyes and set up small 'pockets' or nests of infested tissue before there is any sign on the surface. On peeling an affected tuber at this early stage, these first pockets of infestation can be seen as small off-white spots in the otherwise healthy flesh. Later they enlarge, darkening a little and showing a woolly texture, and are often slightly hollow at the centre. The surrounding tissue rots as secondary organisms gain entry, and the skin dries, shrinks and cracks to give the typical appearance described above. The secondary invaders commonly include harmless nematodes that feed on fungi or bacteria; these nematodes may easily be confused with potato tuber nematode.

It is difficult, if not impossible, to diagnose potato tuber nematode with certainty from the external appearance: sample tubers should be cut, or peeled, and examined for the characteristic whitish pockets where most of the nematodes are found. Fig. 85.1 shows these pockets, groups of which have coalesced in places; their presence distinguishes the effects of tuber nematode from those of blight, with which it is sometimes confused. Microscopic examination by a specialist is necessary to confirm the presence of the nematode.

Potatoes in store

When stored dry, potatoes already infested gradually dry up and become mummified, so that they float if placed in water. In Lincolnshire such potatoes are called 'dry blights'. There is little or no spread from tuber to tuber during storage, provided the potatoes do not become wet. In moist conditions, however, especially in clamps that have not been properly covered, a much wetter rot may ensue owing to the rapid development of bacteria. If a large number of only slightly infested potatoes are present initially, the entire clamp may collapse.

Effects on bulbous iris

As on potato, tuber nematode on bulbous iris causes poor growth but does not produce other above-ground symptoms on the growing crop.

Nematodes enter the bulbs through the root initials. The first symptom of attack is a

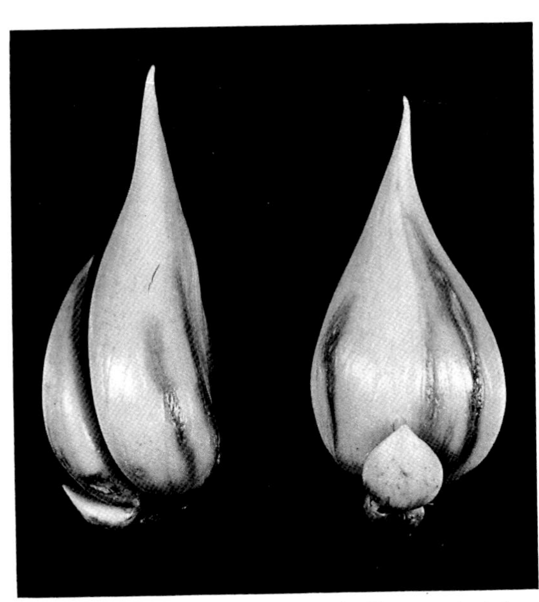

Fig.85.3 English iris bulbs attacked by potato tuber nematode. The scales have been removed to show the dark streaks spreading upwards from the base

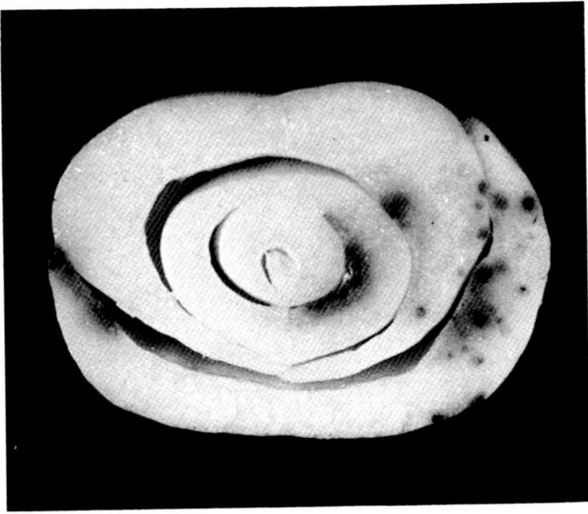

Fig.85.4 Cross section of English iris bulb infested with potato tuber nematode, showing blackening of vascular bundles

blackening of the vascular bundles, visible only when the bulb is cut across near the base (Fig. 85.4). The infestation may die out at this stage or the blackened area may spread. This stage of attack can be seen in the intact bulb, after the brown scales have been removed, as a series of shadowy dark streaks running from the base towards the tip (Fig. 85.3). Later the dark streaks may coalesce to form an extensive area of blackening, and secondary fungal attack, especially by *Penicillium*, may occur. In the past many nematode-infested stocks of iris have probably been considered as attacked only by fungi.

When infested bulbs are planted there is usually uneven development of roots (often only on part of the base plate) and consequent poor growth. Some plants may fail to emerge, or die soon after emergence, especially where there are secondary attacks by fungi. However, under good growing conditions there may be as much as 95% flowering when flowering-sized bulbs are planted.

During the growing season the original bulb withers away and nematode infestation may spread to the daughter bulbs. Depending on soil conditions, the proportion of infested bulbs in a stock may increase or decrease considerably from one season to the next. Infestation usually spreads mostly to the earlier-formed, and therefore larger, bulbs, while the very small bulbs are usually almost free from attack.

The long-term effect of nematode attack is to reduce the rate of multiplication of the stock. Where a clean stock may give 250–300% increase in a single year, the rate of increase of infested stocks may be from 0 to 20% in a year.

Control on potatoes

At planting

Potatoes from a crop known to have been affected by tuber nematode should not be used for seed. For every tuber visibly affected by this pest there may be many more containing light infestations with few or no external symptoms. For planting purposes it is, therefore, useless to attempt to remove nematode-infested tubers by sorting.

No potato variety is known to be resistant to tuber nematode, although King Edward seems to be rather less susceptible than other varieties.

In the field

Good control of weeds (see page 433) helps to eliminate potato tuber nematode from fields. Mechanical cleaning operations are particularly important because the nematode's weed hosts are difficult to control with herbicides. The most practical cultivation method for the control of corn mint or field sowthistle is by repeated cultivations aimed at exhausting the plants' rootstocks.

After lifting

Infested potato crops should not be used for seed. Lightly infested ware crops should preferably be sold at once. (If they must be stored, well-ventilated indoor storage is better than clamping.) Infested potatoes should not be fed to animals or poultry unless they are cooked first, otherwise uneaten pieces containing live nematodes will be trodden in and subsequently spread about the farm with manure.

Control on bulbous iris

Satisfactory, though not complete, control of nematode infestaton can be achieved by hot-water treatment. The recommendation is to warm-store the bulbs at 30 °C for 1–2 weeks and then soak in water plus formalin and a non-ionic wetter for three hours at 44.5 °C.

Treatment of an infested stock may not cause any improvement in the first year. It may, therefore, not be justified where bulbs are being forced for one year and then discarded, but the long-term advantage of clean bulbs in producing greater bulb and flower yields in subsequent years is considerable.

86
Root-knot nematodes on protected crops

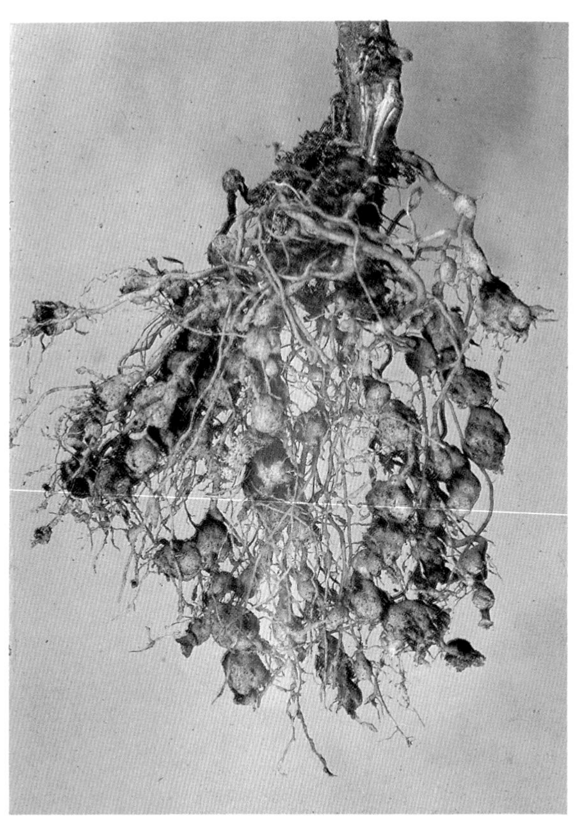

Fig.86.1 Tomato root showing galls caused by feeding of root-knot nematode

Root-knot nematodes (*Meloidogyne* spp.) cause galls to be formed on the roots of tomato (Fig. 86.1), cucumber, and sometimes lettuce, dwarf French bean, chrysanthemum and various ornamental plants grown under protection. These swellings, which vary from the size of a pin-head to that of a walnut, are the plant's reaction to feeding by the nematodes in the root tissues. The galls often merge with one another, resulting in completely deformed root-systems that have few active fibrous roots. With improved methods of crop protection, attacks occur much less frequently than in the past.

Potato cyst nematodes are also pests of tomatoes under glass (see page 428). They cause slight local thickenings of the roots but never large galls.

Certain other gall-like swellings on roots may be confused with galls initiated by root-knot nematodes, e.g. the so-called 'hybridization nodules' which are apparently harmless and contain no parasites, the nitrogen nodules on the root of beans and other legumes, and the corky thickenings on tomato roots caused by a fungus (*Spongospora subterranea*).

Root-knot nematodes, of which many species are known, are destructive pests of a large number of important field crops in tropical climates. In Britain, galls caused by several of the more hardy species of root-knot nematode occur in the open on cereals, carrot, parsnip, red beet, pyrethrum and various other field crops, as well as on shrubs and other ornamental plants in nurseries; damage is occasionally severe. The root-knot nematodes infesting outdoor plants are different species from those usually found under protection.

Description and life history

Female root-knot nematodes are only just visible without magnification. They are glistening white, pear-shaped bodies and occur inside the galled roots. One female may lay from 300 to 1000 or more eggs, which adhere to her body in a gelatinous mass. Most species of root-knot nematode are parthenogenetic, i.e. males play no part in reproduction although they may be present.

On hatching, some of the tiny larvae make their way deeper into the plant tissue; others come out into the surrounding soil, in which they are usually found at depths of 30–40 cm although they may occur at one metre or more. These larvae can live in dry soil for several weeks and in moist soil for several months. Thus they can survive for a considerable time until conditions are suitable for them to start new infestations on successive crops. About a month after entering the roots or the underground portions of the stem of a host plant, the nematodes become mature and a new generation begins.

Symptoms of attack

The first noticeable symptom of attack is a paleness of the lower foliage. Later, there is a strong tendency for the upper foliage to wilt during periods of sunshine. On tomato plants these symptoms become evident at the setting of the fourth or fifth truss; subsequent growth is retarded.

When infestation is severe the deeper roots fail to function. The plants respond by growing additional roots from the base of the stem and these are partially exposed to view on the soil surface.

Although infested plants are not often killed and sometimes produce good crops, losses are considerable on lighter glasshouse soils and in borders in which cucumbers are grown; here the nematodes increase much more than in heavier clay soils.

Resistant tomatoes

Some commercial varieties of tomato are available which are resistant to the root-knot nematode species commonly occurring in glasshouses in the UK. Tomato varieties are chosen by growers on the basis of many factors, particularly yield, earliness and fruit quality. If nematode resistance is also available in the preferred varieties, it is an additional benefit and could avoid the need for specific control measures against the pest.

Resistant rootstocks Where resistant varieties are unsuitable for local conditions, other, non-resistant commercial varieties can be grafted on to rootstocks into which disease resistance has been bred. All the available rootstocks are resistant to corky root caused by *Pyrenochaeta lycopersici*. Rootstock KN also incorporates resistance to some species of root-knot nematode; KNVF is similar, with resistance to verticillium wilt and fusarium wilt in addition.

Prevention and control

The pest may be brought into the glasshouse or nursery with infested soil or by the purchase of infested plants. New soil introduced from any doubtful source should be sterilized with steam or chemicals before use. For this purpose steam sterilization provides the best safeguard. If it is necessary to buy in young plants, growers should buy only those which have been grown in sterilized soil or soil-free composts.

Root-knot nematodes can be spread from house to house with transplants, e.g. lettuce seedlings raised in an infested house.

Before steam-sterilizing or chemically treating soil for control of root-knot nematodes, as many as possible of the infested roots should be removed and burnt.

Steam sterilization

In theory, steam sterilization can be an efficient method of controlling root-knot nematodes. (These pests are easier to kill by heat treatment than are potato cyst nematodes (see page 430).) In practice, however, steaming often fails to give satisfactory control of root-knot nematodes, especially in cucumber houses where conditions favour the rapid build-up of infestations from survivors in the deeper layers of the soil.

When attempting to control nematode pests in glasshouses by steaming, it is especially important to be thorough and to heat not only the soil but also the concrete foundations in the glasshouse and other underground structures. The period during which the plants will remain free, or reasonably free, after sterilization depends on the thoroughness with which the work is done and the depth of soil that is heated.

Chemical control

Several nematicides are now available for control of nematodes in glasshouses, but only volatile soil-fumigant nematicides are recommended for control of root-knot nematodes.

Cultural control

Isolation from infested border soils Tomatoes, cucumbers and other protected crops can be grown satisfactorily in a variety of composts, in peat bags, on straw bales or rockwool slabs as alternatives to glasshouse border soils. Nutrient-film techniques (NFT) can also be used for tomatoes. Where root-knot nematode is present and the use of steam or chemicals is not possible or convenient, a useful alternative is to provide an isolating barrier on top of which the bags, bales, etc. can be arranged conveniently. Various materials can be used to provide such a barrier, but thin-film plastic will be the popular choice, especially where straw-bale culture is envisaged. Thin-film plastics may be used as flat sheets or made into crop containers or troughs. The latter may be placed either on the soil surface or set into the border. Peat or peat/sand composts are the usual choice of media for filling the containers or troughs, or for making shallow growing beds on top of flat plastic sheets. Several types of compost-filled plastic containers or modules of proprietary manufacture are available and give reliable results.

Given skill in supplying water and nutrients to plants growing in the chosen medium, a good crop can be produced from a range of production systems.

Use of peat A layer of peat on the surface of soil in which seriously infested tomato plants are growing is of great value as a temporary measure. Sufficient peat to cover the soil with a layer about 2.5 cm thick should be thoroughly wetted on a clean floor and then spread over the glasshouse soil. It must be kept moist, but not too wet, by damping every second or third day. No attempt should be made to wet the soil beneath the peat layer, otherwise the peat will be washed away. The plants themselves should be kept damp until new roots have been formed; this helps them to withstand the effects of nematode attack and facilitates rooting. New roots are formed in the peat within about eight days. Growth continues so rapidly that the peat is soon filled with clean white roots and, although these are attacked by nematodes in due course, the plants recover their upward growth and bear good crops.

87
Stem nematode on arable and forage crops

Fig.87.1 Oat plants infested with stem nematode, showing 'tulip root' symptoms

Stem nematode (*Ditylenchus dipsaci* (Kühn) Filipjev) is a destructive pest of many different plants. It occurs as 'races' or 'strains' which differ in the range of host plants that they attack. One or more races infesting oats can also attack rye, sugar beet, fodder beet, mangel, field bean, pea, vetch and, occasionally, potato among agricultural crops. Certain horticultural crops, such as onion, broad bean, runner and dwarf French beans, parsnip, carrot, rhubarb and strawberry can also be attacked (see pages 458 and 413). Different races of stem nematode infest clovers and lucerne.

Description and life history

The adult nematode is about 1.2 mm long but, being colourless and very slender, it cannot be seen without magnification. Each female lays many eggs. The newly hatched nematodes, although smaller than the adults, closely resemble them in appearance.

The nematodes live as parasites in the plant tissues where they breed continuously. They can move through moist soil to invade other plants. If infested plants dry up, the nematodes become dormant; in this condition they often survive for several years, becoming active again in the presence of moisture.

In moist soil in the absence of a host plant, stem nematodes can live for a year or more according to the type of soil. They are especially persistent in soils with a high clay content and in areas of heavier soil within infested fields.

Symptoms of attack

On oats

The effects of infestation are generally seen in the earlier stages of plant growth, well before the ears emerge. Symptoms usually appear about 6–8 weeks after sowing. A badly infested crop looks uneven and patchy, often with rather pale foliage. Most plants within the affected patches are stunted and characteristically swollen at the bases of the main shoots and tillers, which are often small, pale and contorted (Fig. 87.1). Common names for this disorder include 'tulip root' and, in the north of England, 'segging'. The infested tissues have a characteristic spongy texture. Badly infested plants tend to rot at ground level and are easily pulled away from the soil, leaving most of the roots behind. Those less severely damaged may produce an ear, but this is usually poor and the yield of grain is small.

Stem nematode attacks on oats are seasonal and are encouraged by mild, wet conditions in the spring. The incidence of attacks on oats has decreased following the decline in the growing of this crop and the introduction of resistant varieties.

Symptoms resembling 'tulip root' can sometimes be seen in oat plants attacked by frit fly (see page 260). Such plants appear swollen at the base owing to excessive tillering, but only the central leaf turns yellow and a small white maggot can usually be found feeding at its base. The two pests sometimes occur together.

On rye

The symptoms of stem nematode on rye resemble those on oats. Infestation of this crop is uncommon in Britain, but heavy losses often occur in countries where rye is grown more extensively.

On sugar beet, fodder beet and mangel

These crops may be attacked in both spring and autumn. In spring the nematode can attack the seedlings, causing swellings, and the tissue becomes characteristically spongy. The swellings affect the stalk, midrib and main veins of the leaves and sometimes form distinct galls (Fig. 87.2). The growing point may be malformed or killed; if it is killed, fresh growing points develop, leading to a multiple-crowned or 'many-necked' condition. During the summer, healthier growth is produced and the damage becomes less apparent. Plants attacked in the spring may develop a rotting or canker of the crowns in the autumn, or fresh plants may be attacked at that time.

The first sign of infestation in the autumn is usually cracking of the surface skin of the roots at or above soil level. The tissues just under the skin become soft and brownish, and a dry spongy rot gradually extends inwards. This is often followed by a darker and wetter rot due to bacteria, until the inside of the root breaks down completely. In advanced stages of attack

Fig.87.2 Young fodder beet plants showing swellings caused by stem nematode

the crown and leaves can often be pulled away from the roots.

A deficiency of the trace element boron in the soil produces two distinct symptoms in sugar beet and mangels, i.e. death of the growing point and crown canker. Either of these symptoms can be confused with the effects of stem nematode. Microscopic examination is thus essential to confirm the presence of the nematode.

Damage to sugar beet by stem nematode should not be confused with that caused by beet cyst nematode, which forms cysts on the fibrous roots underground but not in the crowns or foliage (see page 396).

On field bean

Both the oat-onion race and a so-called 'giant race' of stem nematode infest winter and spring varieties of field bean; infested seed is sometimes produced. Seed-borne infestations usually have little effect on the vigour of the first bean crop, but they may be important in introducing the nematode to uninfested land. Even after an interval of several years, subsequent bean crops can become sufficiently infested from the soil to cause loss of crop.

Susceptible oat varieties following beans infested with the oat-onion race of stem nematode are liable to be severely attacked. Beans following heavily infested oats are also badly affected. Similarly, strawberries have suffered severe damage by stem nematode when planted after beans.

Fig.87.3 Field bean stems showing reddish-brown discoloration caused by stem nematode infestation

Symptoms on the plant The stems of established bean plants infested with stem nematode show a reddish-brown discoloration that darkens with age, starting at the base and sometimes extending to the pod-bearing region (Fig. 87.3). However, such discoloration in leguminous plants can be caused by other agents including disease organisms. Symptoms are most readily seen in July, while pods are still green, but they eventually become masked by the effects of chocolate spot or the general browning due to ripening of the crop.

Stems infested with the oat-onion race of stem nematode are seldom distorted but are often thinner and shorter than those of uninfested plants. Those infested by the 'giant race' are more severely damaged, often being twisted and blistered; infestations in the pod-bearing regions are more common and leaf petioles and pods may be distorted. Secondary rotting at the stem base sometimes causes the collapse of the whole plant.

Symptoms on the seed Ripe,infested pods contain many dried resistant nematode larvae attached to the seeds, especially in the slit of the hilum. Sometimes the nematodes are massed together to form 'nematode wool'. When the seed coat is removed from heavily infested seeds, a discoloured patch or lesion containing many nematodes can often be seen either side of the radicle in the depressions on the cotyledons. More than 10 000 stem nematodes have been recovered from an individual seed.

On pea and vetch

Stem nematode on field peas is frequently associated with vertical cracking at the base of the stem at about soil level, often without discoloration or serious checking of growth. Sometimes, however, there is stunting and malformation. The type of symptom may depend on which race of the nematode is involved; for example, stunting and malformation can occur on pea seedlings following an infested onion crop (see page 459).

Vetches are occasionally found to be

Fig.87.4 Vetches (right) attacked by stem nematode compared with normal plant (left)

infested (Fig. 87.4), sometimes with oats in a mixed crop: the oats show typical 'tulip root' symptoms and the vetches in the same patches are severely stunted.

On red clover

Stem nematode is one of the causes of a condition commonly known as 'clover sickness' and is responsible for serious losses in red clover crops. The same term has also been applied to the effects of a fungus disease which is more often called clover rot.

Stem nematode may be introduced into fields on infested seed or into a clean crop by contamination with infested hay, plant debris, soil or water. The effects of attack may be seen as early as immediately after germination, when a characteristic swelling can appear just below the two seed leaves. The seedlings may be killed and bare or thin patches may develop in the crop, but damage is often not noticed until later.

The effects of stem nematode infestation in older plants are apparent mainly in the spring and autumn. In spring, thin patches appear in the crop. Plants around the patches are often stunted and the bases of shoots swollen (Fig. 87.5) and characteristically spongy. Leaves are frequently thickened and twisted and sometimes have brown patches. These symptoms may be masked by the rapid growth of plants later in the season, when the only signs of attack may be the bare or weedy patches where badly damaged plants have died. In late summer and autumn, swellings and other distortions may develop on the side shoots and below the flower heads, which are often distorted and small. The nematodes may invade the flower heads. Although contaminated seed is an important means of spread, the nematodes usually occur in particles of crop debris rather than on or in the seeds themselves.

Plants affected by clover rot do not show the swellings characteristic of nematode attack.

The red clover race of stem nematode can also severely damage kidney vetch and strawberry. Alsike, lucerne, sainfoin, trefoil and trifolium (crimson clover) can be lightly infested but show little or no sign of damage; for practical purposes they can be considered immune.

On white clover

The race of stem nematode which attacks white clover is distinct from the race attacking red clover. Alsike is the only other cultivated plant attacked to any significant extent; the white clover race does not attack red clover.

The symptoms and effects of attack are similar to those shown by red clover, allowing for the differences in the structure of the plants. Characteristic swellings may develop just below the flower heads and on the young shoots produced on the stolons.

The white clover race can be seed-borne, though in practice infested seed has rarely been reported.

Fig.87.5 Red clover infested with stem nematode. Left: plant showing basal swelling of shoots and leaf distortion. Right: severely damaged shoot

On lucerne

Lucerne can be considered to be immune to the clover races of stem nematode, although it can be lightly infested by the white clover race without significant damage. There is, however, a separate race of stem nematode which is very damaging to lucerne. Alsike is also attacked by the lucerne race of stem nematode, but symptoms are slight and it is generally an inefficient host.

The symptoms of attack in lucerne are similar to those that occur in clover. The stems and foliage become stunted and buds, shoots and stem bases become swollen and distorted. The roots are sometimes infested; internal cavities or outgrowths may develop around the crown and sub-crown parts of the tap roots. Infection of the crop appears as patches of poor, stunted plants. Stem nematode appears to interact with bacterial wilt of lucerne (*Corynebacterium insidiosum*) and may help to reduce the resistance to the disease shown by some varieties.

Stem nematode infestation is seed-borne in lucerne, and infested seed is an important means of spread and introduction to new areas. As with red clover, the nematodes occur in particles of crop debris with the seed, rather than on the seeds themselves.

On teasel

The scientific name of the stem nematode, *Ditylenchus dipsaci*, is derived from that of the fuller's teasel (*Dipsacus fullonum*) — the plant on which it was first discovered. Teasels are still grown commercially for the wollen industry on a few farms in Somerset.

Stem nematode causes severe stunting of this crop and the race of nematode responsible is probably one that affects oats. Infested plants produce either small malformed flower heads or none; heads from such plants are soft and useless for dressing cloth. Large numbers of nematodes move into the flower heads and infest the seed. The growers usually save their own seed and most of this is infested.

Weed hosts

Stem nematode may infest common chickweed, mouse-ear chickweed, cleavers (goosegrass), scarlet pimpernel, wild oats and many other weeds. These weeds enable the nematode to persist after an attacked crop in sufficient numbers to initiate a damaging attack on the next host crop.

Control measures

Crop rotation helps to prevent nematode damage. Where oats, rye, onion, sugar beet, mangel, beans or vetch have been attacked, all susceptible crops should be excluded from the rotation for at least two years. Wheat and barley can be grown safely on such land.

Neither red clover nor kidney vetch should be grown for several years in fields where either of these crops has been attacked by stem nematode. Lucerne and white clover may be useful alternative forage crops for such fields. Similarly, fields infested with the lucerne race of stem nematode should be given several years' break from lucerne.

As stem nematode can live in common field weeds, it is important to maintain good weed control. Wild oats are probably the most troublesome, and also the most costly, annual weeds of arable land. The fact that they are hosts of stem nematode makes their control even more important.

It is also important to ensure that infested material, such as oat straw or mangels, is not carted to other fields. Stem nematode can quickly be spread about a farm with infested straw in farmyard manure. If manure has become contaminated in this way, the heap should be well composted and turned at intervals, so as to heat all parts thoroughly, before spreading it on arable land. Because of the risk of spreading the infestation, infested clover crops should never be cut for hay; instead they should be fed off the field.

Expert examination is usually necessary to confirm the presence of nematodes in a crop.

Several chemicals are available for treating soil to control nematodes, but only those recommended for the control of stem nematode in arable or forage crops should be used.

Resistant varieties

Where appropriate, varieties resistant to stem nematode can be grown. The list of recommended varieties issued by the National Institute of Agricultural Botany should be consulted for information on varieties of arable crops having nematode resistance.

On field bean

Field bean crops should be checked during July for signs of stem nematode. Seed for sowing should be saved only from clean crops. Seed samples can be checked in the laboratory for the presence of stem nematode. Field beans may introduce this pest to uninfested land and are therefore a potential source of infestation for other susceptible crops such as oats, onion, pea, sugar beet, mangel, rhubarb and strawberry. If oats or other susceptible crops are to be grown within three or four years of an infested field bean crop, a resistant variety should be used.

Fumigation of seed

Red clover Stem nematode in red clover seed can be controlled by fumigation. Growers are recommended to buy fumigated seed. Under the Fodder Plant Seeds Regulations 1985, if stem nematode is detected during field inspection of a red clover crop grown for Basic or Higher Voluntary Standard grade, the seed must be fumigated effectively or downgraded to Certified (minimum standard).

Lucerne The Plant Health Order (Great Britain) 1987 requires either that imported lucerne seed is fumigated prior to export or that the seed was taken from a crop grown where no symptoms of stem nematode have been observed and was also sampled and found to be free from stem nematode in laboratory tests. Virtually no lucerne seed is grown in the UK.

88
Stem nematode on narcissus

Fig.88.1 Narcissus foliage affected by stem nematode: centre leaf shows typical spickels

Stem nematode (*Ditylenchus dipsaci* (Kühn) Filipjev) is a common pest of bulbs and is well known to bulb growers in Britain. Because many countries, especially those within the European Community, strictly prohibit its entry, this nematode is of particular concern to those who grow narcissus and tulip bulbs for export. It is also a pest of many other plants, including various farm crops (see page 440), vegetables (page 458) and strawberry (page 413).

Several apparently distinct 'races' or 'strains' of stem nematode occur on different crops. Some races breed readily on a wide range of plants, while others are more specialized; sometimes the host ranges of two or more races overlap. Races can be distinguished only by their ability to multiply on various host plants and by their effects on these plants. However, their attacks are not limited to their most suitable hosts; thus stem nematodes of any race may enter and survive in, or even breed slowly in, many kinds of plant. There is no guarantee that plants considered 'safe' in rotation with bulbs will remain so.

At least three races are known specifically to

infest bulbs in Britain, i.e. the narcissus, tulip and hyacinth races. The **narcissus race** attacks narcissus and tulip but does not usually breed in tulip although it can damage it. The narcissus race is common and very important commercially in both eastern and south-west England. The **tulip race** (see page 454) can infest tulip, narcissus and hyacinth and causes severe symptoms and damage in narcissus. It occurs mainly in eastern England. The **hyacinth race** is rare in Britain; it has occurred occasionally in the south-west of England. It does not affect narcissus or tulip. Each of these races can attack certain other kinds of bulb.

Description and life history

The adult nematodes are tiny, thread-like and transparent; they are about 1.2 mm long and cannot normally be seen without magnification. Both males and females live as parasites within the plant tissues, where they often occur in enormous numbers. The females lay many eggs. The newly hatched nematodes, apart from their small size, closely resemble the adults in appearance. Breeding is rapid and continuous as long as the host plant remains alive. Nematodes frequently leave infested bulbs and migrate to adjacent ones through the soil. This migration may continue if the bulb dies, but many of the nematodes die in the decaying remains.

Stem nematodes can also breed in stored bulbs. Heavily infested bulbs are often killed and become completely rotten; great numbers of nematodes then find themselves without any further food supply. Under these conditions many of them do not develop beyond the pre-adult stage, when they are particularly resistant to unfavourable conditions. Thousands of these pre-adult larvae, forming glistening off-white masses easily visible without magnification, may sometimes be found oozing out at the base of a rotten narcissus bulb. These masses of nematodes gradually dry to form pieces of dirty-white or buff-coloured material known as 'nematode wool' ('eelworm wool'). Mycelium of *Fusarium*, which also occurs frequently round the base of rotted bulbs, differs from 'nematode wool' in being pure white and more fluffy in appearance (Fig. 88.2).

In this dormant condition the nematodes can remain alive for several years, becoming active again only when moistened. Small particles of infested bulb tissue blowing about from bulb stores or carried into fields with bulbs can quickly start fresh infestations. Infested fragments can similarly contaminate cleaning, grading or other machinery and thus spread infestation to clean bulbs. Pieces of leaf infested with dormant nematodes can also be a means of spread, especially when tops are 'flailed' off. The 'wool' stage is more resistant to hot-water treatment than the active stages.

Fig.88.2 Basal parts of two bulbs showing piece of buff-coloured 'nematode wool' on upper bulb and white mycelium of *Fusarium* on lower one

This dormant or quiescent stage in the life history is characteristic of all races of stem nematode and makes it the more persistent and insidious. However, the nematodes seldom remain dry for long in the field and they cannot live in moist soil for much more than a year without host plants.

Symptoms of attack

Stem nematodes usually enter the bulbs from the soil in the region of the neck, the initial infestation often being caused by only a few nematodes. They invade the young leaf tissue and some are carried upwards by growth of the leaves while others move down into the leaf bases.

The establishment of small breeding colonies in the leaves leads to the formation of small local swellings called spickels (Fig. 88.1). These are usually conspicuous by their unhealthy pale yellow colour. With slight infestations, however, they may be more easily recognized by the unevenness felt when the leaf is run between finger and thumb. In more severe cases, the spickels may be large and show brown areas of dead tissue in the centre of the swelling. When leaves are heavily infested, the spickels tend to run together and the whole leaf becomes twisted, distorted and discoloured.

Flower stalks are affected in the same way as the leaves: they often have spickels and in severe cases are misshapen and stunted. Nematode-infested bulbs often flower late, so that a late-flowering patch is sometimes a sign of nematode attack.

Although stem nematodes may invade bulbs at any time during the growing season when conditions are suitable, invasion occurs mostly during the cooler months and often before the leaves emerge above ground. When the attack occurs late or is slight, spickels may not appear until after flowering or may not show at all during the first year of growth. Thus the absence of symptoms, especially in one-year-down bulbs, cannot be taken as a sure sign that the stock is nematode-free. Bulbs selected as apparently healthy from a stock in which some are infested are likely to harbour nematodes. A stock in which stem nematode has been found should be regarded as infested and treated as such.

As the season progresses, the nematodes multiply in the leaf tissue, particularly in the basal parts of the leaves, but when the foliage begins to senesce, they move downwards into the scale leaves of the bulb. This initial infestation of the bulb usually leads to orange-brown or greyish-brown areas in the infested scale leaves. If a bulb at this stage is cut across just below the neck, only a small patch of discoloured tissue may be visible in one scale where the nematodes are feeding. Later, the infestation will spread right round an affected scale, the nematodes taking the line of least resistance and spreading as far as possible before moving to an adjacent scale. Spread from scale to scale is usually via the base plate of the bulb: after destroying one scale, the nematodes enter the base plate and then work upwards into the adjacent scale. This behaviour accounts for the characteristic brown rings seen when heavily infested bulbs are cut across (Fig. 88.3) and the loosening of the base plate to form a plug of infested tissue which often remains in the soil at lifting.

Infested bulbs tend to become soft in store, particularly round the neck, and usually appear dull by contrast with the lustrous appearance of healthy bulbs. They are often secondarily attacked by bulb mites and small narcissus flies (see page 292); presence of the latter is frequently an indication of stem nematode attack.

Other plants attacked

Bulbs The narcissus race of the nematode which occurs in the south-west of England can infest and cause damage to snowdrops, bluebell (wild hyacinth) and scillas. On snowdrops it produces spickels similar to those found on narcissus; on bluebell and scillas it causes stunting and twisting of leaves and flower

Fig.88.3 Narcissus bulb cut across to show 'brown-ring' effect of stem nematode infestation

stalks, with pale streaks and lesions on the leaves. The effects of an attack of stem nematode on hyacinth are rather similar to those on narcissus, except that definite spickels are not usually formed on the leaves.

Other crops Stem nematodes from narcissus may also breed and persist in onion, broad bean, dwarf French bean, runner bean, pea and strawberry. The nematode can sometimes breed in brassica seedlings but seems not to persist in them; brassica transplants may spread stem nematodes and should not be propagated on infested land. Cereals and grasses seem unable to maintain the nematode.

Weeds Stem nematode has been found in most common weeds growing in fields where bulbs have been infested. Examples are bindweed, chickweed, cleavers (goosegrass), groundsel, knotgrass, rayless mayweed, scarlet pimpernel and speedwells. Some of these weeds may maintain the nematodes in the absence of host crops, though there is little evidence of their practical importance.

Means of spread

The most common way in which the bulb races of stem nematode are spread is by the distribution and planting of infested bulbs. Nematodes can also be spread about farms or gardens in pieces of dead foliage, in bulb debris and in soil, which may be blown about or carried on implements and vehicles.

Within fields, infestations generally extend slowly in the soil so that a small infested patch gradually increases in size. This kind of spread is increased in the direction of cultivation and may be greatly accelerated by movement of surface water — an effect which is sometimes seen clearly on a sloping field, where nematode-infested patches are elongated in the direction of the slope. Spread by cultivations can be very serious. On sloping fields cross-cultivation makes the situation worse by extending infestation sideways and starting new concentrations, which may then be spread downhill by surface water.

Control measures

Cultural control

As with other pests and diseases, sound crop rotations, hygiene and use of healthy, good quality planting material do much to prevent crop losses caused by stem nematode.

Inspection Growers are advised to walk their stocks during the growing season, keeping a sharp look-out for suspect plants and consulting their local adviser when necessary. Symptoms of nematode attack are far more easily detected on the foliage than in the dry bulbs. When inspecting stocks of narcissi, it is advisable to avoid periods of bright sunshine because contrasting shadows make it more difficult to see the spickels. In lightly infested stocks late-developing spickels can easily be overlooked at about ground level. The best time to detect stem nematode is just after flowering.

Roguing Spread can be reduced by roguing out

infested plants as soon as these are discovered, together with adjoining plants in the same and adjacent rows, instead of leaving them to spread the pest further in the stock. It must be emphasized, however, that this is only a check and stocks should be hot-water treated soon after lifting (see below). Removal of all soft bulbs before hot-water treatment will improve the efficiency of the treatment.

Rotation Bulbs left behind as self-sets should be removed when they appear in the following spring; only then should the rotation period be considered as having started. It is recommended that at least four years should elapse between bulb crops. Where the narcissus or tulip strain of the nematode is known to be present, narcissus rotations should preferably not be closer than one in seven years, and certainly not closer than one in four, ensuring wherever possible that no other host crop is grown in the interval. Weeds should also be kept to a minimum.

Hot-water treatment

Hot-water treatment is a long established and efficient method for control of nematodes and other pests in narcissus bulbs. It was originally introduced to control the large narcissus fly (see page 293).

The present recommendation for stem nematode control is to soak the bulbs for three hours in water at 44.4 °C during the period between lifting and replanting. The treatment should be timed from the moment the water temperature reaches 44.4 °C after addition of the bulbs.

Because of the risk of damage to the bulbs, particular attention should be paid to the stage of development of the bulbs at which treatment is applied (see below). The extent of hot-water damage is also greatly influenced by the storage temperature during the 2–3 weeks before treatment. When treatment is applied at the correct stage of internal bud development (Fig. 88.4), serious damage to bulbs or foliage is unlikely provided the bulbs have been stored at 18 °C before treatment, although flowers in the year following may be slightly distorted. If storage is below 16 °C, yields can be decreased considerably.

Fig.88.4 Optimum development stage of narcissus flower for hot-water treatment. Petals have been removed to show carpels, stamens and trumpet initials (× 18)

Bulbs for forcing or for sale should not be hot-water treated.

Warm storage of bulbs at 30 °C for seven days immediately before hot-water treatment considerably reduces subsequent damage to flowers and foliage and, correctly applied, often enables perfect first-year flowers to be marketed. However, pre-warming tends to increase the resistance to hot-water treatment of any nematodes that may be present, while a prolonged period of warming and drying encourages the formation of 'nematode wool', which is highly resistant to treatment. Therefore, a higher water temperature in conjunction with a pre-soak (see below) is essential if high-temperature storage is used. Storage at 30 °C for seven days makes possible safe hot-water treatment of bulbs at higher temperatures, namely 46.7 °C for three hours. This warm storage treatment should not be given to stocks known to be infested as it makes subsequent control much more difficult; such stocks are also sources of infestation during storage.

Pre-soaking entails soaking the bulbs for three hours or overnight in cold water (with a wetter and formalin added) immediately before hot-water treatment. This process helps to drive out air bubbles that might otherwise insulate nematodes; it also wets and reactivates 'nematode wool', making control a little easier. However, pre-soaking, especially with a wetter added to the water, increases the damage caused by subsequent hot-water treatment unless the bulbs have been warm stored as above.

Timing of treatment The timing primarily affects the risk of damage to the subsequent crop, but it can also influence the degree of nematode control. Bulbs from infested stocks should be lifted before the foliage has died and then treated as soon as possible to destroy the nematodes before they reach the resistant 'wool' stage. The subsequent flower crop will almost certainly be damaged by this early treatment, but stem nematode is such a serious menace that growers should be prepared to sacrifice a flower crop to ensure clean bulbs.

Precautionary treatment of apparently uninfested stocks of bulbs lifted at the normal time can be started as soon as the bulbs have been cleaned and graded, but it must be completed by late August before the root initials have started to grow. Early treatment damages the flowers but does not affect bulb yield. If it is intended to crop flowers in the first year after replanting, hot-water treatment should be applied as soon as the flower buds are fully developed within the bulb (Fig. 88.4). Sample bulbs must be cut to determine the flower-development stage. This is not difficult using a ×8 hand lens, but expert advice should be sought if in doubt.

Pale blotching of the upper parts of the leaves caused by late treatment is sometimes confused with the effects of the virus disease 'mosaic' or 'stripe', but the markings due to this disease are lines rather than blobs and are more evenly distributed.

The order in which different varieties flower does not indicate the order in which they should be hot-water treated, i.e. the late poeticus types should be treated first, followed by short cup, large cup and trumpet varieties in that order. Doubles should be treated according to their origin, e.g. double whites, being poeticus sports, should be treated early.

Equipment and operation Many instances of unsatisfactory hot-water treatment can be traced to inefficient hot-water tanks, incorrect use of the installations, or to lack of hygiene leading to reinfestation. The following are some of the most important requirements.

1. Efficient forced circulation of the water, but avoiding foaming. Where foaming occurs an anti-foam agent should be used.
2. Good heat insulation, including a lid, over the whole tank.
3. Accurate temperature control (preferably by an automatic regulator).
4. An accurate thermometer of the mercury-in-glass type (*not* one encased in metal or other material) or an accurate electronic thermometer.
5. A clean, draught-free building to house the equipment.
6. A good, non-ionic wetting agent added to the water.
7. A fungicide added to the water to prevent bulb rots and other fungus troubles; commercial formalin at a strength of 1 in 200 or a suitable proprietary material may be used. Formalin also helps to kill nematodes present in the water and should always be used whether or not another fungicide is added.
8. Where bulk-handling systems are not used, rigid open-mesh containers, bulb trays or nets for the bulbs are satisfactory alternatives, but close-woven sacks are *not*. Plastic or cotton nets are widely used in commerce. Nets stacked on pallets should be filled tightly and cross-stacked to ensure adequate circulation of water between the nets. However, where nets are loaded into bins through which water is pumped, they should be loosely filled to ensure a more uniform distribution of heat.

9. With bulk-handling systems — using for example 500 kg wooden boxes having slatted floors — water circulation, temperature control and post-treatment drying all require special attention.
10. Tanks must not be overloaded: too many bulbs impede circulation and make temperature control more difficult. A ratio of bulbs to water of 1:2 should never be exceeded; a ratio of 1:3 or more is ideal.
11. Strict attention to hygiene. Frequent hosing down of the building, a 'one-way traffic' system and new or sterilized containers for treated bulbs are recommended. However, with front loading (and some bulk-bin treatment tanks) a one-way system is not possible. With bulk handling, treated bulbs remain in the same containers throughout, so the containers are treated with the bulbs.

All growers wishing to install new hot-water tanks or to improve their existing ones should obtain more detailed advice from their local advisers.

After treatment, bulbs should be allowed to cool in a well-ventilated area. Where bulk bins have been used, clean air at ambient temperature should be blown through the bulbs, otherwise prolonged heating and exposure to moisture will cause deterioration. Ideally the bulbs should be planted at once, but if this is not possible they must be stored in a 'clean' area well away from sources of reinfestation.

Treated bulbs should, of course, never be planted on infested land. Long rotations and clean cultivation reduce the risks of reinfestation (see page 451).

Chemical treatment of soil

Treatment of the soil with fumigants can be effective in controlling stem nematode but is not economic for large areas. However, it may sometimes be worth treating small patches to control a slight infestation before it spreads. Treatment should be applied only after lifting and destroying the affected bulbs together with a margin of surrounding healthy ones. Fumigation is less effective in the surface layers of the soil, so it is essential to achieve a good seal at the soil surface. Plastic sheeting can be used to seal small areas; for larger areas, a machine incorporating a powered roller that smears the soil surface is preferable.

89
Stem nematode on tulip

Fig.89.1 Tulip flower (left) infested with stem nematode. Note lesions on stem and persistence of green colour on petal nearest site of infestation

Attacks of stem nematode (*Ditylenchus dipsaci* (Kühn) Filipjev) on narcissus have been recorded since the beginning of the century, but it was not until 1943 that infested tulips were discovered in England. During the following 10 years tulip stem nematode (tulip eelworm) spread quickly in the important bulb-growing areas of south Lincolnshire and became one of the most serious problems of the tulip industry. Much progress has been made in its control, but it remains a potential problem requiring strict attention.

Although the stem nematode that damages tulip is a separate race from the narcissus race (see page 447), it can damage narcissus at least as severely as the narcissus race can. The tulip race appears to be particularly virulent and to have a wider host range than the other races known to infest bulbs. Races other than the tulip race may invade tulip bulbs or plants and cause lesions, but damage is not usually obvious and infestations do not persist in tulips.

Description, life history and spread

The nematodes are tiny, thread-like and transparent; when adult they are about 1.2 mm long. They live and breed as parasites in the plant tissues, where adults, eggs and young occur together.

Nematodes can migrate from infested to healthy tulips through the soil. If the bulb is killed and rots, this migration may continue, but many of the nematodes will probably die in the decaying remains. Heavily infested bulbs may die in storage; such bulbs usually become dried and shrivelled. Under these conditions a large proportion of the nematodes are likely to be at the pre-adult stage,when they are particularly resistant to unfavourable conditions and can survive drying for several years. These nematodes can also dry up and become dormant in pieces of infested stem, leaf or flower, so that any infested plant fragments, in bulb stores or blowing about the field, could be a means of spread.

Symptoms of attack

The effects of stem nematode infestation on tulips are best seen at, or just before, flowering. With slight attacks, the first sign is usually the appearance of a pale whitish or purplish streak on one side of the stem just below the flower (Fig. 89.1). This streak develops into a lesion which increases in size as the flower develops and colours; the surface tissues often become blistered and split. The flower is frequently bent over towards the side where the damage has occurred; such bent flowers are a good indication of the presence of the nematode in a stock (Fig. 89.2). The damage often spreads up from the stem on to the petals, and the affected parts of the flower generally fail to colour (Fig. 89.1). In severe attacks the outer petals may be severely malformed and devoid of colour, while the inner petals seldom show more than a small lesion. The stems may show similar streaks or lesions lower down, particularly near junctions with leaves; with very severe infestations, the stem may be stunted and variously bent or twisted. The leaves are sometimes affected in a similar way, especially in forced stocks. They may show pale streaks and ragged splits and are easily torn lengthwise.

The effects of nematode attack on the dry bulbs are not so well marked, or so easily distinguished from other troubles, as with narcissus bulbs. Attack most commonly starts on one side just above the base plate, seldom at the neck as with narcissus, although the bulbs may be attacked at any point. Slightly infested bulbs show little or no sign of trouble, even when cut across. The damage soon increases in storage and can be recognized by the appearance on the outer scale of glossy, greyish or brownish patches which merge indistinctly into the healthy tissue and do not have firm margins (Fig. 89.3). These patches generally feel soft and spongy when pressed by the thumb, and become more so as the nematodes increase in the bulb. When cut across, infested bulbs may show a patchy brown or grey discoloration, but this seldom shows as distinct rings as in narcissus bulbs. Badly infested bulbs finally rot and die, usually becoming dry and shrivelled. These dead bulbs are sometimes called 'corks'.

As slight infestations are difficult to detect in dormant bulbs, stocks of tulips are best

Fig.89.2 Tulips more severely infested with stem nematode than in Fig.89.1, showing characteristic bend in flower stalks. Uneven coloration of petals is again apparent

Fig.89.3 Tulip bulbs infested with stem nematode

examined for stem nematode at flowering time — the stem lesions found just under the flower are a very sensitive criterion. As with narcissi, it is best not to examine tulips in strong sunlight.

Expert examination is usually needed to confirm the presence of tulip stem nematode and, when it is suspected, growers are advised to consult their local adviser.

Other plants attacked

Besides tulip and narcissus, the tulip race of stem nematode from eastern England has been shown experimentally to infest and cause damage to hyacinth, bluebell, *Scilla siberica*, phlox, onion, pea, broad bean, dwarf French bean, runner bean and strawberry. However, none of these plants has so far been found attacked by this race in the field, although some of them are known as hosts of other races of stem nematode.

Stem nematode has been found in various common weeds growing in infested tulip fields. Some of these weeds probably maintain infestations in the absence of tulips.

Control measures

Cultural control

Bulbs that appear to be clean and sound but have come from an infested stock are likely to be slightly infested and may introduce the nematode to clean land. For this reason growers should regard the *stock* as the infested unit and should make it a regular practice at flowering time, before heading or cropping, to walk each stock and search carefully for signs of nematode attack. All visibly infested plants should be rogued out as soon as they are detected, placed at once in containers, such as empty fertilizer bags, and burnt. It is sound practice also to remove and destroy apparently clean bulbs growing within one metre of an infested plant.

The stock on each side of an infested one must be considered contaminated even when showing no obvious sign of nematode. Such stocks should be examined carefully in the following two seasons to check whether they are infested or not. Those that prove to be lightly infested may be used for forcing, but such bulbs will contaminate soil and boxes and could introduce the nematode into glasshouses.

Spread from one variety to another will be minimized if stocks are planted in the same order each year; then, if trouble occurs unnoticed, only the same neighbours are likely to be contaminated each time.

Hot-water treatment

Early experiments on the control of stem nematode in tulip showed that good commercial control could be obtained by hot-water treatment, but the bulbs were often damaged with subsequent loss of flowers and bulb yield. More recently it has been found that the damaging effects of the treatment can be minimized by warm storage of the bulbs beforehand.

The following procedure is suggested for cleaning up a stock of tulips by hot-water treatment. The stock must be rogued very thoroughly: all infested plants and neighbouring plants within one metre of them should be removed from the field and burnt. The bulbs should be lifted before the end of June, stored at 34 °C for four days, soaked overnight (for 12 hours) in cold water containing wetter and formalin, and hot-water treated at 45.5 °C with wetter and formalin for three hours.

Hot-water treatment affects the final yield of almost all varieties. Some, however, are more sensitive than others: for example, Bonanza, Coriolan, Orange Monarch and Rose Copeland should not be hot-water treated.

Buying new stocks

Nearly all new cases of stem nematode in tulips are caused by buying infested stocks. All possible precautions should be taken to ensure that bulbs are healthy before buying them. The only way to be certain is to examine them in the field while they are flowering in isolation from other stocks. If this is impossible, stocks should be bought only from a reputable grower who can be relied on to sell clean bulbs. Tulip stem nematode is a very serious pest and it is far easier to keep it out than to get rid of it once it has gained a foothold.

90
Stem nematode on vegetables

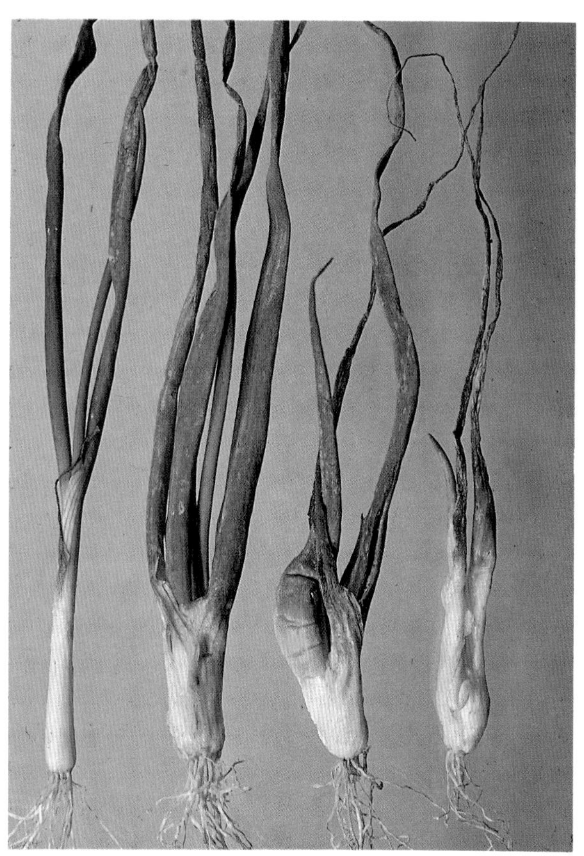

Fig.90.1 Three onion seedlings (right) showing characteristic 'bloat' caused by stem nematode compared with uninfested seedling (extreme left)

Stem nematode (*Ditylenchus dipsaci* (Kühn) Filipjev) is a destructive pest of many different plants. Within this species many races or strains have been recognized according to the range of plants attacked, but the races are difficult to define clearly and some seem able to adapt to new hosts. Onion is the most seriously and widely affected vegetable crop, but carrots, parsnips, beans, peas and rhubarb are also affected. At least some of these crops are attacked by a race which also attacks strawberries (see page 413), oats and some other crops (see page 440).

Description and life history

The adult nematodes are about 1.2 mm long but, being very slender and transparent, they cannot normally be seen without magnification. They live as parasites within the plant tissues, where both males and females often occur in enormous numbers. The females lay many eggs and breeding is continuous. The newly hatched nematodes are smaller than the adults but closely resemble them in appearance.

Throughout the life of an infested plant, nematodes enter the soil and, under suitable conditions, invade fresh plants. If the plant dies, migration into the soil may continue for a short time, but most of the nematodes die when the plant tissues decay. If pieces of infested plant tissue dry up, the nematodes become dormant; in this condition they can survive for several years, only becoming active again when moistened. In moist soil the nematodes can remain alive in the absence of a host plant for a year or longer, according to soil type: they are more persistent in heavy soils.

Symptoms of attack

The effects of stem nematode on its host plants are often striking and characteristic: the usual symptoms are swelling and distortion of stems or leaves and necrosis or rotting of stems, stem bases or crowns. However, symptomless but infested plants also occur, e.g. in bean, pea and onion crops, probably more often than is realised. Expert examination is usually necessary to confirm the presence of the nematode.

On onion

Onion plants may be attacked at any time after germination. Infested seedlings or young plants become greatly swollen at the base and have malformed and twisted leaves (Fig. 90.1). The infested tissues have a characteristic loose, puffy texture and the skin is generally dull in appearance. This condition is known to growers as 'bloat' — a name which is very descriptive of the appearance of the plants. Later on, rotting often occurs at soil level, so that badly infested plants are easily pulled away, leaving most or all of the roots behind. Eventually the whole plant rots and dies.

Slight infestations may pass unnoticed at lifting but may increase during storage, especially if the temperature is not kept low. Bulbs which look sound at harvest may deteriorate during prolonged storage, but spread of nematode from bulb to bulb should be minimized if modern refrigerated stores are used and the bulbs are dry and not sweating.

When onions are grown for seed, nematodes may be carried upwards with the growing flower stem and infest the flowers. Seed from such plants is often heavily infested with nematodes; it is probably by this means that the nematodes are most commonly introduced into fields.

Shallot, **chive**, **garlic** and **leek** may be infested, with effects similar to those on onion.

On parsnip and carrot

The 'onion race' of stem nematode also attacks parsnip and carrot, causing a dry mealy rotting and splitting of the crown (Fig. 90.2) with swelling and splitting of the leaf bases. The symptoms may be confused by the presence of other pests or diseases and there may be secondary attacks by fungi or soft-rot bacteria.

On beans and peas

Heavy attacks on these crops early in the season can cause stunting, malformation or even death of plants. The main shoot of broad bean may be killed and replaced by lateral shoots. Badly attacked dwarf French bean plants are severely stunted with the leaves

Fig.90.2 Carrot infested with stem nematode

Fig.90.3 Dwarf French bean plant (left) severely infested with stem nematode; distorted pod on right

bunched around the growing point (Fig. 90.3).However, damage usually shows as staining of the stem: reddish-black in broad bean (Fig. 90.4), reddish-brown in runner bean and greyish-black in pea. Staining of the stems of beans or peas can also be caused by other organisms or by frost. The effects of a nematode attack often do not become obvious until close to harvest. When the attack is severe the stems are weakened; broad beans collapse and are difficult to harvest, runner bean vines become brittle and break, whereas peas may wither and collapse during flowering.

Stem nematode can be seed-borne in broad bean. Thousands of nematodes have been found in small lesions in the cotyledons just under the seed coat.

On rhubarb

Rhubarb is affected by a disease called 'crown rot' in which the rootstocks are badly damaged by a rotting of the tissues. The damage appears to start on the crowns at about soil level and extends down into the fang roots and sometimes up into the buds and leaf stalks. Two parasites are associated with this condition: stem nematode and a bacterium (*Erwinia rhapontici*). The affected crown tissue is often a dark chocolate-brown colour, but areas of paler discoloured tissue are generally present as well. Large numbers of nematodes may be found in the affected crowns, although few are present in the most rotten parts. Nematodes are also found in buds and leaf stalks. Affected stalks become swollen at the base, then split and soon begin to rot. The bacterium alone can cause symptoms very similar to crown rot, whereas the nematode alone can infest and seriously damage the young crowns and leaves of rhubarb seedlings. The race attacking rhubarb is the same one that attacks oats and onion.

On other vegetables

Stem nematode associated with stunted and twisted foliage has been found in leaf beet and spinach. In swede it has caused malformation of seedlings and a multiple-crown effect. Such attacks are, however, rare.

Stem nematode can also infest potato haulms. It causes stunting, puckering of leaves, swelling of stems and leaf stalks, or sometimes just breakdown of the lower stem tissues leading to collapse of the haulm. The tubers may or may not be attacked but, when they are, the symptoms superficially resemble those of potato tuber nematode (see page 433). It should be noted that the effects of stem nematode on the potato plant differ from those of potato cyst nematodes, which infest the roots (see page 423). Attacks on potato by stem nematode are rare in Britain, but they occur more frequently in several other European countries.

Weed hosts

Stem nematode can live in many common weeds, thus enabling the infestation to persist in the absence of host crops. The nematodes that infest onion and most of the other vegetables mentioned here can live in bindweed, chickweed, cleavers (goosegrass), knotgrass, mayweed, scarlet pimpernel, speedwell and several other weeds.

Means of spread

Stem nematode is commonly introduced into fields with the seed of many crops, e.g. onion, broad bean, field bean, pea, carrot and red beet. A slight seed infestation is likely to affect only individual plants or small patches in the first year, but the patches will probably increase with successive host crops. The rate of spread is generally quicker in the direction of cultivation and may be greatly increased if the field slopes. A slow spread can still occur within a field when non-host crops are grown, but this can be limited if weeds are kept down.

Fig.90.4 Broad bean plant infested with stem nematode, showing lesion in stem on left and healthy growth on right

Stem nematode also occurs in onion and garlic sets. Badly infested sets are likely to rot in store, but they can contaminate sound ones. There is also a danger of contamination from lightly infested sets which show no signs of attack but will give rise to infested plants later.

Within a field or farm, stem nematodes are readily spread by movement of infested debris, soil and surface water.

Control measures

As with other pests and diseases, high standards of cultivation and hygiene, good drainage and soil structure, sound crop rotations and the use of healthy planting material do much to prevent crop losses from stem nematode. It is important to control weeds rigorously.

Hygiene

Care should be taken to remove as much debris as possible from an infested crop. Since the nematode can survive for long periods in dried debris, all crop residues should be destroyed, preferably by burning.

Careful attention to hygiene in onion stores is also important. When stores are cleared, all trash and soil left behind should be destroyed and *never* carted back to the land.

Rhubarb. When rhubarb is followed by other crops, any groundkeepers should be removed. If a field is left down to rhubarb for several years, any diseased roots should be removed and burnt as soon as they are found. When rootstocks are split for propagation, any that show signs of disease should be discarded. It is not sufficient to cut off bad parts: all diseased roots should be rejected completely.

Cultural control

Crop rotations help to limit the build up of stem nematode populations, although damage to onion has occurred in fields where host crops have been absent for many years. However, stem nematode may be introduced with the seed of crops that do not themselves show damage. Most seed merchants take steps to ensure that their onion seed comes from crops apparently free from nematodes on the basis of field and laboratory tests. However, such tests cannot guarantee complete freedom from nematodes and, if in doubt, growers are recommended to use fumigated onion seed. Seed of other crops is not normally examined for stem nematode.

Leaf brassicas and lettuce are not attacked by stem nematode and are, therefore, in addition to wheat and barley, valuable break crops on infested land.

Chemical control

Some granular nematicides are recommended for the control of stem nematode in maincrop onions. The granules should be applied in a narrow band immediately in front of the seed coulters so that the granules are mixed with the soil as the seed is sown. They should not be drilled alongside onion seed as this may cause damage to germinating seedlings. Stem nematode in silverskin onions can be controlled by an overall application before sowing.

Growers planting out onions raised in peat blocks or smaller modules should consult their local adviser for information on methods of applying nematicides.

The use of these nematicides usually controls stem nematode sufficiently well to ensure an acceptable yield of marketable onions from infested land. However, by harvest time some onions will probably contain light infestations of nematode and these may lead to deterioration of the crop during prolonged storage. It is, therefore, better not to grow onions on land known to be infested with stem nematode.

Disinfesting seeds and sets

Seed fumigation. Suitable techniques have been developed for fumigating both large and small quantities of seed, but special equipment is needed. For successful treatment, the percentage germination of the seed must be high, the moisture content should not exceed 12%, and the concentration of gas to which the seed is exposed must be measured precisely. The seed may be damaged if the correct conditions are not fulfilled.

Hot-water treatment. Nematodes in shallot or onion sets can be controlled by hot-water treatment at 44 °C for two hours. Treatment may be applied in autumn or spring. However, when hot-water treatment precedes winter storage, thorough drying is important to avoid risk of rotting. Autumn-treated onion sets have been shown to yield slightly more than those treated in spring, although autumn treatment is not usually practicable with imported onion sets.

Infestations in garlic can be controlled by soaking individual cloves in water containing one per cent formaldehyde for three to four hours at room temperature, followed by hot-water treatment at 45.6 °C for two hours. Treatment applied when the cloves are dormant is likely to give the best results.

MOLLUSCS

Mollusca

91
Slugs and snails

Fig.91.1 Field slug on damaged wheat plants showing severing of stems and leaf shredding

Slugs

Slugs are perhaps the most widely known of all the pests that attack cultivated plants. They cause serious damage on farms, in gardens and on allotments. Although few crops escape attack, autumn-sown cereals and maincrop potatoes are the worst affected. Other farm crops often severely damaged are peas, clover, reseeded ryegrass, and root crops, especially in the seedling stages. Many kinds of vegetables and flowers are also damaged.

Description and habits

Slugs are active and feed throughout the year whenever the temperature and moisture conditions are suitable. During very dry weather, and often when it is frosty, they stop feeding and move down into the soil or shelter under debris. They are most active on still nights when the soil is wet and the atmosphere humid; wind and heavy rain decrease their activity.

Slugs lay their eggs in clusters of up to 50 in the soil or in decaying organic matter (Fig. 91.2). The eggs are quickly killed by drought or frost unless they are protected by the warmth and damp provided by rotting vegetation and soil. The length of time before hatching varies greatly: under warm spring conditions hatching may be complete within three weeks, whereas eggs laid in late autumn may not hatch until the following spring.

Apart from their size and lighter colour, the young slugs resemble the adults. On hatching they first feed on organic matter and humus in the soil, but later their feeding habits vary. Most species will feed both underground and on the surface on a great variety of materials, including living and decayed parts of plants.

The species most harmful to crops are described below.

The field slug (*Deroceras reticulatum* (Müller)) (Fig. 91.1) is very variable in colour but is usually light grey or fawn with milky mucus (slime). The adult is about 35–50 mm long when extended; the hind end tapers to a point. This slug is probably the most common and injurious throughout Britain; it is active at low temperatures and feeds above ground on the aerial parts of plants, at the surface and also underground. There is a peak of breeding in April–May and another in September–October.

Fig.91.2 Egg cluster of a slug (× 3)

Other* Deroceras *species Three species closely related to the field slug are the marsh slug (*Deroceras laeve* (Müller)), *Deroceras agreste* (L.) and the chestnut slug (*Deroceras panormitanum* (Lessona & Pollonera)). The marsh slug is similar to the field slug but is darker and smaller and produces clear mucus; it has been found in significant numbers with the field slug in slug-damaged cereal crops. *D. agreste* is a pale uniform oatmeal colour and is never speckled. In northern Britain it also occurs with the field slug in significant numbers, especially in leys and meadows. The chestnut slug can be distinguished most easily from the field slug by its greater activity and faster movement. It occurs in gardens and glasshouses and can be very damaging.

The garden slugs (*Arion distinctus* (Mabille) and *Arion hortensis* Férussac (Fig. 91.3)) are small, tough-skinned species which vary from grey to black on the upper surface and are yellow below. They secrete yellow-tinged mucus. The adults are usually slightly smaller than the adult field slug; the hind end is rounded. In some areas these slugs are nearly as common as the field slug and are important pests, feeding above and below ground. They have one generation a year and breed mainly in mid to late summer.

Fig.91.3 Garden slug (about life size)

The white-soled slugs (*Arion circumscriptus* Johnston, *Arion fasciatus* (Nilsson) and *Arion silvaticus* Lohmander (Fig. 91.4)) are similar in size to the field slug, have a flattened appearance and are usually dark grey with a strikingly white sole. They are often found in fields where winter wheat or seeded grass has been damaged soon after drilling.

Other* Arion *species The black slug (*Arion ater ater* (L.)) (Fig. 91.5) is usually much larger than any of the other species described here, being up to 200 mm long. The mature form is covered with prominent tubercles and is black or brown with a grey sole. The immature form is yellowish with dark tentacles. This species is not important on agricultural land, though the young have been found with other slugs feeding on cereal seeds. It causes some damage in gardens and is a common species in market gardens.

A reddish-brown to yellow sub-species (*Arion ater rufus* (L.)) also occurs and is more common in southern England. Hybrids of the two sub-species are also found.

Other species responsible for crop damage are the dusky slug (*Arion subfuscus* (Draparnaud)), which is a medium-sized red-brown slug with a smooth skin, and the hedgehog slug (*Arion intermedius* Normand), which is a very small grey slug with conspicuous spiky tubercles when hunched.

The keeled slugs (*Milax gagates* (Draparnaud) and *Tandonia* spp.) are grey, dark brown or black with a ridge or keel down the centre of the back. The adults are larger than those of the *Deroceras* and most *Arion* species described above, being 50–75 mm long. They are mainly subterranean though they also feed above ground. Most serious damage to potatoes is caused by *Tandonia budapestensis* (Hazay) (Fig. 91.6), which is black or brown with a dirty-yellow or orange keel; the sole has a dark central and paler lateral areas. This species is very slender when extended. *M. gagates* is grey with a darker keel and a uniformly pale sole; the mucus is colourless. This slug is more common in south-west England. It also damages potatoes. *Tandonia sowerbyi* (Férussac) is brown with darker speckles but a pale keel and a broad, uniformly pale sole; the mucus is yellow. The keel of this species is very prominent.

The worm-eating slugs (*Testacella* spp.) are rather uncommon. They are predatory animals feeding mainly on earthworms and are peculiar in having a small external shell at the hind end of the body.

Snails

Snails differ from slugs in possessing a shell into which they can withdraw completely. They are much less important pests than slugs but are troublesome in gardens and black currant plantations. They tend to be restricted to calcareous soils. Like slugs, they feed on plant material, mainly at night, and are particularly active in moist conditions. They do not usually feed during the winter but hibernate in clusters in relatively dry and sheltered places. They

Fig.91.4 White-soled slug (*Arion silvaticus*) (× 1¼)

Fig.91.5 Black slug (*Arion ater ater*) (× ¼)

emerge in the spring and eggs are laid in the soil during the late summer or early autumn. Snails mature in one to two years, depending on the species; adults may live for several seasons.

The garden snail (*Helix aspersa Müller*) is the most common and most widely distributed species of snail in gardens. It is easily distinguished by its large greyish-brown shell with paler markings (Fig. 91.7).

The strawberry snail (*Trichia striolata (Pfeiffer)*) is seldom more than 13 mm long; the shell is flatter than that of the garden snail and varies from dirty-grey to reddish-brown or brown. It attacks various garden plants.

The banded snails (*Cepaea nemoralis* (L.) and *Cepaea hortensis* (Müller)) possess shells that may be white, grey, pale yellow, pink or brown with one to five darker bands.

Natural enemies

Slugs are occasionally eaten by birds, including the jackdaw, rook, lapwing and mallard. They are also eaten by frogs, moles, shrews, hedgehogs and various beetles, especially carabids.

Slugs on agricultural crops

Slugs are more active in wet seasons and most damage occurs in the autumn and in mild periods during the winter. Serious damage to cereals does not occur in dry autumns after a dry summer unless conditions at drilling are warm and moist, favouring slug activity, or unless germination is delayed until there is fairly heavy rainfall. On farm land, slugs are more or less confined as serious pests to the heavier soils, being most numerous where drainage is poor and where the soil contains abundant organic matter. During wet seasons slug populations in leafy crops tend to increase rapidly and, even on light land, severe attacks can occur. Surveys of slug populations have shown that most fields contain a mixture of slug species, but the field slug was present in almost every field studied.

Fig.91.6 Keeled slug (*Tandonia budapestensis*) (about life size)

Fig.91.7 Garden snail (× 1¼)

Light mineral soils hold less moisture than heavy soils and, because of their texture and tendency to dry out quickly, are generally unsuitable for slugs. Although peat soils have a high water-holding capacity, there is rarely any trouble from slugs on these soils.

Damage

Most agricultural crops are damaged occasionally, but the most serious losses occur in autumn-sown cereals, especially winter wheat on heavy soils, and in potatoes. Low glucosinolate varieties of oilseed rape are susceptible to slug damage. However, slugs are not generally considered to be major pests of this crop in England.

Cereals. Attacks on cereals are usually worst where the crop follows oilseed rape or other brassica seed crop, clover, ley or meadow grass, cereals, beans or peas, especially after a damp summer and autumn which enable slug numbers to increase. Complete failures of winter wheat can occur under these conditions, but damage is usually patchy over the field. Damage to cereals rarely occurs after sugar beet or potatoes, even when the potatoes have been damaged by slugs. Slug numbers have been shown to increase where cereal straw is incorporated in the soil, but this does not necessarily lead to increased damage. It is only where seedbed conditions enable slugs to move through the soil that damage to seeds and young seedlings is likely.

Serious losses are caused by slugs scraping the germ of winter wheat below ground, often hollowing out the grain before it has had a chance to germinate (Fig. 91.8). Similar damage may occur on winter rye and barley but not on oats. Several species of slug, particularly the field slug, contribute to the attack on grain. On heavy land, when cereals are direct-drilled (see page 472) seed damage is more likely than when conventional drilling is practised, particularly where crops are direct-drilled into straw or stubble.

The shoots of all cereals can be damaged very soon after germination. Shoots are cut through just above the seed or at ground level (Fig. 91.1). Once the shoots thicken they are rarely severed completely but may still be holed at one side. The holes tend to be elongate in contrast to the round holes made by wireworms (see page 218).

Fig.91.8 Wheat grains damaged by slugs

Another kind of damage which may occur is the familiar shredding of seedling leaves after emergence (Fig. 91.1). This is usually less important than grain hollowing or shoot severing, unless the crop is thin and establishing slowly. Leaf shredding can continue in mild, damp weather throughout the life of the crop. The flag leaf may be damaged in wet weather and slugs may feed directly on the ear, especially if the crop lodges.

Leys Serious attacks on young leys may occur soon after germination, the young shoot being cut through at ground level.

Potatoes Slugs often spoil maincrop potatoes by making holes in the tubers (Fig. 91.9); damage can be serious in some seasons, particularly in wet autumns following mild, wet summers. Much smaller numbers of slugs may be important in the potato crop than in cereals. The garden and keeled slugs are the main pests. The field slug, although said to attack mainly damaged tubers, is capable of penetrating the skin of tubers and can contribute considerably to the damage. Slug damage is often confused with damage by cutworms (see page 113). However, cutworm damage is most prevalent on light and medium soils (including peats) after hot, dry summers, whereas slug damage is most prevalent on medium to heavy soils after moist summers and/or autumns, though it can

Fig.91.9 Potato tuber damaged by slugs

occur in dry conditions.

Sugar beet. Slugs may feed on sugar beet plants at any time of year and can readily kill beet seedlings and seriously damage small stecklings. Damage to individual plants is not particularly characteristic and can be confused with that caused by cutworms, leatherjackets (see page 275) or sand weevils. Slime trails are diagnostic but are only rarely visible. The field slug scrapes irregular areas from the leaves and stem and, after one or two nights, only the root remains. White-soled slugs feed below soil level and sever the root so that the plant wilts and dies.

Direct-drilled crops Crops direct-drilled on heavy soils by triple-disc machines are more vulnerable to slug damage than those established by orthodox means or by other types of direct drill. Cereals, grass and brassicas can be severely damaged. When damage is expected, slugs should be controlled before drilling if conditions permit. Alternatively, slug pellets can be admixed with the seed in direct-drilled crops.

Cultural control

Underground attacks to the seed of wheat are most serious in direct-drilled crops and on ploughed land where the seedbed tilth is rough and cloddy. On fields of winter wheat that have failed as a result of slug attack, the headlands are often little damaged; similarly other areas that have been consolidated before or after drilling, or have received additional cultivations, may be little damaged. Additional cultivations are worth considering as a cultural control measure, except on the heaviest soils, where they are virtually impossible in wet seasons. Spring sowings following a winter failure are usually successful, and often an autumn re-drilling succeeds if measures are taken to reduce the risk of further damage.

Damage to maincrop potatoes increases in the autumn when conditions become favourable for slug activity. Onset of damage may also be partly related to the maturation of the tubers and can occur under dry conditions. Damage can be checked by lifting the crop as soon as it has matured, but in a wet summer the damage can be quite severe long before maturity. If damage is likely, it is better to avoid growing varieties that are susceptible to slug damage. Trials on varietal susceptibility showed the order of *increasing* susceptibility to slug damage to be as follows: Pentland Dell and Record; Désirée, King Edward, Pentland Crown, Pentland Squire and Wilja; Kingston; Cara, Estima and Maris Piper; Marfona.

Chemical control

As chemical control measures rely on slug activity on the soil surface, they should be used when slugs are likely to be active above ground, i.e. when the relative humidity is above about 80% and when the soil is damp or wet. Peak activity occurs at about 10 °C if other factors are suitable. Dry, windy weather provides the worst conditions for slug control.

Test-baiting with small quantities of molluscicide pellets under 15-cm square tiles is sometimes used for assessing the need for treatment. These traps should be examined regularly. Test-baiting will give a measure of slug activity on the soil surface and hence an indication of the likely success of broadcasting pellets over the field. However, this technique does not give a reliable prediction of crop damage.

Control of slug damage to winter wheat is best obtained by applying a recommended molluscicide treatment before sowing. This should be done while the slugs are active on the soil surface. The field should not be worked for at least three days following treatment. However, timely sowing in good conditions is more important in reducing slug damage and it is not worth delaying sowing to allow a treatment to be applied. Drilling molluscicide pellets mixed with the seed is less effective but can give useful control when the seedbed is very cloddy as the slugs are more likely to encounter the pellets. In direct-drilled crops, admixture of pellets with the seed is the preferred method of application. However, in most situations, the best practical option is to broadcast pellets to the soil surface at or immediately after drilling.

If damage becomes evident after a cereal crop has emerged, it may be worth applying a molluscicide to prevent further plant loss at this stage, particularly if the crop has already been thinned by hollowing of the grain. Grazing of the leaves is not in itself of much significance but, if plants are grazed down to, or below, ground level during early growth, they may be killed.

Some control of damage to potatoes is possible with molluscicide pellets broadcast over the ridges in late July or August. It is important to apply the pellets when slugs are active on the surface; this usually occurs when about 25 mm of rain falls within a week during this period, or if a similar quantity of irrigation water has been applied. Surface activity can be confirmed by test-baiting as described above. When treating, the soil should be thoroughly moist and there should be a good cover of foliage, thus creating a humid environment in which slugs will be active.

Molluscicide pellets can be applied satisfactorily with many modern types of fertilizer distributor and air-assisted granule applicators. Alternatively, specially designed low-ground-pressure molluscicide distributors can be used. Before using a machine for the first time it is necessary to ensure that the mechanism will handle pellets of the size chosen and that it is capable of dispensing them at the low rate required without breaking them excessively. At the beginning of work, the distribution pattern should be checked and any necessary adjustments made to the machine and, for broadcasters, to the bout width as well. For small areas, pellets can be broadcast by a fiddle drill or by hand, when it is necessary to wear suitable protective clothing, including gloves.

Slugs on horticultural crops

In market gardens and plant nurseries there is often an abundance of organic matter in the soil, as well as cover for slugs throughout the year. The use of green or partly rotted organic manures encourages them. There is, therefore, a tendency for slugs to occur on a wider range of soil types than on farms and they are sometimes quite common on sandy soils. Damage occurs under temporary and permanent cover of all kinds, as well as in the open.

In glasshouses and polythene tunnels slug injury is usually associated with outside sources of infestation, such as boxes and pots of plants waiting to be brought in and adjacent weeds or grass. Baiting and weed control at these sources will usually prevent injury later. On some nurseries baiting of the standing grounds, especially under the boxes, is required to prevent slug or snail injury while the plants are standing out. Damage is often worse around the edges of glasshouses and tunnels.

Damage

Bulbs, corms and outdoor flowers Tulip, iris and lily bulbs may be severely holed. Sometimes the emerging foliage is made ragged (tulip) or is shredded (gladiolus and iris). Narcissus may be grazed when being rooted outside under straw or other cover before forcing under glass. In south-west England, young anemone and ranunculus plants and flowers are sometimes badly grazed. Chrysanthemums, and a range of other plants grown for cut flowers, can be badly damaged, especially soon after planting.

Seedlings of many crops may be eaten by slugs. Examples are brassicas, early carrots, celery, lettuce and herbaceous ornamentals, e.g. *Dianthus*, pansy, primula and sweet pea. Early-sown and newly transplanted crops in the open are particularly susceptible to injury. Dwarf French, broad and runner beans may have the cotyledons, first leaves and the stems injured soon after germination, while still below ground.

Brussels sprout buttons are damaged by slugs in some seasons and marketable yields can be seriously affected. The attack starts on the lower buttons and gradually spreads up the stem. The slugs prefer the buttons and damage to the leaves is minimal. Damage to the buttons is unacceptable when the crop is for the freezing, pre-packing or any other high quality market.

Celery Stalks of self-blanching crops can be badly damaged.

Lettuce Crops can easily be spoilt by slug feeding. Secondary infection with disease often follows slug damage.

Fruit Ripening strawberries are very susceptible to slug damage. Strawberries grown on raised beds can be severely affected because the polythene cover provides a moist environment that is ideal for slugs. Fruits can be holed and hollowed by the slugs. Other pests, such as caterpillars, can cause similar damage.

Control measures

Recommended molluscicides can be used as for slugs on agricultural crops (see page 472). On small plots subject to reinfestation from uncultivated ground, more frequent treatment is needed than on a field scale. Where practicable, pellets should be applied when the ground is bare, so that mildly poisoned slugs are unable to reach shelter where they might recover. When using pellets to protect older plants, the pellets should be placed around the plants rather than on them, since decaying pellets on crops such as lettuce may lead to crop rejection. Produce so contaminated may be regarded as unfit for human consumption and the grower may be liable to prosecution. Decaying pellets may also encourage botrytis. On Brussels sprouts the most effective time to apply pellets is as soon as the crop meets between the rows and when the soil surface is moist. On strawberries not grown on beds, pellets should be applied just before strawing; on raised beds where the soil has not been sterilized, pellets should be applied after bed-forming but before polythene is laid.

Snails in black currant plantations

Large populations of snails (the garden snail and banded snails), often several thousands per hectare, may occur in black currant plantations, particularly if weed control is poor. Although snails do not cause direct damage to the crop, they rest in the bushes and cause problems by contaminating consignments of mechanically harvested fruit. They also seek shelter in fruit trays left standing in the plantations before picking.

Control measures

Control of weeds by the efficient use of herbicides, particularly on headlands, is a sure way of reducing snail populations in black currant plantations. Contamination of fruit containers may be prevented by placing them on a hard standing or by separating them from direct contact with the soil or weeds using, for example, a polythene sheet.

Molluscicide pellets applied at the black currant 'grape' stage and broadcast on the headlands of plantations have a long-term effect in reducing snail populations.

Control of slugs and snails in gardens

Much can be done to check infestations by keeping a tidy garden which offers the minimum of shelter and food to slugs and snails.

Where slugs are a perennial problem the numbers can be decreased in autumn by baiting attractive crops that are not ready for harvesting. The lower leaves of brassica crops falling to, or touching, the ground in autumn provide excellent cover and food for slugs, as well as a good opportunity for effective baiting. Small heaps of molluscicide pellets, about 30 cm apart, can be covered with a tile or leaf to keep off rain and to make the poison less accessible to birds and domestic animals. The poisoned slugs should be collected. Mouldy bait is not attractive to slugs and should be replaced with fresh bait while conditions favour slug activity.

APPENDIX A

Leaflet titles listed alphabetically

APPENDIX B

Leaflet titles listed alphabetically under major host crops

Cereals

Cereal aphids (page 36)
Cereal cyst nematode (page 399)
Frit fly (page 258)
Gout fly (page 264)
Leatherjackets (page 274)
Saddle gall midge (page 306)
Slugs and snails (page 467)
Stem nematode on arable and forage crops (page 440)
Swift moths (page 132)
Wheat blossom midges (page 310)
Wheat bulb fly (page 315)
Wireworms (page 216)

Grass

Cereal aphids (page 36)
Chafer grubs (page 162)
Frit fly (page 258)
Leatherjackets (page 274)
Pests of grass and clover seed crops (page 193)
Slugs and snails (page 467)
Swift moths (page 132)
Wireworms (page 216)

Potatoes

Chafer grubs (page 162)
Colorado beetle (not established in Britain) (page 168)
Cutworms (page 112)
Millepedes and centipedes (page 377)
Potato aphids (page 60)
Potato cyst nematodes (page 422)
Potato tuber nematode (page 432)
Slugs and snails (page 467)
Stem nematode on arable and forage crops (page 440)
Swift moths (page 132)
Wireworms (page 216)

Sugar beet

Beet cyst nematode (page 395)
Beet leaf miner (page 233)
Black bean aphid (page 21)
Cutworms (page 112)
Docking disorder of sugar beet (page 408)
Flea beetles (page 173)
Leatherjackets (page 274)
Millepedes and centipedes (page 377)
Pygmy beetle (page 200)
Slugs and snails (page 467)
Symphylids (page 382)
Wireworms (page 216)

Top fruit

Apple and pear midges (page 225)
Apple and pear suckers (page 11)
Apple aphids (page 16)
Apple blossom weevil (page 159)
Apple sawfly (page 93)
Bryobia mites (page 329)
Capsid bugs on fruit (page 32)
Codling moth (page 108)
Fruit tree red spider mite (page 341)
Pear and cherry slugworm (page 121)
Pear leaf blister mite (page 347)
Plum aphids (page 56)
Plum fruit moths (page 123)
Scale insects on fruit trees (page 66)
Stem-boring caterpillars on fruit plants (page 126)
Tortrix moths on apple (page 136)
Two-spotted spider mite on outdoor crops (page 369)
Wasps (page 321)
Web-forming caterpillars (page 145)
Winter moths (page 153)
Woolly aphid (page 78)

Hops

Plum aphids (page 56)
Swift moths (page 132)
Two-spotted spider mite on outdoor crops (page 369)
Wingless weevils (page 210)

Bush and cane fruits

Bryobia mites (page 329)
Capsid bugs on fruit (page 32)
Caterpillars on currants and gooseberry (page 103)
Currant and gooseberry aphids (page 42)
Raspberry beetle (page 204)
Raspberry cane midge and midge blight (page 301)
Raspberry leaf and bud mite (page 350)
Reversion disease and gall mite of black currant (page 354)
Scale insects on fruit trees (page 66)
Slugs and snails (page 467)
Stem-boring caterpillars on fruit plants (page 126)
Two-spotted spider mite on outdoor crops (page 369)
Wingless weevils (page 210)
Winter moths (page 153)

Strawberry

Capsid bugs on fruit (page 32)
Chafer grubs (page 162)
Leatherjackets (page 274)
Nematodes on strawberry (page 413)

Slugs and snails (page 467)
Strawberry mite (page 366)
Swift moths (page 132)
Tortrix moths on strawberry (page 141)
Two-spotted spider mite on outdoor crops (page 369)
Wingless weevils (page 210)
Wireworms (page 216)

Brassica seed crops including oilseed rape

Cabbage aphid (page 27)
Cabbage caterpillars (page 97)
Cabbage root fly (page 237)
Flea beetles (page 173)
Insect pests of brassica seed crops (page 176)
Insect pests of oilseed rape (page 182)
Turnip gall weevil (page 207)

Peas, beans and other legumes

Bean seed flies (page 230)
Black bean aphid (page 21)
Pea, bean and clover weevils (page 190)
Pea cyst nematode (page 419)
Pea midge (page 297)
Pea moth (page 117)
Pests of grass and clover seed crops (page 193)
Slugs and snails (page 467)
Stem nematode on arable and forage crops (page 440)
Stem nematode on vegetables (page 458)
Thrips on peas (page 88)
Wireworms (page 216)

Vegetables

Cabbage aphid (page 27)
Cabbage caterpillars (page 97)
Cabbage root fly (page 237)
Carrot fly (page 244)
Celery fly (page 252)
Cutworms (page 112)
Flea beetles (page 173)
Leatherjackets (page 274)
Lettuce aphids (page 51)
Onion fly (page 294)
Slugs and snails (page 467)
Stem nematode on vegetables (page 458)
Turnip gall weevil (page 207)
Two-spotted spider mite on outdoor crops (page 369)
Willow–carrot aphid (page 73)
Wireworms (page 216)

Protected edible crops

Glasshouse whitefly (page 46)
Mushroom pests (page 279)
Potato cyst nematodes on tomato (page 428)
Root-knot nematodes on protected crops (page 436)
Spider mites on protected crops (page 360)
Symphylids (page 382)
Wireworms (page 216)
Woodlice (page 387)

Bulbs

Bulb scale mite (page 334)
Gladiolus thrips (page 85)
Narcissus flies (page 290)
Potato tuber nematode (page 432)
Slugs and snails (page 467)
Stem nematode on narcissus (page 447)
Stem nematode on tulip (page 454)

Protected ornamentals

Chrysanthemum gall midge (page 255)
Chrysanthemum nematode (page 404)
Cyclamen mite on glasshouse plants (page 338)
Glasshouse whitefly (page 46)
Leaf miners of chrysanthemum (page 268)
Root-knot nematodes on protected crops (page 436)
Spider mites on protected crops (page 360)
Symphylids (page 382)
Wingless weevils (page 210)

Index of pest and beneficial invertebrates

(also diseases of pests)

Index of plants

Index of plant diseases and pathogens